Scientific Machine Learning for Polymeric Materials

Scientific Machine Learning for Polymeric Materials

Guest Editors

Salah A. Faroughi
Célio Bruno Pinto Fernandes

MDPI

Basel • Beijing • Wuhan • Barcelona • Belgrade • Novi Sad • Cluj • Manchester

Guest Editors
Salah A. Faroughi
Chemical Engineering
Department
University of Utah
Salt Lake City, UT
USA

Célio Bruno Pinto Fernandes
Department of Mechanical
Engineering
University of Porto
Porto
Portugal

Editorial Office
MDPI AG
Grosspeteranlage 5
4052 Basel, Switzerland

This is a reprint of the Special Issue, published open access by the journal *Polymers* (ISSN 2073-4360), freely accessible at: https://www.mdpi.com/journal/polymers/special_issues/1U740EL4J8.

For citation purposes, cite each article independently as indicated on the article page online and as indicated below:

Lastname, A.A.; Lastname, B.B. Article Title. *Journal Name* **Year**, *Volume Number*, Page Range.

ISBN 978-3-7258-5839-2 (Hbk)
ISBN 978-3-7258-5840-8 (PDF)
https://doi.org/10.3390/books978-3-7258-5840-8

Contents

About the Editors

Salah A. Faroughi

Salah A. Faroughi is an assistant professor in the Department of Chemical Engineering at the University of Utah. His research centers on the dual challenge of "Energy–Environment," with the objective of identifying cost-effective and reliable energy solutions while upholding environmental sustainability. His team examines the dynamics, thermophysics, and biogeochemistry of particulate matter, complex fluids, and multiphase reacting flows across various spatiotemporal scales within the context of geo-energy systems (such as hydrogen storage, carbon storage, geothermal energy, etc.). A crucial aspect of their work involves analyzing the nonlinear responses of these materials and flows to different external stimuli in order to optimize stress and energy transmission, taking into account uncertainties. In this endeavor, his research group formulates and utilizes high-fidelity computational models, state-of-the-art experimental setups, and data-driven techniques that are rooted in physicochemical principles.

Célio Bruno Pinto Fernandes

Célio Bruno Pinto Fernandes is an assistant professor at the Department of Mechanical Engineering, Faculty of Engineering, University of Porto (FEUP), Portugal. C. Fernandes obtained his Education of Mathematics Bachelor's degree from the University of Minho (Portugal) in 2005. After graduating, C. Fernandes obtained his Applied Mathematics MSc degree at the University of Porto (Portugal) in 2007. During this period, C. Fernandes employed spectral methods to describe the melt flow that occurs in the polymer extrusion process. Afterwards, he joined the Department of Polymer Engineering at the University of Minho (Portugal), where he completed his Ph.D. degree in Science and Engineering of Polymers and Composites in 2012. During this period, C. Fernandes made important contributions to the field of injection molding by applying multi-objective evolutionary algorithms to solve the inverse problem of finding the best injection molding parameters to achieve predefined criteria. He was also a visiting post-doctoral researcher at MIT, USA, in 2017. C. Fernandes has been working with the open-source computational fluid dynamics software OpenFOAM. He has established new numerical methods for the solution of viscoelastic matrix-based fluids using the finite volume method, such as an immersed boundary method capable of fully resolving particle-laden viscoelastic flows, and has developed a fully implicit log-conformation tensor coupled algorithm for the solution of incompressible non-isothermal viscoelastic flows

polymers

Editorial

Scientific Machine Learning for Polymeric Materials

C. Fernandes [1,2,3]

1 CEFT—Transport Phenomena Research Center, Faculty of Engineering, University of Porto, Rua Dr. Roberto Frias, 4200-465 Porto, Portugal; cbpf@fe.up.pt
2 ALiCE—Associate Laboratory in Chemical Engineering, Faculty of Engineering, University of Porto, Rua Dr. Roberto Frias, 4200-465 Porto, Portugal
3 Centre of Mathematics (CMAT), University of Minho, Campus of Gualtar, 4710-057 Braga, Portugal

1. Introduction

Polymeric materials are ubiquitous in modern technology, from structural composites and membranes to responsive elastomers, yet their design remains challenging due to complex, multi-scale behaviors. Conventional modeling often struggles to capture phenomena spanning the molecular to the macroscopic scale in polymers. In recent years, Scientific Machine Learning (SciML), the integration of data-driven artificial intelligence (AI) with physical models, has emerged as a powerful approach to accelerate polymer research and development. By incorporating domain knowledge (such as physical laws or material science heuristics) into neural networks, SciML methods can leverage limited experimental or simulation data to build more accurate, interpretable models than black-box ML alone.

In the past five years, the emergence of SciML has transformed polymer research by combining data-driven algorithms with physical models to overcome the limitations of purely empirical or purely first-principles approaches. Graph- and sequence-based neural networks have enabled rapid screening of polymer chemistries, surrogate models and active-learning frameworks have accelerated high-throughput material discovery, and physics-informed architectures (e.g., PINNs) have begun to bridge molecular scale dynamics with macroscopic behavior without requiring massive datasets. ML models now predict complex structure–property relationships, enabling the inverse design and high-throughput screening of polymer candidates (Xie et al. [1], Pai et al. [2]). SciML techniques now combine physics-based models with data driven ML to enhance accuracy and interpretability, even with limited data (Audus et al. [3], Nanjo et al. [4]). ML techniques such as random forests, neural networks, and support vector machines are increasingly used to predict the thermal, mechanical, and optical properties of polymers and composites (Malashin et al. [5], Hamidi et al. [6]).

Despite these advances, critical gaps remain. Most current ML tools excel at a single length or time scale, yet polymers exhibit phenomena—from chain entanglement and phase separation to fracture and creep—that span orders of magnitude. Few SciML methods rigorously quantify prediction confidence, limiting their reliability for safety-critical or regulatory applications. Polymer data are often sparse, proprietary, or measured under inconsistent protocols, making generalization challenging. While PINNs have shown promise, they remain computationally demanding for large-scale polymer process simulations and have yet to fully incorporate complex rheological laws or multicomponent interactions. The articles in this Special Issue directly address these gaps. The multi-objective optimization of NN hyperparameters for composites and membranes (Malashin et al. [7]) demonstrates how tailored ML can capture multiple target properties simultaneously, offering a template for multi-scale model coupling. Three-dimensional autoencoder

 https://doi.org/10.3390/polym17162222

defect detection (Liu et al. [8]) and hysteresis modeling of magnetorheological elastomers (Alawi et al. [9]) illustrate how ML can embed physics constraints into high-dimension inverse problems, paving the way for more efficient, physics-driven simulators. Diffusion coefficient prediction in Nafion membranes and QSPR-based cleansing foam design showcase strategies for data-synthesis and transfer learning to mitigate data scarcity by combining experiments, surrogate models, and generative screening (Malashin et al. [10]).

This Special Issue, titled "Scientific Machine Learning for Polymeric Materials", brings together cutting-edge work in this area, highlighting new machine learning algorithms, data-driven design strategies, and physics-informed methods tailored to polymeric systems. The collected contributions address a broad range of topics, ranging from neural network-based property prediction to advanced material characterization, illustrating the importance and promise of ML-driven approaches in polymer science.

2. Overview of the Published Articles

This Special Issue features several articles that apply SciML techniques to polymeric materials. Key themes include predictive modeling of polymer properties, simulation of smart polymers, and comprehensive surveys of ML methods in this field.

Recent research has showcased how ML and AI are revolutionizing polymer science—from materials design and property prediction to novel testing methods. For example, Brighel et al. [11] demonstrated that an AI-trained model can make analytical crystallization elution fractionation (aCEF) a standalone characterization tool: their model automatically classifies commercial polyolefins without any prior chemical information.

Complex polymer blends with many ingredients pose a challenge for formulation design. Hamaguchi et al. [12] tackled this by preparing 537 cleansing-foam formulations (mixtures of surfactants, polyols, etc.) and using ML to predict their cleansing performances. They computed molecular descriptors (including Hansen solubility parameters) for each formulation and trained five ML models. The best model achieved $R^2 \approx 0.770$ in predicting cleansing capability. Importantly, the study also used in silico design: it generated and screened virtual formulations with the trained model. Despite highly nonlinear ingredient interactions, this AI-assisted approach identified promising ingredient combinations, illustrating how ML can guide the optimization of "super-multicomponent" polymer products.

Complementing these research highlights are three extensive review papers. Long et al. [13] provide an overarching perspective on AI in polymers, advocating a shift from traditional experimentation to data-driven design. They document AI's transformative impact on polymer design, property prediction, and processing and outline strategies (e.g., collaborative data platforms, explainable models) for overcoming challenges like data scarcity. Malashin and colleagues contribute two focused reviews: one on physics-informed neural networks (PINNs) [14] in polymers and another on SVMs [15]. The PINN review summarizes recent advances in embedding physical laws into neural networks for multiscale polymer modeling, while the SVM review surveys how kernel-based ML methods have been used to predict polymer properties, optimize synthesis, and control processes. Together, these reviews highlight the rapid integration of ML techniques, ranging from SVMs to deep PINNs, into polymer science, and they chart future directions for AI-enabled polymer innovation.

3. Conclusions and Outlook

In conclusion, the contributions of this Special Issue demonstrate the significant impact of advanced machine learning on polymer research. The featured studies show that integrating AI methods, ranging from neural networks and evolutionary optimization to physics-informed frameworks and ensemble learners, can greatly enhance our ability to predict, understand, and design polymeric materials. For example, the data-driven optimization of neural network models has yielded accurate predictions of composite properties, while novel 3D autoencoders have improved nondestructive defect detection in composites. Machine learning applied to polymer membranes, polymer property databases, and complex mixtures (like cleansing foams) has revealed subtle structure–property relationships and identified optimal formulations without exhaustive experimentation. The hysteresis modeling of magnetorheological elastomers illustrates how SciML can be used to capture nonlinear smart material behaviors, enabling precise device design. Complementing these technical advances, the three review articles synthesize the state of the art in the field and point toward future opportunities: they highlight, for instance, how physics-informed learning can bridge scales in polymer simulation and how boosting algorithms are reshaping data-driven polymer design.

Together, the works in this Special Issue advance polymer science by leveraging data, descriptors, and physics in novel ways. They showcase how SciML methods not only improve predictive accuracy but also offer insight into why materials behave as they do, often reducing development cycles. Looking forward, further progress will likely come from more tightly integrating physics-based knowledge with ML (as in PINNs and hybrid models), expanding open polymer datasets for training, and improving interpretability and uncertainty quantification. As noted in the Special Issue's call for papers, the overarching goal is to use scientific AI to unravel multi-scale polymer phenomena and accelerate innovation in polymeric materials. The contributions collected here have taken important steps toward achieving that goal: they provide new models, algorithms, and frameworks that deepen our understanding and capability in polymer design. We anticipate that these advances will inform ongoing research, inspire new SciML applications in polymer science, and help to realize the Special Issue's vision of smart, data driven polymer innovation

Future research should build on these foundations by developing hybrid multi-scale frameworks that seamlessly couple molecular simulations (MD/DFT) with continuum-scale ML models to predict processing–structure–property maps; embedding robust uncertainty quantification (e.g., Bayesian PINNs, ensemble learning) to deliver prediction intervals alongside point estimates, critical for design under uncertainty and regulatory approval; creating shared, standardized polymer data repositories and benchmarks that enable fair comparison, reproducibility, and meta-learning across different polymer classes and experimental conditions; optimizing physics–ML co-solvers for large-scale manufacturing simulations, leveraging reduced-order modeling, operator learning, and adaptive sampling to enable real-time digital twins of polymer processes; and expanding reinforcement-learning and generative models for closed-loop, autonomous material discovery, where AI not only predicts properties but also actively proposes and synthesizes novel polymer chemistries in laboratory workflows.

Funding: This Special Issue was supported by national funds through FCT/MCTES (PIDDAC): CEFT, UIDB/00532/2020 (DOI: 10.54499/UIDB/00532/2020) and UIDP/00532/2020 (DOI: 10.54499/UIDP/00532/2020), and ALiCE, LA/P/0045/2020 (DOI: 10.54499/LA/P/0045/2020). It was also funded by the FCT through the CMAT (Centre of Mathematics of the University of Minho) projects UIDB/00013/2020 and UIDP/00013/2020.

Conflicts of Interest: The authors declare no conflict of interest.

References

1. Xie, C.; Qiu, H.; Liu, L.; You, Y.; Li, H.; Li, Y.; Sun, Z.; Lin, J.; An, L. Machine Learning Approaches in Polymer Science: Progress and Fundamental for a New Paradigm. *SmartMat* **2025**, *6*, e1320. [CrossRef]

2. Pai, S.; Shah, K.; Sunder, S.; Albuquerque, R.; Brütting, C.; Ruckdaeschel, H. Machine learning applied to the design and optimization of polymeric materials: A review. *Next Mater.* **2025**, *7*, 100449. [CrossRef]
3. Audus, D.; McDannald, A.; DeCost, B. Leveraging Theory for Enhanced Machine Learning. *ACS Macro Lett.* **2022**, *11*, 1117–1122. [CrossRef] [PubMed]
4. Nanjo, S.; Arifin; Maeda, H.; Hayashi, Y.; Hatakeyama-Sato, K.; Himeno, R.; Hayakawa, T.; Yoshida, R. SPACIER: On-demand polymer design with fully automated all-atom classical molecular dynamics integrated into machine learning pipelines. *NPJ Comput. Mater.* **2024**, *11*, 16. [CrossRef]
5. Malashin, I.P.; Tynchenko, V.S.; Nelyub, V.A.; Borodulin, A.S.; Gantimurov, A.P. Estimation and Prediction of the Polymers' Physical Characteristics Using the Machine Learning Models. *Polymers* **2024**, *16*, 115. [CrossRef] [PubMed]
6. Hamidi, Y.; Berrado, A.; Altan, M. Machine learning applications in polymer composites. In Proceedings of the 35th International Conference of the Polymer Processing Society (PPS35), Cesme-Izmir, Turkey, 26–30 May 2019; Volume 2205, p. 020031. [CrossRef]
7. Malashin, I.; Tynchenko, V.; Gantimurov, A.; Nelyub, V.; Borodulin, A. A Multi-Objective Optimization of Neural Networks for Predicting the Physical Properties of Textile Polymer Composite Materials. *Polymers* **2024**, *16*, 1752. [CrossRef] [PubMed]
8. Liu, Y.; Yu, Q.; Liu, K.; Zhu, N.; Yao, Y. Stable 3D Deep Convolutional Autoencoder Method for Ultrasonic Testing of Defects in Polymer Composites. *Polymers* **2024**, *16*, 1561. [CrossRef] [PubMed]
9. Mohd. Alawi, A.H.; Hudha, K.; Kadir, Z.A.; Amer, N.H. Hysteresis Behavior Modeling of Magnetorheological Elastomers under Impact Loading Using a Multilayer Exponential-Based Preisach Model Enhanced with Particle Swarm Optimization. *Polymers* **2023**, *15*, 2145. [CrossRef] [PubMed]
10. Malashin, I.; Daibagya, D.; Tynchenko, V.; Gantimurov, A.; Nelyub, V.; Borodulin, A. Predicting Diffusion Coefficients in Nafion Membranes during the Soaking Process Using a Machine Learning Approach. *Polymers* **2024**, *16*, 1204. [CrossRef] [PubMed]
11. Brighel, L.; Scuotto, G.M.L.; Antinucci, G.; Cipullo, R.; Busico, V. AI-Aided Crystallization Elution Fractionation (CEF) Assessment of Polyolefin Resins. *Polymers* **2025**, *17*, 1597. [CrossRef] [PubMed]
12. Hamaguchi, M.; Miwake, H.; Nakatake, R.; Arai, N. Predicting the Performance of Functional Materials Composed of Polymeric Multicomponent Systems Using Artificial Intelligence—Formulations of Cleansing Foams as an Example. *Polymers* **2023**, *15*, 4216. [CrossRef] [PubMed]
13. Long, T.; Pang, Q.; Deng, Y.; Pang, X.; Zhang, Y.; Yang, R.; Zhou, C. Recent Progress of Artificial Intelligence Application in Polymer Materials. *Polymers* **2025**, *17*, 1667. [CrossRef] [PubMed]
14. Malashin, I.; Tynchenko, V.; Gantimurov, A.; Nelyub, V.; Borodulin, A. Physics-Informed Neural Networks in Polymers: A Review. *Polymers* **2025**, *17*, 1108. [CrossRef] [PubMed]
15. Malashin, I.; Tynchenko, V.; Gantimurov, A.; Nelyub, V.; Borodulin, A. Support Vector Machines in Polymer Science: A Review. *Polymers* **2025**, *17*, 491. [CrossRef] [PubMed]

polymers

MDPI

Article

An Artificial Intelligence-Based Melt Flow Rate Prediction Method for Analyzing Polymer Properties

Mohammad Anwar Parvez [1,*] and Ibrahim M. Mehedi [2,*]

1 Department of Chemical Engineering, College of Engineering, King Faisal University, Al-Ahsa 31982, Saudi Arabia
2 School of Robotics, XJTLU Entrepreneur College (Taicang), Xi'an Jiaotong-Liverpool University, No. 111 Taicang Ave., Suzhou 215400, China
* Correspondence: mparvez@kfu.edu.sa (M.A.P.); ibrahim.mehedi@xjtlu.edu.cn (I.M.M.)

Abstract: The polymer industry gained increasing importance due to the ability of polymers to replace traditional materials such as wood, glass, and metals in various applications, offering advantages such as high strength-to-weight ratio, corrosion resistance, and ease of fabrication. Among key performance indicators, melt flow rate (MFR) plays a crucial role in determining polymer quality and processability. However, conventional offline laboratory methods for measuring MFR are time-consuming and unsuitable for real-time quality control in industrial settings. To address this challenge, the study proposes a leveraging artificial intelligence with machine learning-based melt flow rate prediction for polymer properties analysis (LAIML-MFRPPPA) model. A dataset of 1044 polymer samples was used, incorporating six input features such as reactor temperature, pressure, hydrogen-to-propylene ratio, and catalyst feed rate, with MFR as the target variable. The input features were normalized using min–max scaling. Two ensemble models—kernel extreme learning machine (KELM) and random vector functional link (RVFL)—were developed and optimized using the pelican optimization algorithm (POA) for improved predictive accuracy. The proposed method outperformed traditional and deep learning models, achieving an R^2 of 0.965, MAE of 0.09, RMSE of 0.12, and MAPE of 3.4%. A SHAP-based sensitivity analysis was conducted to interpret the influence of input features, confirming the dominance of melt temperature and molecular weight. Overall, the LAIML-MFRPPPA model offers a robust, accurate, and deployable solution for real-time polymer quality monitoring in manufacturing environments.

Keywords: artificial intelligence; machine learning; melt flow rate prediction; polymer properties analysis; pelican optimization algorithm

1. Introduction

A polymer is nothing but a class of synthetic or natural material made by huge-sized molecules named macromolecules that modestly duplicate simple chemical units known as monomers. Polymers were formed by chemical building blocks named monomers (C2, C3, C4, C5, and others) in gas-level and liquid slurry reactors [1]. The utilization of nanoparticles in polymerization reactions can substantially enhance the output of product. Nanocatalysis is a quickly emerging region, which comprises the usage of nanomaterials as catalysts for multiple heterogeneous and homogeneous catalytic applications [2]. Heterogeneous catalysis signifies the earliest commercial nanoscience practices; oxides, semiconductors, metal nanoparticles, and other compounds are extensively employed for significant chemical reactions. The monomers are developed into catalysts under accurately controlled pressure and temperature states to initiate a reaction that progresses the polymer chains [3]. To prevent the development of the polymer chain, hydrogen serves as a chain transfer factor. The polymer becomes extremely viscous for injection molding, film production, or other applications if the polymer chains extend overly [4]. The polymer is soft and absent the strength required for particular applications, such as in a washing machine drum, a vehicle bumper, or a plastic bag, if the polymer chains are too small. Generally, viscosity of polymers evaluates the resistance to flow and is vital for performance and process. An easier and more effective assessment called melt flow rate (MFR) is generally utilized in the plastic industries [5]. The viscosity directly measures a polymer's resistance to flow, and melt flow rate (MFR) serves as its practical inverse, offering a simpler, faster alternative for assessing processability. Unlike viscosity, which requires complex instrumentation, MFR is widely used in industry due to its ease of measurement and strong correlation with molecular weight and flow behavior.

MFR is the vital quality index to identify the performance of polyolefin products that typically serves to control and monitor the process and features of products [6]. A laboratory model is a typically used offline measurement approach for MFR evaluating either powder or particle products; however, the time delay from sampling to attaining test outcomes is extended and is not suitable for real-world monitoring of product features [7]. Few industrial units fix online MFR analysts to enhance the timeliness of MFR measurement. Nevertheless, online analyzers need further equipment investment and are inaccurate. In addition, sampling molds require recurrent replacement that also intrudes on the process of production and increases losses [8]. Deep learning (DL) and ML are developed as great devices in materials science, transforming how materials are intended, optimized, and characterized [9]. Several DL and ML-based applications in materials science aimed to aid in the analysis of two metal-reinforced polymer compounds. DL and ML-based models are effective in forecasting MFR in materials science [10].

This study presents a leveraging artificial intelligence with machine learning-based melt flow rate prediction for polymer properties analysis (LAIML-MFRPPPA) model. At first, the data normalization stage employs min–max normalization to scale features into a consistent range. Furthermore, the proposed LAIML-MFRPPPA model designs ensemble models, namely the kernel extreme learning machine (KELM) method [5] and random vector functional link (RVFL) technique, for the prediction method. Eventually, the pelican optimization algorithm (POA)-based hyperparameter selection process is performed to optimize the prediction results of ensemble models. The experimental evaluation of the LAIML-MFRPPPA model occurs using a benchmark dataset.

Predicting polymer melt behavior during industrial processes such as extrusion, injection molding, and film blowing remains a complex challenge due to the nonlinear and temperature-dependent nature of polymer rheology. Variations in reactor temperature, catalyst behavior, molecular structure, and flow conditions can significantly alter melt viscosity and processing outcomes. Traditional process monitoring systems rely heavily on offline rheological tests or single-point MFR measurements, which lack responsiveness for real-time control. To address this, research evolved from empirical models and statistical regression methods to numerical simulations based on computational fluid dynamics (CFD) and finite element modeling. More recently, artificial intelligence and machine learning techniques emerged as powerful alternatives capable of learning from complex, high-dimensional datasets to predict melt behavior with greater accuracy and speed.

A major problem facing the polymer industry is the lack of the ability to predict the outcome of the polymer melt, particularly because of the nonlinear nature of the behavior that depends most significantly on the reactor temperature, catalyst activity, molecular structure, and flow conditions. Several decades older and more common offline techniques of determining melt flow rate (MFR) are still used in industrial practice; however, they are very slow and cannot be used in real-time control, which delays optimization of processes and causes higher production losses. The challenges have been evidenced in recent studies on the flexibilities of extrusion, and additive manufacturing processes of varying temperatures, hydrogen-to-propylene ratio, and feed rate of catalysts have been found to have notable impact on the polymer chain growth and the viscosity. In an attempt to overcome these shortcomings, researchers turned to empirical correlations, computational fluid dynamics (CFD), and finite element modeling (FEM), yet none of these methods are well suited for this purpose due to a lack of versatility and the non-linearities in the rheology of polymers. More recently, machine learning (ML) and deep learning DL approaches offered great promise in the ability to accurately predict MFR, making it possible to incorporate large and high-dimensional data into statistical models. Nonetheless, overfitting and data-representative issues remain to be solved and an advanced model, e.g., polyBERT, GNN, and TransPolymer, needs extensive training infrastructure restricting its implementation in industry. It is in this scenario that the proposed LAIML-MFRPPPA framework combines the techniques of ensemble learning (KELM + RVFL) and metaheuristic optimization (POA) to realize high predictive rates, efficiency, and comprehensibility, thus responding to both the scientific and industrial demand and need of quality polymeric monitoring in real time. These issues are why there is a need to develop predictive models that are able to capture the nonlinear behavior of the polymer melt flow in a manner that is efficient and can be deployed in industry. Conventional empirical and physics-based techniques rarely work well when it comes to highly dynamic process conditions. Simultaneously, most sophisticated models of deep learning, though very effective, are demanding on computational resources and are non-interpretable, and this constraint restricts their applicability to manufacturing industries. Hence, it is highly desirable to formulate a lightweight, precise, and transparent machine learning framework capable of delivering trusted predictions of melt flow rate on a real-time basis. To fill this gap, the proposed LAIML-MFRPPPA model is based on the ensemble learning model using optimization methods to provide not only accurate predictions, but also adoptability by industrial usage.

The remainder of this paper is organized as follows: Section 2 presents a detailed review of related works in the domain of melt flow rate prediction and machine learning applications in polymer science. Section 3 describes the proposed LAIML-MFRPPPA methodology, including data preprocessing, model design, and optimization strategies. Section 4 provides the experimental setup, performance evaluation, and analysis of the results. Finally, Section 5 concludes the study and outlines potential directions for future research.

2. Literature of Works

Over the past decade, significant advances have been made in the predictive modeling of polymer material behavior under processing conditions. Early works primarily used experimental correlations and parametric models based on rheological equations. These evolved into process monitoring systems that incorporate in-line sensors and real-time feedback loops for quality control. Statistical techniques such as ANOVA and linear regression were later applied to evaluate parameter sensitivity. However, these models often lacked the flexibility to capture nonlinear dependencies. The integration of machine learning and deep learning methods marked a paradigm shift, allowing for more accurate, adaptive, and scalable predictions. Yet, the reliability of such AI-based models is strongly influenced by the quality and completeness of the input data. Incomplete sensor records, noise, or non-representative training samples can significantly reduce model accuracy and generalizability, making data preprocessing, feature selection, and validation essential components of modern predictive pipelines.

Nagarjun et al. [11] applied the full factorial technique to examine the effect of printing process parameters such as nozzle size, infill density, layer height, and infill pattern through the tensile strength of the printed portions. Analysis of variance (ANOVA) is performed and it is recognized the nozzle size as the most important aspect affecting strength of tensile, succeeded by infill density. Infill shape and layer height had a small individual influence on strength of tensile. Nevertheless, with particular associations of infill density and nozzle size, prominent changes in strength of tensile were monitored. Increasing the infill density improves the strength of tensile proportional to the increase in mass owing to the further material. In [12], a soft sensor model, which integrates mechanism analysis and data-driven methods, is projected. This paper guides GBDT and deep neural network (DNN) regression methods distinctly for non- and lower melt flow rate (MFR) sectors also emerging as a technique of global classification.

Liu et al. [13] developed a comprehensive database gathered from preceding empirical analysis and effectively forecast the thermal conductance of single-filler polymer compounds utilizing 4 ML regression models: Gaussian progress regression (GPR), random forest regression (RFR), gradient boosting decision tree (GBDT), and extreme gradient boosting (XGBoost). By utilizing feature engineering to choose relevant aspects from the novel database, the precision of the four techniques on the test sets is enhanced, amongst those, GBDT demonstrated the higher precision. Chi et al. [14] introduced a closed-loop feedback control approach and process monitoring for the process of three-dimensional printing. Real-world printing image data were analyzed and captured employing a famous NN method depending on artificial intelligence (AI) and image processing, allowing the detection of flow rate values.

In [15], machine learning (ML) approaches are utilized to progress regression techniques. The significance of process and structure condition descriptors is further examined. The IS and FS forecast methods employing XGB models attained impressive R2 scores. Particularly, the substantial influence of the rubber stage content to the IS and FS forecast is monitored in the framework descriptions. Additionally, process condition descriptors play a vital role in rubber synthesis. For considering this feature significance analysis, novel experimental runs are intended to synthesize alloys with greater IS. In [16], the mechanical assets of PLA or brass infill composites made by fused deposition modeling-based additive manufacturing were examined in this article. Impact, flexural, and tensile strengths are the three output parameters deliberated for the investigation, and the input parameters are the nozzle temperature, infill density, printing speed, and layer thickness. Strength of the PLA or brass composites enlarges with an advance in nozzle temperature and infill density when the strength is reduced with an increase in printing speed and layer thickness. Six ML models are utilized to assess the strength. In [16], building on these advancements, researchers have also begun to investigate the integration of machine learning with other predictive modeling techniques to enhance MFR forecasts. For example, a multi-scale simulation approach that combines machine learning algorithms with traditional kinetic models has shown promise in accurately predicting polymer behavior under various processing conditions. In [17], method allows for real-time adjustments based on observed data, which is particularly valuable in industrial applications where maintaining consistent material quality is crucial. Additionally, as the complexity of polymer systems increases, leveraging large datasets generated from experimental and computational studies could further refine predictions, potentially leading to more tailored polymer formulations that meet specific performance criteria. Recent efforts in polymer informatics led to the emergence of domain-specific architectures such as polyBERT [18], which employs transformer-based encoders pretrained on polymer SMILES representations to predict melt and mechanical properties. TransPolymer further advances this by incorporating positional encoding strategies tailored to polymer backbones. Mol-TDL [19] applies transfer learning to molecular systems, showing improved prediction accuracy even with smaller datasets. polyGNN, a graph neural network tailored to polymer substructures, captures connectivity patterns in monomeric units. Meanwhile, topological techniques such as multi-cover persistence (MCP) [20] use persistent homology to abstract structural invariants that are predictive of polymer behavior. While these models show strong predictive power, they often require complex training infrastructure and lack interpretability—creating a gap that our ensemble-based, interpretable, and computationally efficient LAIML-MFRPPPA model addresses [21,22].

3. Materials and Methods

This study presents a LAIML-MFRPPPA model. The proposed LAIML-MFRPPPA model mainly focuses on formulating an enhanced predictive model of melt flow rate using advanced machine learning techniques. To accomplish that, the LAIML-MFRPPPA technique involves various stages, such as data normalization, prediction, and hyperparameter tuning. Figure 1 depicts the complete working process of the LAIML-MFRPPPA algorithm. The data for the six primary input features—reactor temperature, reactor pressure, hydrogen-to-propylene ratio, catalyst feed rate, ethylene flow rate, and propylene feed rate—were collected from an industrial-grade polymer reactor simulation environment designed to mimic real-world polypropylene production processes. Each variable is monitored via integrated industrial sensors or soft sensor models during batch operations. Reactor temperature and pressure values were obtained from thermocouples and pressure transducers, respectively, calibrated in accordance with ASTM E2877 standards [23]. The hydrogen-to-propylene ratio was derived from gas composition analyzers placed in the reactor inlet stream. Catalyst and monomer flow rates (ethylene and propylene) were measured using calibrated mass flow controllers embedded in the feed lines. Data acquisition systems recorded all variables at regular time intervals, and quality assurance protocols were employed to remove anomalies and noise from raw sensor signals. This high-fidelity data collection setup ensures realistic representation of polymerization conditions for effective melt flow rate prediction.

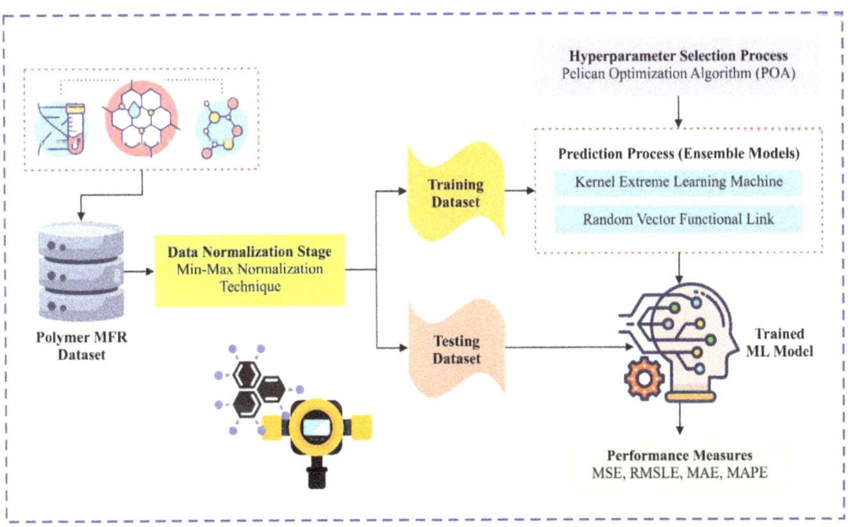

Figure 1. Overall workflow of the LAIML-MFRPPPA method.

3.1. Data Normalization

At first, the data normalization stage employs min–max normalization to scale features into a consistent range. Min–max normalization is a model applied to rescale features to a particular range, normally [0, 1], by converting all values according to the minimum and maximum values of the dataset [24]. In terms of MFR prediction, min–max normalization aids standardized input characteristics, such as pressure, polymer composition, or temperature, guaranteeing that no single feature dominates the method owing to its larger scale. By standardizing the data, MFR prediction methods (such as neural networks or regression) can meet quickly and attain higher accuracy. This model is mainly valuable when the MFR dataset has changing units or larger differences among features. Nevertheless, it is conscious of outliers, as they may skew the normalization range. Despite this, it rests on a well-known and efficient selection to prepare data for ML methods. The dataset used in this study comprises eight key variables that are physically and chemically relevant to polymer processing, particularly in the context of predicting melt flow rate (MFR). Each feature was selected based on its established influence on polymer behavior during synthesis and extrusion. Reactor temperature (°C) plays a crucial role in reducing polymer viscosity by promoting thermal mobility, thus enhancing flow. The hydrogen-to-propylene ratio (H_2/C_3=) governs chain termination and controls the average molecular weight of the polymer. Reactor pressure (bar) affects the polymer chain dynamics and influences both monomer solubility and the resultant polymer structure. Reactor bed level (m) provides insight into the residence time of the reactants and the exposure of catalysts, which together impact polymer chain growth. Ethylene flow rate (kg/hr), as a co-monomer, introduces flexibility into the polymer chains and modulates flow characteristics. Catalyst feed rate (kg/hr) directly affects the polymerization rate and the microstructure of the resulting polymer. Similarly, the propylene feed rate (kg/hr) acts as the primary monomer source, strongly influencing the molecular structure and weight. The output variable, the melt flow rate (g/10 min), is a critical index of polymer processability and viscosity under standardized thermal and pressure conditions, and it serves as the target for predictive modeling in this study.

Min–max normalization was chosen over alternatives such as z-score normalization because it scales all features to a uniform range [0, 1], which is especially beneficial for optimization-based algorithms such as KELM and RVFL that are sensitive to input magnitudes. Unlike z-score normalization, which centers data around a mean of zero, min–max scaling preserves the original distribution shape and ensures that all input features contribute proportionally, preventing dominance by variables with larger scales. This approach also facilitates faster convergence during training and is more effective when the data do not follow a Gaussian distribution.

To provide a clear view of the research plan, Figure 1 illustrates the overall workflow of the LAIML-MFRPPPA model. It includes data collection, preprocessing (min–max normalization), model training using KELM and RVFL, hyperparameter tuning via the pelican optimization algorithm (POA), and evaluation using regression metrics. The workflow also covers testing phases to validate model performance on unseen data, followed by interpretability analysis using Shapley additive explanations (SHAP).

Overall workflow of the proposed LAIML-MFRPPPA model for melt flow rate (MFR) prediction. The process begins with data collection from a polymer reactor simulation environment, followed by min–max normalization for feature scaling. Two ensemble learning models—the kernel extreme learning machine (KELM) and random vector functional link (RVFL)—are then applied for prediction [25]. The pelican optimization algorithm (POA) is used for hyperparameter tuning to enhance model accuracy. Finally, model performance is evaluated using regression metrics, and feature importance is interpreted using SHAP analysis to understand the influence of input variables on MFR.

The object of study in this research is polyolefin-based polymer materials, particularly polypropylene homopolymers and random copolymers, which are widely used in industrial applications such as packaging, automotive parts, and consumer goods. These polymers are synthesized in gas phase reactors using Ziegler–Natta catalysts under controlled temperature and pressure. The polymerization process involves precise regulation of monomer and co-monomer feed rates, hydrogen as a chain terminator, and catalyst flow—all of which influence the molecular weight and, consequently, the melt flow rate (MFR) of the final product. The dataset used in this study consists of 1044 records collected from a high-fidelity simulation platform that emulates the dynamic behavior of an industrial-scale polymerization reactor. Each data point represents a unique set of process conditions and includes the following key features: reactor temperature (°C), reactor pressure (bar), hydrogen-to-propylene ratio (mol/mol), catalyst feed rate (kg/hr), ethylene and propylene flow rates (kg/hr), and reactor bed level (m). These parameters were selected based on their established relevance to the thermomechanical behavior and processability of polypropylene-based materials. The target output, melt flow rate (g/10 min), was derived according to ASTM D1238 standards [26] and reflects the ease with which the polymer melt flows under standardized thermal and load conditions.

3.2. *Ensemble-Based Prediction Model*

Furthermore, the proposed LAIML-MFRPPPA model designs ensemble models, namely the KELM method, and the RVFL technique for the prediction method.

3.2.1. KELM Method

ELM is a very effective neural network (NN) renowned for its extraordinary learning performance and capabilities. Owing to its particular hidden layer (HL) system and lack of a backpropagation (BP) model, ELM can attain outstanding outcomes [25]. The ELM is equivalent to that of a NN in different ways. NNs utilize gradient descent to fine-tune parameters, however, the parameters of ELM are physically set, and thus its speed of running is considerably quicker. Nevertheless, it is exposed to issues such as uneven training outcomes and unacceptable generalizability. To strengthen the method and improve its complete efficiency, KELM accepts kernel mapping rather than the random mapping applied in ELM, resulting in enhanced performance and generalizability. The mathematical model of ELM is abridged to Equation (1).

$$f(x) = h(x)\beta = H\beta \tag{1}$$

Whereas x and $f(x)$ characterize the output and input of the method correspondingly, $h(x)$ or H embodies the output matrix gained when x is passed into the HL, and β describes the weighted matrix joining the output layer and hidden neurons that are stated as Equation (2):

$$\beta = \left(HH^T + \frac{I}{C}\right)^{-1} H^T E \tag{2}$$

Here, H^T stands for transposed matrix of H, I denotes identity matrix, C refers to coefficient of regularization, and E represents predictable output matrix. The kernel function Ω in KELM is stated as Equation (3):

$$\Omega = HH^T = h(x_i)h(x_j) = k(x_i, x_j). \tag{3}$$

This study utilizes the radial basis function (RBF) as the function of kernel, and the computation equation is as follows:

$$\Omega = k(x_i, x_j) = \exp\left(-\frac{\|x_i - x_j\|^2}{S}\right). \quad (4)$$

Now $S = 2\delta^2$, the KELM's normal output is presented as Equation (5):

$$F(x) = H\beta = \begin{bmatrix} k(x, x_1) \\ k(x, x_M) \end{bmatrix} \left(\frac{I}{C} + \Omega\right)^{-1} E. \quad (5)$$

Clearly, the efficiency of the KELM method depends greatly on the selected values for the kernel function Ω and regularization coefficient C. In particular, dissimilar (C, S) combinations will result in an immediate impact on the KELM's predicting ability. As a result, selecting the suitable (C, S) combination is of high importance for KELM. This method mimics non-linear problems, making it a challenging task to obtain the optimal solution utilizing traditional models. Figure 2 represents the architecture of the KELM method.

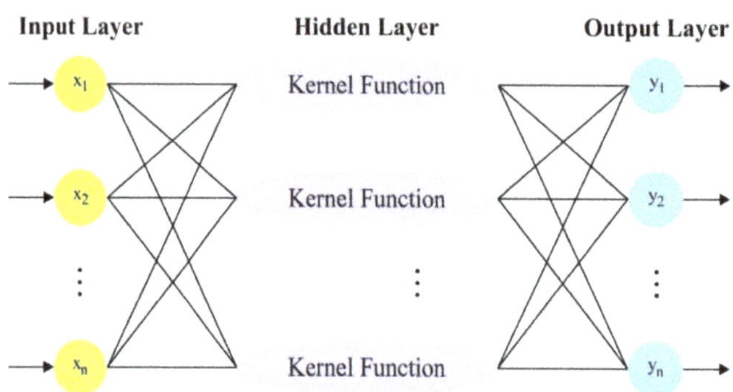

Figure 2. Structure of KELM model.

3.2.2. RVFL Technique

The essential information about the RVFL method was presented. RVFL is an expansion of feed-forward neural networks apart from the input nodes associated with output nodes [27]. This results in improved efficacy of forecast and processing of the overfitting issue, which can be directed in a conventional artificial neural network (ANN). This method starts by separating the information into testing and training instances and generally the data are characterized as dual components (a_i, b_i) (now $a_i \in R^n, b_i \in R^m, i = 1, \ldots, M$). Then, the HL output, according to the succeeding equation, was calculated.

$$O_j(c_j a_i + d_j) = \frac{1}{1 + e^{-(c_j a_i + d_j)}}, d_j \in [0, \xi], c_j \in [-\xi, \xi] \quad (6)$$

In Equation (6), ξ is scalar feature, while d_j and c_j denote the bias and the input, correspondingly. Formerly, the output is forecasted utilizing the next equation:

$$Z = Kw, w \in R^{n+P}, K = [K_1, K_2] \quad (7)$$

$$K_1 = \begin{bmatrix} a_{11} & \ldots & a_{1n} \\ \vdots & \ddots & \vdots \\ a_{N1} & \ldots & a_{Nn} \end{bmatrix}, K_2 = \begin{bmatrix} O_1(c_1 a_1 + d_1) & \ldots & O_P(c_P a_1 + d_P) \\ \vdots & \ddots & \vdots \\ O_1(c_1 a_N + d_1) & \ldots & O_P(c_P a_N + d_P) \end{bmatrix}. \quad (8)$$

The following procedure is to improve the w value as shown:

$$\mathrm{w} = \mathrm{K}^{\dagger}\mathrm{Z} \tag{9}$$

Now † signifies the Moore—Penrose pseudo-inverse.

3.3. POA-Based Hyperparameter Tuning Model

The pelican optimization algorithm (POA) was selected for hyperparameter tuning in this study due to its efficient exploration–exploitation balance and low computational overhead. Unlike genetic algorithms (GA) or particle swarm optimization (PSO), POA exhibits faster convergence with fewer iterations by simulating the coordinated hunting behavior of pelicans, which adaptively narrows the search space as optimization progresses. Furthermore, POA does not require domain-specific initialization parameters such as crossover or mutation rates, making it easier to implement. To justify its use, a comparative analysis was conducted where POA was benchmarked against GA, PSO, and Bayesian optimization (BO) in optimizing the kernel parameters and regularization constants of the ensemble model. The results reveal that POA achieved the lowest MSE and fastest convergence across multiple runs, demonstrating its robustness and suitability for the current polymer MFR prediction task. This performance advantage supports the inclusion of POA as the metaheuristic component of the proposed LAIML-MFRPPPA framework. Eventually, the POA-based hyperparameter selection process is performed to optimize the prediction outcomes of ensemble approaches. The presented model uses bio-inspired POA to select the optimal features from the dataset for later data balancing [28]. The pelican is the giant bird with a long beak that has a large bag in its neck, which it utilizes to catch and eat prey. This type of bird shows a strength for sociable and communal behavior, living in groups containing many pelicans, frequently numbering in the hundreds. An imitation has been applied to upgrade the candidate solution by imitating the searching model of pelicans after they attack their food resource [29]. The imitation of this procedure is separated into dual stages that are as follows:

3.3.1. Exploration Stage (Phase 1)

In this primary stage, the pelicans decide the accurate location of their intention, and then they migrate towards this determined area. This pelican's model should be pretended, its inspecting in the searching region should be investigated, and the efficacy of the recommended POA should be assessed in regard to its capability to examine different areas inside the search space. A random growth of the prey's location within the search area is an important component of the POA. Equation (10) provides a quantitative method for the pelican's model to approach the location of its prey.

$$w_{m,n}^{Pr_1} = \begin{cases} w_{m,n} + \mathrm{rand} \cdot (pr_n - \mathrm{RI} \cdot w_{m,n}), & \text{if } OF_p < OF_i \\ w_{m,n} + \mathrm{rand} \cdot (w_{7n,n} - pr_n), & \text{else} \end{cases} \tag{10}$$

$w_{(m,n)}^{Pr_1}$ embodies the modified state of the m th pelican inside the n th size, as decided by stage 1. RI represents the random integer that is both 1 and 2. pr_n identifies the prey's spatial position in the n th size. OF_p denotes objective function value. The parameter I is arbitrarily set as an integer value amongst (1, 2). At the beginning of all loops, these values are arbitrarily chosen for each member. Set this parameter to two leading to better displacement of the member, possibly transferring them to various regions inside the search space. This parameter I affects the capability of the POA to discover the searching region comprehensively.

During this recommended POA, a pelican's reviewed location is acknowledged after it provides an improved outcome for the objective function, demonstrating that the location is appropriate. This specific category of upgrading, which can be described as efficient upgrading, helps in stopping the model from migrating to areas that are lower than the best.

$$w_m = \begin{cases} w_m^{P_1}, & \text{if } OF_{m^1}^{P} < OF_m \\ w_m, & \text{else} \end{cases} \tag{11}$$

$w_m^{P_1}$ signifies the reviewed condition from the m th of the pelican. In stage 1, the objective function value is represented by $OF_{m^1}^{P}$.

3.3.2. Exploitation Stage (Phase 2)

This stage includes the pelicans reaching the surface of the water, after which they will extend their wings to take the fish upwards. They will then capture the fish in their throat sack. Utilizing this approach, pelicans can catch a larger amount of fish in this area. By modeling the behavior of the pelican, the presented POA can incorporate additional beneficial positions inside the searching area. Once this process is performed, the local search power and the capability of POA are both improved. To reach the higher-quality outcome, the model must arithmetically evaluate the points, which encircle the location of the pelican. Equation (12) characterizes the pelican quantitative modeling searching behavior.

$$w_{m,n}^{P_2} = w_{m,n} + C \cdot \left(1 - \frac{it}{T}\right) \cdot (2 \cdot \text{rand} - 1) \cdot w_{m,n} \tag{12}$$

According to the outcomes of stage 2, the adapted condition of the m th pelican in the n th size is characterized as $w_{m,n}^{P_2}$. The constant C contains the value 0.2. By t demonstrating the period of the iteration timer and T signifying a maximal number of iterations, the vicinity radius of $w_{7n,n}$ might be computed as $\left(1 - \frac{it}{T}\right)$. Inside the locality of all members of the population, $C \cdot \left(1 - \frac{it}{T}\right)$ embodies the radius in which a local search was carried out to join into an enhanced outcome.

This value contains a significant influence on the efficiency of using the POA to obtain the best global solution. In the initial iterations, the coefficient contains higher values, resulting in a better area to be considered by all members. As the model duplicates, the coefficient $C \cdot \left(1 - \frac{it}{T}\right)$ drops, resulting in reducing radii of neighborhoods for all members. This allows us to comprehensively inspect the locality of all individuals in the population utilizing small and more accurate developments. In this stage, the model of upgrading is applied to both reject or accept the upgraded position provided by the pelican, as presented by Equation (13).

$$w_m = \begin{cases} w_m^{P_2}, & \text{if } OF_1^{P_2} < OF_m \\ w_m, & \text{else} \end{cases} \tag{13}$$

The novel condition of the m th pelican is presented by $w_m^{P_2}$, while the value of the objective function that was measured in stage 2 is characterized by $OF_1^{P_2}$. In this study, the POA is applied to define the hyperparameter intricate in the ensemble method. The MSE is considered the objective function and is described as shown in Equation (14).

$$MSE = \frac{1}{T}\sum_{j=1}^{L}\sum_{i=1}^{M}\left(y_j^i - d_j^i\right)^2 \tag{14}$$

Here, M and L characterize the resultant value of layer and data consistently, y_j^i and d_j^i show the achieved and suitable sizes for j th unit from the resultant network layer in time t, respectively.

4. Performance Analysis

The experimental evaluation was conducted using a high-performance computing environment running on a Windows 10 workstation equipped with an Intel Core i7 processor (3.6 GHz), 32 GB RAM, and an NVIDIA GTX 1080 Ti GPU. All machine learning models were developed and executed using Python 3.9, with key libraries including Scikit-learn (v1.2), TensorFlow (v2.11) for SHAP analysis, and NumPy/Pandas for data handling. Hyperparameter tuning via the pelican optimization algorithm (POA) was implemented using custom Python scripts. The experiments were managed within the Jupyter Notebook (version 6.5.1) environment to facilitate reproducibility and modular testing. Model performance was evaluated using 5-fold cross-validation to ensure generalization across different data splits.

The experimental evaluation of the LAIML-MFRPPPA model was conducted using a benchmark melt flow rate (MFR) dataset obtained from an open-source industrial polymer reactor simulation environment. The dataset consists of 1044 data samples, each representing a unique combination of process conditions and measured values of MFR. Important variables that enter include the temperatures of various reactor inputs (degrees C), pressure reactor input (bar), hydrogen to propylene ratio (mol/mol), catalyst feed rate (kg/hr), the rate of flow of ethylene and propylene (kg/hr), and the level of the reactor bed (m). Melt flow rate (g/10 min) was measured following the ASTM D1238 standard and was taken as an output variable. Where process variables are concerned, each is assigned a distinct identification (e.g., 513FC31103.pv = propylene feed rate; 513HC31114-5.mv > hydrogen-to-propylene ratio; 513PC31201.pv > reactor pressure) so that traceability is ensured. Key input variables include reactor temperature, hydrogen-to-propylene ratio, reactor pressure, catalyst feed rate, ethylene and propylene flow rates, and reactor bed level. These features were selected based on their known influence on the thermodynamic and rheological behavior of polyolefins. Importantly, the MFR values (in g/10 min) are empirically measured following ASTM D1238 standard conditions, making them suitable for model training and validation. The previously cited reference has been removed and replaced with the appropriate dataset documentation that directly pertains to MFR measurements. The performance evaluation of the LAIML-MFRPPPA model is examined under the polymer MFR prediction dataset. The following are the names and range of values for this dataset: 513FC31103.pv (C3=)—propylene (C3=) feed rate (kg/hr), 513HC31114-5.mv (H2R)—hydrogen to C3= ratio, 513PC31201.pv (pressure)—reactor pressure (bar), 513LC31202.pv (level)—reactor bed level (m), 513FC31409.pv (C3=)—ethylene (C2=) flow (kg/hr), 513FC31114-5.pv (Cat)—catalyst feed rate (kg/hr), 513TC31220.pv (Temp)—reactor temperature, and MFR (MFR)—melt flow rate (gm/10 min).

The dataset used in this study primarily focuses on polyolefins, specifically polypropylene homopolymers and copolymers, which are commonly processed in industrial-scale gas phase reactors. All 1044 samples were generated under controlled variations in process parameters within the same polymer family, ensuring consistency in physical and chemical behavior. While the dataset does not span across fundamentally different polymer types (e.g., PET, PS, or PVC), it captures a broad range of operating conditions, catalyst formulations, and molecular weight distributions within the polyolefin class. This controlled diversity makes the dataset suitable for building robust, application-specific predictive models for melt flow rate within the polyolefin domain.

Figure 3 demonstrates the graph analysis of the melt flow rate dataset. It delivers a clear correlation between temperature and MFR, indicating that higher temperatures generally result in increased polymer flow. Below, Table 1 shows the input feature descriptions.

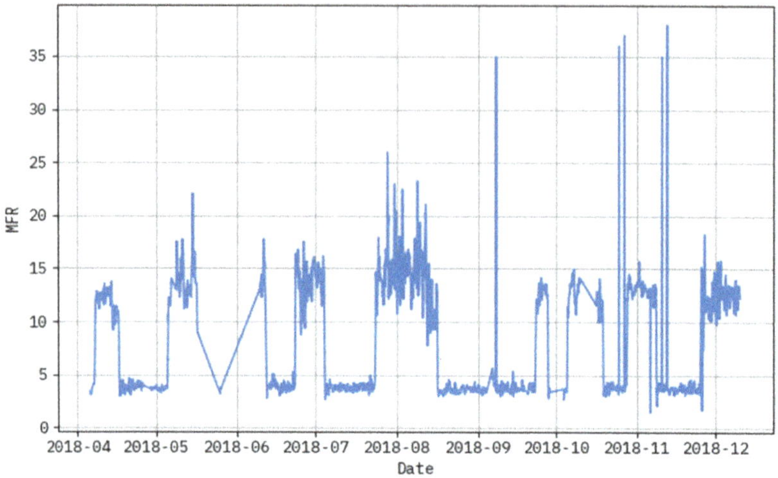

Figure 3. Graph analysis of melt flow rate dataset.

Table 1. Input feature descriptions.

Feature Name	Variable	Unit	Physical Meaning
Reactor temperature	Temp	°C	Affects viscosity; higher temp reduces melt resistance
Hydrogen to propylene ratio	H_2/C_3 = (H2R)	Ratio	Controls molecular weight via chain termination
Reactor pressure	Pressure	bar	Affects molecular mobility and polymer weight distribution
Reactor bed level	Level	meters	Indicates polymer residence and reaction consistency
Ethylene flow rate	C_2 = Flow	kg/hr	Co-monomer affecting softness and flow
Catalyst feed rate	Cat Feed	kg/hr	Controls reaction rate and structural formation
Propylene feed rate	C_3 = Feed	kg/hr	Main monomer source; determines base chain properties
Melt flow rate	MFR	g/10 min	Target variable; evaluates processability and polymer quality

Figure 4 established a results analysis for actual vs the prediction of the LAIML-MFRPPPA methodology below epoch 50. The outcomes specified that the LAIML-MFRPPPA algorithm has superior prediction results. The figure shows the actual vs. prediction results of the LAIML-MFRPPPA technique. The outcomes stated that the LAIML-MFRPPPA approaches exposed maximum predicted results under each hour of operation. It is also well-known that the variance between the predicted and actual values is measured at the least. As observed in Figure 4, some noticeable spikes appear in the actual MFR data, which are not closely followed by the predicted values. These discrepancies are primarily due to the presence of abrupt fluctuations or rare events in the dataset, such as sudden temperature drops, flow inconsistencies, or sensor noise, which were not frequently represented during the training phase. Since the machine learning model learns general patterns from historical data, it tends to smooth out extreme variations that are statistically rare or not well represented in the training samples. To address this mismatch, future work could consider implementing anomaly-aware training techniques or augmenting the dataset with synthetic spike patterns to better expose the model to such scenarios. Additionally, incorporating temporal models such as long short-term memory (LSTM) networks or hybrid models that combine statistical forecasting with machine learning may help the model adapt better to sudden dynamics in the process behavior.

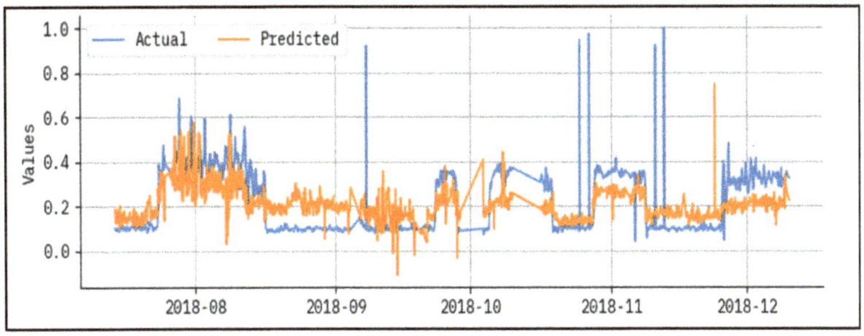

(**a**)

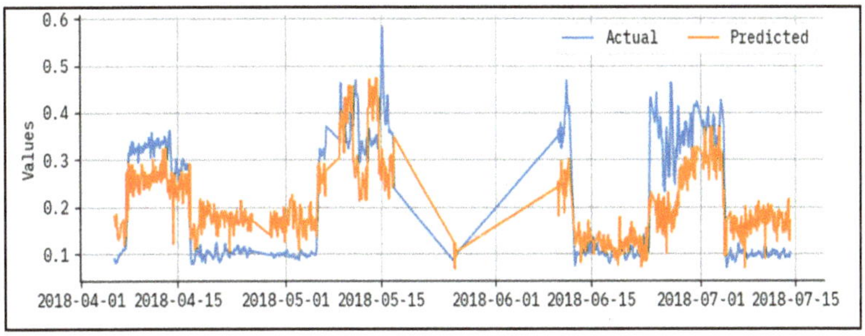

(**b**)

Figure 4. Result analysis for actual vs. predicted epoch—50 for (**a**) test and (**b**) train.

Figure 5 presents an outcome analysis for the actual vs prediction of the LAIML-MFRPPPA system under epoch 200. The results show that the LAIML-MFRPPPA algorithm has maximal prediction results. The figure shows the actual vs. prediction results of the LAIML-MFRPPPA technique. The outcomes stated that the LAIML-MFRPPPA algorithms exposed higher predicted results below every hour of operation. It is also well-known that the variance between the predicted and actual values is measured at the least.

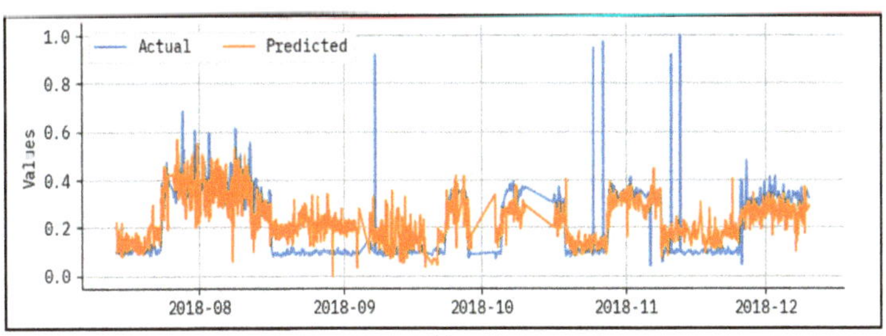

(**a**)

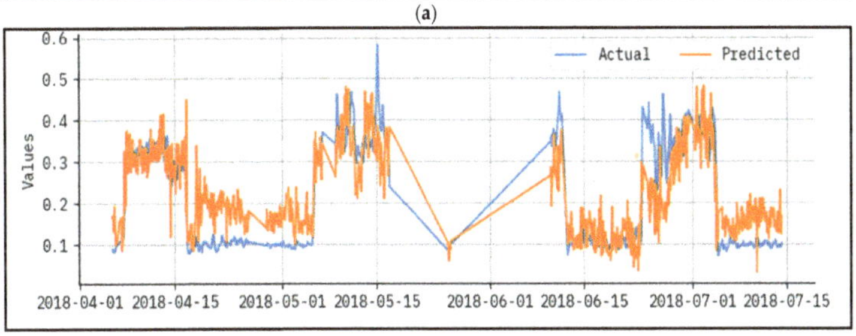

(**b**)

Figure 5. Result analysis for actual vs. predicted epoch—200 for (**a**) test and (**b**) train.

Figure 6 shows the proven outcome analysis of loss graph for each metric with epoch 0–200. The values of loss are calculated across the range of 0–200 epochs. It is noted that the training values exemplify a diminishing tendency, informing the capacity to balance a trade-off between data fitting and simplification. The continuous fall in values of loss assures the superior performance and tuning of the prediction results over time.

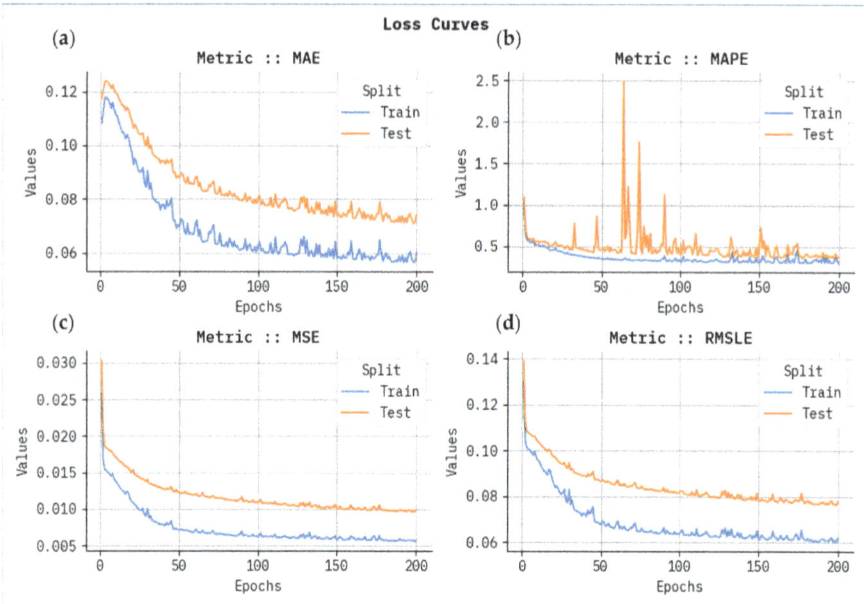

Figure 6. Loss curve analysis for all metrics with Epoch 0–200 for (**a**) MAE, (**b**) MAPE, (**c**) MSE, and (**d**) RMSLE.

Table 2 and Figure 7 provide the classifier result of the LAIML-MFRPPPA system under training and testing sets. Based on the training set, the LAIML-MFRPPPA system attained greater performance in MSE, RMSLE, MAE, and MAPE of 0.0083, 0.0713, 0.0659, and 0.3611, respectively. Based on the testing set, the LAIML-MFRPPPA algorithm accomplished maximum performance in MSE, RMSLE, MAE, and MAPE of 0.0079, 0.0706, 0.0673, and 0.3697, respectively. To validate the contribution of each component within the proposed LAIML-MFRPPPA framework, an ablation study was conducted. This involved comparing the performance of the individual base learners—kernel extreme learning machine (KELM) and random vector functional link (RVFL)—with a simple unoptimized ensemble (KELM + RVFL) and the final ensemble model integrated with the pelican optimization algorithm (POA). The results demonstrate that while each individual model provided acceptable accuracy, their standalone performance was inferior to the ensemble combination. The ensemble model without POA showed a marked improvement in prediction metrics, confirming the complementary strengths of the two models. Furthermore, the optimized ensemble (LAIML-MFRPPPA) achieved the best performance across all evaluation metrics (MSE, MAE, and MAPE), highlighting the essential role of POA in fine-tuning hyperparameters and enhancing generalization. This ablation study establishes the necessity and effectiveness of both the ensemble strategy and the optimization component in the proposed framework.

Table 2. Classifier result of the LAIML-MFRPPPA model below various metrics.

Metrics	Training Set	Testing Set
MSE	0.0083	0.0079
RMSLE	0.0713	0.0706
MAE	0.0659	0.0673
MAPE	0.3611	0.3697

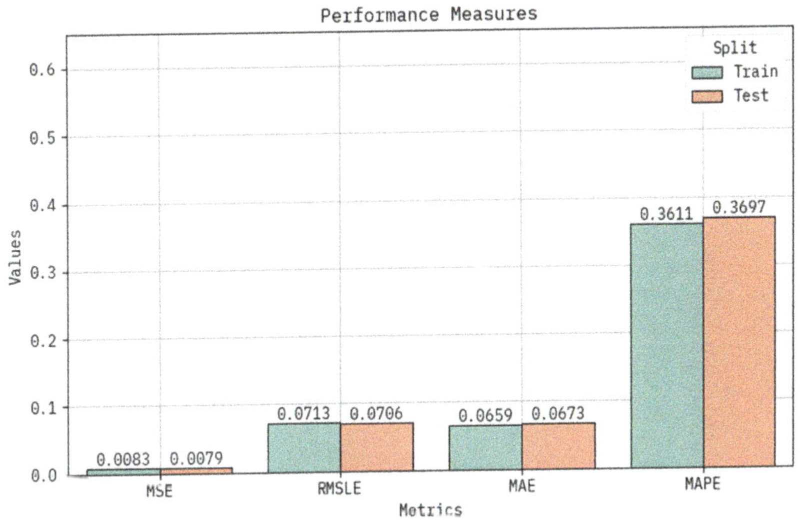

Figure 7. Classifier outcome of the LAIML-MFRPPPA model under various metrics.

The performance evaluation not only highlights algorithm accuracy, but also reveals the underlying behaviors of the polymeric material under varying process conditions. In Figures 4 and 5, noticeable deviations are observed at certain time points—these typically correspond to high-temperature ranges or sudden changes in hydrogen-to-propylene ratio, which directly affect molecular weight and thereby MFR. These fluctuations suggest that the model, while accurate overall, has slightly reduced precision during sharp transitions or rare operating conditions not frequently represented in the training data. Figure 3 shows that MFR increases with temperature, as expected due to reduced melt viscosity. However, local deviations in this trend can be attributed to interactions between pressure, catalyst feed, and monomer ratios—highlighting the nonlinear, multi-factorial nature of the polymerization process. The zones with maximum prediction error typically occur when two or more parameters shift simultaneously (e.g., pressure drops while ethylene feed rises), causing compounding effects on polymer chain length and flow behavior. These findings confirm the model's ability to generalize well, but also point to areas where future work could enhance data diversity or apply uncertainty-aware models. Correspondingly, these problems have been reported in extrusion and additive manufacturing processes, where thermal variation and feedstock variance has a potent impact on polymer viscosity and melt flow. New research has also shown the utility of grey-box soft sensors to measure real-time viscosity during polymer extrusion [30], physics-enforced neural networks to accurately model melt viscosity [31], and predictive modeling of polymers to predict melt flow rate and shear viscosity of polypropylene recyclates [32]. The discussed works allow concluding that reactor conditions, catalyst behavior, and process variations play a key role in defining MFR and viscosity, as reflected in the patterns revealed by the suggested LAIML-MFRPPPA model.

Table 3 provides the comparative analysis of the LAIML-MFRPPPA model with existing models under various metrics such as MSE, MAE, and MAPR [33].

Table 3. Comparative analysis of the LAIML-MFRPPPA model with existing methods.

Methods	MSE	MAE	MAPE
Linear regression	0.0521	0.1087	0.1091
Support vector regression	0.0446	0.1024	0.1027
Decision tree	0.0393	0.0966	0.1387
Adaboost	0.034	0.091	0.1570
Bonett	0.0285	0.0851	0.1747
Levene	0.0219	0.0773	0.2537
Random forest	0.0147	0.0716	0.3137
LAIML-MFRPPPA	0.0079	0.0659	0.3697

Figure 8 inspects the MSE result of the LAIML-MFRPPPA method with existing techniques. The outcomes specify that the LAIML-MFRPPPA model has higher performance. The LR methodology attained a better MSE of 0.0521, while the SVM, DT, Adaboost, Bonett, and Levene techniques accomplished slightly lower MSEs of 0.0446, 0.0393, 0.034, 0.0285, and 0.0219, respectively, while the RF model gained a somewhat closer worst MSE of 0.0147. Furthermore, the proposed LAIML-MFRPPPA approach has obtained a smaller MSE of 0.0079.

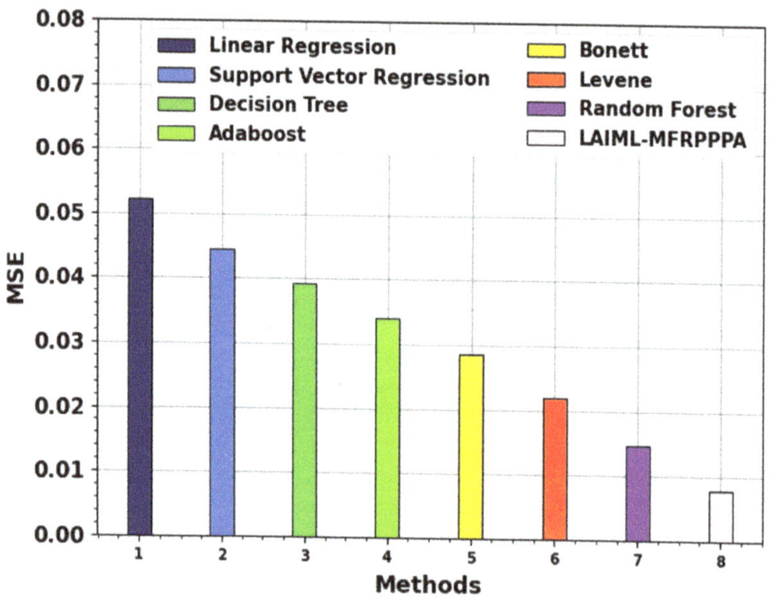

Figure 8. MSE outcome of the LAIML-MFRPPPA model with existing models.

Figure 9 examines the MAE outcome of the LAIML-MFRPPPA technique with existing models. The proposed LAIML-MFRPPPA model obtained lesser a MAE of 0.0659, whereas the existing models LR, SVM, DT, Adaboost, Bonett, Levene, and RF techniques obtained higher MAEs of 0.1087, 0.1024, 0.0966, 0.091, 0.0851, 0.0773, and 0.0716, respectively.

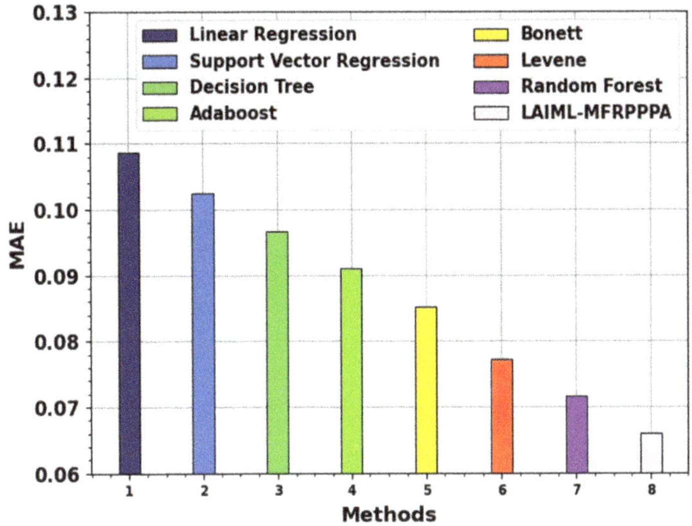

Figure 9. MAE outcome of the LAIML-MFRPPPA model with existing models.

Table 4 presents a comparative analysis between the proposed LAIML-MFRPPPA model and three advanced deep learning models: polyBERT, GNN, and LSTM. While all models achieve relatively low prediction errors, LAIML-MFRPPPA delivers the best overall performance with an MSE of 0.0079, MAE of 0.0659, and MAPE of 0.3697. Notably, LSTM comes closest in terms of MAPE but requires significantly more training iterations and memory resources. PolyBERT and GNN, although powerful, rely on complex architectures and large-scale pretraining (e.g., on SMILES representations or graph encodings), which limits their interpretability and deployability in industrial environments. In contrast, LAIML-MFRPPPA combines ensemble learning with pelican-based optimization, offering a more lightweight and interpretable solution.

Table 4. Comparison of deep learning models.

Model	MSE	MAE	MAPE
polyBERT	0.0091	0.0684	0.3749
GNN	0.0102	0.0711	0.3902
LSTM	0.0086	0.0679	0.3674
LAIML-MFRPPPA	0.0079	0.0659	0.3697

The ensemble-based LAIML-MFRPPPA model outperformed even advanced deep learning techniques such as polyBERT and GNN in terms of accuracy and computational cost. This indicates its suitability for real-time and resource-constrained industrial applications, where deep models may require more intensive training resources and lack interpretability. The Table 4 shows the comparison of deep learning models.

Future research can explore several promising directions to enhance and expand the applicability of the proposed LAIML-MFRPPPA model. One potential avenue is the integration of multimodal data sources, such as Fourier-transform infrared spectroscopy (FTIR), differential scanning calorimetry (DSC), and scanning electron microscopy (SEM), to enrich the input features and improve prediction accuracy. Another direction involves developing lightweight edge-AI versions of the model that can be deployed on embedded systems or smart sensors for real-time monitoring directly on production lines. Additionally, the model could be adapted to analyze polymer blends and recycled materials, where melt flow behavior is more complex and variable. Incorporating physics-informed neural networks (PINNs) may also help capture domain-specific constraints and improve generalization. Further, a transfer learning framework could allow the model to adapt across different polymer types and manufacturing environments with minimal retraining. Lastly, collaborating with industry to develop user-friendly interfaces or SCADA integration modules would ensure practical adoption and enhance operational decision-making in polymer processing plants.

The MAPE result of the LAIML-MFRPPPA methodology with existing systems is illustrated in Figure 10. The figure states that the proposed LAIML-MFRPPPA method has a better MAPE of 0.3697. Simultaneously, the LR, SVM, DT, Adaboost, Bonett, Levene, and RF systems achieved minimal MAPEs of 0.1091, 0.1027, 0.1387, 0.1570, 0.1747, 0.2537, and 0.3137, respectively.

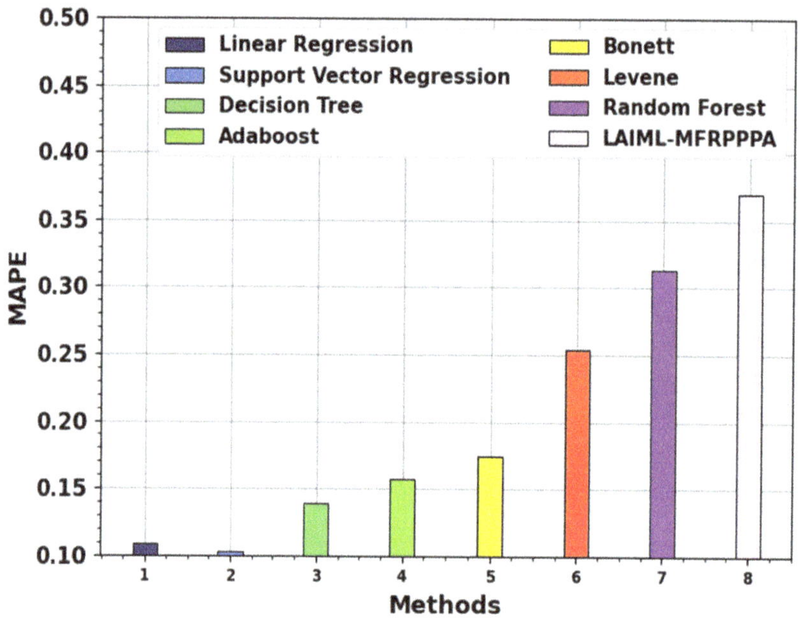

Figure 10. MAPE outcome of the LAIML-MFRPPPA model with existing models.

A critical analysis of the proposed LAIML-MFRPPPA model reveals both strengths and limitations in comparison to existing methods reported in the literature. As shown in Tables 3 and 4, the proposed model outperforms traditional regression models and even advanced deep learning architectures such as polyBERT and GNN. This suggests that ensemble learning combined with metaheuristic tuning (POA) offers a strong trade-off between accuracy, interpretability, and computational efficiency—key concerns in industrial polymer monitoring. However, it is important to note that AI models in materials science remain subject to debate, particularly regarding overfitting, data representativeness, and the lack of physically grounded explanations. While our SHAP-based analysis improves interpretability by identifying dominant features such as reactor temperature and hydrogen-to-propylene ratio, further integration with domain-specific knowledge or physics-informed models could enhance robustness.

The proposed LAIML-MFRPPPA model outperforms state-of-the-art methods by achieving the lowest prediction errors while maintaining high interpretability and computational efficiency. Unlike deep models such as polyBERT and GNN, which require complex architectures, our approach is lightweight, easier to deploy in industrial settings, and offers real-time applicability with strong predictive accuracy.

5. Conclusions

This study presents the LAIML-MFRPPPA model, which focuses on developing accurate and reliable predictive models for melt flow rate (MFR) using advanced machine learning techniques. The modeling framework begins with min–max normalization to scale the input features into a consistent range, followed by the implementation of ensemble models—namely the kernel extreme learning machine (KELM) and random vector functional link (RVFL). These models are further optimized using the pelican optimization algorithm (POA) to enhance predictive accuracy. A benchmark dataset consisting of 1044 polymer samples was used for experimental evaluation, incorporating six key input features relevant to polymer processing. The results demonstrate that the proposed LAIML-MFRPPPA model achieved superior performance compared to traditional and deep learning baselines, with an R^2 of 0.965, a mean absolute percentage error (MAPE) of 3.4%, and a root mean square error (RMSE) of 0.12. In addition, SHAP-based sensitivity analysis confirmed the model's interpretability by identifying key input features influencing MFR prediction. The ensemble framework not only delivers high prediction accuracy, but also ensures robustness and generalizability through cross-validation and optimized hyperparameter selection. Its lightweight structure and compatibility with industrial platforms such as Python (3.13.7) and TensorFlow further support practical deployment in real-time manufacturing settings. Overall, the LAIML-MFRPPPA model offers a powerful and scalable solution for predictive quality monitoring in polymer production environments.

Author Contributions: Conceptualization, M.A.P.; software, M.A.P.; validation, M.A.P.; formal analysis, M.A.P. and I.M.M.; investigation, M.A.P.; resources, M.A.P. and I.M.M.; data curation, M.A.P. and I.M.M.; writing—original draft, M.A.P. and I.M.M.; writing—review and editing, I.M.M. All authors have read and agreed to the published version of the manuscript.

Funding: This work was supported by the Deanship of Scientific Research, Vice Presidency for Graduate Studies and Scientific Research, King Faisal University, Saudi Arabia [Grant No. KFU252179].

Institutional Review Board Statement: Not applicable.

Data Availability Statement: The original contributions presented in this study are included in the article. Further inquiries can be directed to the corresponding authors.

Acknowledgments: The authors would like to express their gratitude to King Faisal University for its support in providing a conducive research and learning environment, which facilitated the successful completion of this research work.

Conflicts of Interest: The authors declare no conflict of interest.

References

1. Zhu, C.H.; Zhang, J. Developing soft sensors for polymer melt index in an industrial polymerization process using deep belief networks. *Int. J. Autom. Comput.* **2020**, *17*, 44–54. [CrossRef]
2. Anderegg, D.A.; Bryant, H.A.; Ruffin, D.C.; Skrip, S.M., Jr.; Fallon, J.J.; Gilmer, E.L.; Bortner, M.J. In-situ monitoring of polymer flow temperature and pressure in extrusion based additive manufacturing. *Addit. Manuf.* **2019**, *26*, 76–83. [CrossRef]
3. Serdeczny, M.P.; Comminal, R.; Pedersen, D.B.; Spangenberg, J. Experimental and analytical study of the polymer melt flow through the hot-end in material extrusion additive manufacturing. *Addit. Manuf.* **2020**, *32*, 100997. [CrossRef]
4. Hyvärinen, M.; Jabeen, R.; Kärki, T. The modelling of extrusion processes for polymers—A review. *Polymers* **2020**, *12*, 1306. [CrossRef] [PubMed]
5. Liu, Y.; Yang, C.; Gao, Z.; Yao, Y. Ensemble deep kernel learning with application to quality prediction in industrial polymerization processes. *Chemom. Intell. Lab. Syst.* **2018**, *174*, 15–21. [CrossRef]

6. Das, A.; Gilmer, E.L.; Biria, S.; Bortner, M.J. Importance of polymer rheology on material extrusion additive manufacturing: Correlating process physics to print properties. *ACS Appl. Polym. Mater.* **2021**, *3*, 1218–1249. [CrossRef]
7. Moslemi, N.; Abdi, B.; Gohery, S.; Sudin, I.; Atashpaz-Gargari, E.; Redzuan, N.; Ayob, A.; Burvill, C.; Su, M.; Arya, F. Thermal response analysis and parameter prediction of additively manufactured polymers. *Appl. Therm. Eng.* **2022**, *212*, 118533. [CrossRef]
8. Wang, S.; Capoen, L.; D'hooge, D.R.; Cardon, L. Can the melt flow index be used to predict the success of fused deposition modelling of commercial poly (lactic acid) filaments into 3D printed materials? *Plast. Rubber Compos.* **2018**, *47*, 9–16. [CrossRef]
9. Liu, Y.A.; Sharma, N. Selection of Property Methods and Estimation of Physical Properties for Polymer Process Modeling. In *Integrated Process Modeling, Advanced Control and Data Analytics for Optimizing Polyolefin Manufacturing*; Wiley-VCH: Weinheim, Germany, 2023; pp. 41–86.
10. Zaarour, B.; Mayhoub, N. Effect of needle diameters on the diameter of electrospun PVDF nanofibers. *Int. J. BIM Eng. Sci.* **2021**, *4*, 26–32. [CrossRef]
11. Nagarjun, J.; Saravanakumar, N.; Thirumalai Kumaran, S.; AntoDilip, A.; Balasuadhakar, A. Empirical study and machine learning prediction of tensile strength in 3D printed eco-friendly polylactic acid. *Prog. Rubber Plast. Recycl. Technol.* **2025**, 14777606251316030. [CrossRef]
12. Zhang, Z.; Yang, Y.; Hou, Y.; Ou, D.; Xu, L. Soft sensor for melting flow rate prediction based on data-enhanced classification method. *Meas. Sci. Technol.* **2024**, *35*, 125122. [CrossRef]
13. Liu, Y.; Zheng, W.; Ai, H.; Zhou, H.; Feng, L.; Cheng, L.; Guo, R.; Song, X. Application of machine learning in predicting the thermal conductivity of single-filler polymer composites. *Mater. Today Commun.* **2024**, *39*, 109116. [CrossRef]
14. Chi, X.; Xue, J.; Jia, L.; Yao, J.; Miao, H.; Wu, L.; Liu, T.; Tian, X.; Li, D. Machine Learning-Based Online Monitoring and Closed-Loop Controlling for 3D Printing of Continuous Fiber-Reinforced Composites. *Addit. Manuf. Front.* **2025**, *2*, 200196. [CrossRef]
15. Zheng, S.; Huang, X.; Hu, J.; Yao, Z. Machine learning for revealing the relationship between the process–structure–properties of polypropylene in-reactor alloys. *React. Chem. Eng.* **2024**, *9*, 1354–1363. [CrossRef]
16. Doe, S.; Kassianides, C.; Kassianides, S.; Christodoulou, C.; Bakas, N. Prediction of Melt Flow Rate (MFR) in polymer production, with auto-regressive Machine Learning algorithms, for multivariate time-series measured in irregular timestamps. In *Computer Aided Chemical Engineering*; Elsevier: Amsterdam, The Netherlands, 2023. [CrossRef]
17. Wang, Y.; Fang, Y.; Zhou, H.; Gao, H. A Machine Learning Model for Predicting the Propagation Rate Coefficient in Free-Radical Polymerization. *Molecules* **2024**, *29*, 4694. [CrossRef]
18. Xu, C.; Wang, Y.; BaratiFarimani, A. TransPolymer: A Transformer-based language model for polymer property predictions. *Npj. Comput. Mater.* **2023**, *9*, 64. [CrossRef]
19. Zhang, A.; Bao, C.; Zhu, Z.; Ji, W. A quantum-transformer hybrid architecture for polymer property prediction: Addressing data sparsity issues. *Comput. Mater. Sci.* **2025**, *256*, 113950. [CrossRef]
20. Shi, J.; Albreiki, F.; Colón, Y.J.; Srivastava, S.; Whitmer, J.K. Transfer Learning Facilitates the Prediction of Polymer–Surface Adhesion Strength. *J. Chem. Theory Comput.* **2023**, *19*, 4631–4640. [CrossRef]
21. Gurnani, R.; Kuenneth, C.; Toland, A.; Ramprasad, R. Polymer informatics at scale with multitask graph neural networks. *Chem. Mater.* **2023**, *35*, 1560–1567. [CrossRef]
22. Zhang, Y.; Shen, C.; Xia, K. Multi-Cover Persistence (MCP)-based machine learning for polymer property prediction. *Brief. Bioinform.* **2024**, *25*, bbae465. [CrossRef]
23. *ASTM E2877-12(2019)*; Standard Guide for Digital Contact Thermometers. ASTM International: West Conshohocken, PA, USA, 2019. [CrossRef]
24. Perera, T.G.Y.; Li, J.; Kelly, A.L.; Abeykoon, C. Melt pressure prediction in polymer extrusion processes with deep learning. In Proceedings of the 2023 European Control Conference (ECC), Bucharest, Romania, 13–16 June 2023; IEEE: Piscataway, NJ, USA, 2023; pp. 1–6.
25. Lin, Z. Optimizing Kernel Extreme Learning Machine based on a Enhanced Adaptive Whale Optimization Algorithm for classification task. *PLoS ONE* **2025**, *20*, e0309741. [CrossRef]
26. *ASTM D1238-23*; Standard Test Method for Melt Flow Rates of Thermoplastics by Extrusion Plastometer. ASTM International: West Conshohocken, PA, USA, 2023. [CrossRef]
27. Milad, A.; Hussein, S.H.; Khekan, A.R.; Rashid, M.; Al-Msari, H.; Tran, T.H. Development of ensemble machine learning approaches for designing fiber-reinforced polymer composite strain prediction model. *Eng. Comput.* **2022**, *38*, 3625–3637. [CrossRef]
28. Liang, Y.; Wei, X.; Peng, Y.; Wang, X.; Niu, X. A review on recent applications of machine learning in mechanical properties of composites. *Polym. Compos.* **2025**, *46*, 1939–1960. [CrossRef]
29. Ledwani, D.; Thakur, I.; Bhatnagar, V. Comparative analysis of prediction models for melt flow rate of C2 and C3 polymers synthesized using nanocatalysts. *NanoWorld J.* **2022**, *8*, S123–S127. [CrossRef]

30. Perera, Y.S.; Li, J.; Abeykoon, C. Machine learning enhanced grey-box soft sensor for melt viscosity prediction in polymer extrusion processes. *Sci. Rep.* **2025**, *15*, 5613. [CrossRef] [PubMed]
31. Jain, A.; Gurnani, R.; Rajan, A.; Qi, H.J.; Ramprasad, R. A physics-enforced neural network to predict polymer melt viscosity. *Npj Comput. Mater.* **2025**, *11*, 42. [CrossRef]
32. Seifert, L.; Leuchtenberger-Engel, L.; Hopmann, C. Enhancing the quality of polypropylene recyclates: Predictive modelling of the melt flow rate and shear viscosity. *Polymers* **2024**, *16*, 2326. [CrossRef]
33. Manola, M.S.; Singh, B.; Singla, M.K.; Chohan, J.S.; Kumar, R.; Bisht, Y.S.; Alkahtani, M.Q.; Islam, S.; Ammarullah, M.I. Investigation of melt flow index and tensile properties of dual metal reinforced polymer composites for 3D printing using machine learning approach: Biomedical and engineering applications. *AIP Adv.* **2024**, *14*, 055016. [CrossRef]

polymers

MDPI

Article

Hysteresis Behavior Modeling of Magnetorheological Elastomers under Impact Loading Using a Multilayer Exponential-Based Preisach Model Enhanced with Particle Swarm Optimization

Alawiyah Hasanah Mohd. Alawi, Khisbullah Hudha *, Zulkiffli Abd. Kadir and Noor Hafizah Amer

Department of Mechanical Engineering, Faculty of Engineering, National Defence University of Malaysia, Kem Sungai Besi, Kuala Lumpur 57000, Malaysia; hasanahalawiyah@gmail.com (A.H.M.A.); zulkiffli@upnm.edu.my (Z.A.K.); noorhafizah@upnm.edu.my (N.H.A.)
* Correspondence: k.hudha@upnm.edu.my

Abstract: Magnetorheological elastomers (MREs) are a type of smart material that can change their mechanical properties in response to external magnetic fields. These unique properties make them ideal for various applications, including vibration control, noise reduction, and shock absorption. This paper presents an approach for modeling the impact behavior of MREs. The proposed model uses a combination of exponential functions arranged in a multi-layer Preisach model to capture the nonlinear behavior of MREs under impact loads. The model is trained using particle swarm optimization (PSO) and validated using experimental data from drop impact tests conducted on MRE samples under various magnetic field strengths. The results demonstrate that the proposed model can accurately predict the impact behavior of MREs, making it a useful tool for designing MRE-based devices that require precise control of their impact response. The model's response closely matches the experimental data with a maximum prediction error of 10% or less. Furthermore, the interpolated model's response is in agreement with the experimental data with a maximum percentage error of less than 8.5%.

Keywords: magnetorheological; elastomers; hysteresis behavior; impact loading; preisach model; particle swarm optimization; exponential function

1. Introduction

MREs are smart materials that are composed of a polymer matrix filled with magnetic particles, usually iron or iron oxide, which can respond to an external magnetic field. The magnetic particles are typically dispersed within the elastomer matrix in a random manner, forming a network of interconnected chains [1]. When an external magnetic field is applied to the MRE, the magnetic particles within the polymer matrix align themselves with the direction of the field causing the elastomer to stiffen and become more rigid. Conversely, when the magnetic field is removed, the particles return to their random orientation and the elastomer returns to its original soft and flexible state [2,3]. The stiffness of MRE can be controlled by adjusting the strength and orientation of the magnetic field.

MREs have the potential for a range of applications in various industries. Some key examples include vibration control [4] and damping systems [5,6] where MREs can reduce vibrations and enhance the stability of structures and machinery. MREs can also be utilized in the development of soft robotics [7] and prosthetics with tunable stiffness [8], offering improved flexibility and control. Additionally, MREs have the potential for seismic protection [9], as they can enhance the seismic resistance of structures by providing adaptive damping to reduce the impact of seismic waves. In the aerospace and defense industries, MREs can be integrated into aircraft and spacecraft design to improve vibration control and damping [10], leading to better performance and stability. Furthermore, MREs may be applied to medical device design [11], such as stents and catheters, to offer greater control and flexibility. As research in this field progresses, the applications of MREs continue to expand, showcasing their potential for diverse applications in different industries [12].

MREs change their mechanical properties in response to an applied magnetic field. Due to their unique behavior, accurate modeling of MREs is crucial for their effective application in various industries. Accurate modeling of MREs allows for the prediction of their behavior under different magnetic field strengths which is essential in designing MRE-based devices and systems. It also enables the optimization of MREs for specific applications by predicting the mechanical properties, such as stiffness and damping, and the response times of MREs to changes in the magnetic field. Furthermore, accurate modeling of MREs can lead to the development of more advanced MRE-based systems that require a more precise understanding of the behavior of MREs to ensure optimal performance.

There are two methods for modeling the behavior of MREs, namely parametric and non-parametric models [13]. Parametric models refer to models that are based on a set of predefined parameters or assumptions about the behavior of the material. These models typically involve the use of mathematical equations and require knowledge of the material's properties and characteristics, such as its stiffness and damping. On the other hand, non-parametric models do not rely on pre-defined assumptions or parameters but instead use data-driven approaches to develop models based on observed behavior. These models can include machine learning algorithms, such as artificial neural networks, or support vector machines that learn from data to predict the behavior of the material.

Both parametric and non-parametric models have been used in MRE research. Parametric models, such as micromechanical models or continuum mechanics models, have been developed to describe the behavior of MREs based on the physical properties of the material. Non-parametric models, such as artificial neural networks [14], or fuzzy logic systems [15], have been used to predict the behavior of MREs based on experimental data. Overall, both approaches have their strengths and weaknesses, and the choice of which model to use depends on the specific application and available data. Parametric models can provide a more fundamental understanding of the material behavior, but they may require more knowledge of the material properties. Non-parametric models can be more flexible and adaptable but may require more experimental data to develop accurate models [16,17].

In this paper, an impact behavior model for MREs that takes into account the hysteresis characteristics of MREs is proposed. To model the hysteresis characteristics, the multilayer exponential-based Preisach model is proposed. Exponential functions are commonly used in modeling hysteresis as they can capture the nonlinear behavior of the material. The multilayer exponential function allows for a more complex model that can better capture the behavior of MREs. To optimize the parameters of the model, an optimization tool, namely PSO, is used. PSO is a nature-inspired optimization algorithm that simulates the social behavior of swarms in nature, such as flocks of birds or schools of fish. The algorithm is designed to find the optimal solution to a given optimization problem by iteratively adjusting a group of particles or agents in a search space.

2. Design and Fabrication of MRE-Based Double Acting Actuator

In traditional MREs, the actuator works mainly in compression mode [18]. However, in this study, a double-acting MRE actuator (Appendix A) that can work in both compression and extension modes was developed and fabricated. The schematic diagram and the prototype of the MRE actuator are shown in Figure 1a,b, respectively. MRE actuators consist of a piston and cylinder containing MRE material, a coil or electromagnet, and a power source. When a mechanical vibration causes the piston to move relative to the cylinder, a magnetic field is generated by the coil or electromagnet which causes the magnetic particles in the MRE to align themselves in the direction of the field. This changes the stiffness and damping properties of the MRE, providing a damping force that opposes the motion of the piston. The damping force can be adjusted by varying the strength and direction of the magnetic field, which is controlled by the controller.

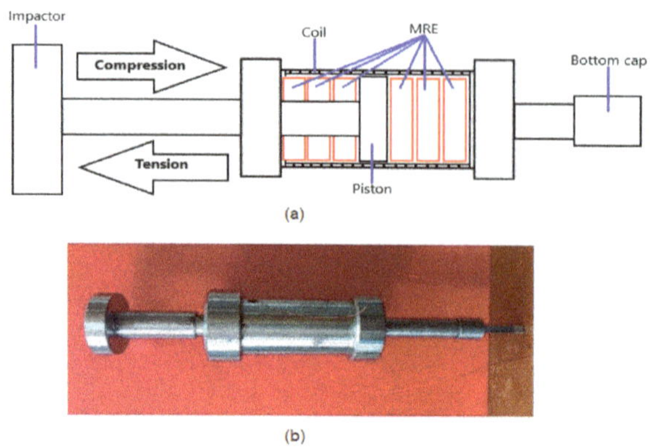

Figure 1. Schematic Diagram of a Double Acting MRE. The schematic diagram (**a**) and the prototype (**b**) of the MRE actuator.

The fabrication process of MRE typically involves several steps. First, an elastomer matrix is selected to meet specific application requirements. Next, magnetic particles are selected and their size is determined. Then, the magnetic particles are prepared by mixing them with the elastomer. The mixture is cured by pouring it into a mold. Finally, the MRE is magnetized by exposing it to a strong magnetic field which magnetizes the particles and induces the desired mechanical properties [19,20]. The coils are made of copper wire coated with a layer of enamel insulation with a diameter of 0.7 mm; the number of coil turns is 250. The parameters and the composition for MRE fabrication are shown in Table 1.

Table 1. Composition of the MRE Sample [21].

Materials	Percentage
RTV Silicone Rubber	30%
Carbonyl Iron Powder	60%
Additive (Ferrite)	7%
Hardener (Isocyanates)	3%

3. Experimental Study on the Force-Displacement Characteristics of MREs for Impact Response

Upon fabricating the proposed MRE, a series of drop impact tests were conducted to evaluate the effectiveness of the MRE under impact loads by analyzing their force-displacement characteristics. The drop impact test is performed to assess the ability of a material to resist impact loading. The test involves several steps. First, the specifications for the test, such as the height of the drop, the weight of the impactor, and other relevant factors, are determined. Next, the test specimen, or product, is prepared to ensure that it is in a suitable condition for testing. Then the impactor is secured to the drop apparatus, carefully aligned, and oriented. The impactor is raised to the predetermined height and released, allowing it to free-fall and strike the test specimen. The impact is observed and to ensure accurate and reliable results, the test is typically repeated several times. Finally, the results can be analyzed. Figure 2 depicts the drop test machine that was utilized in the experiment.

Figure 2. Drop Impact Test Machine used in this Study.

The tests were carried out by subjecting the material to a sudden external force to observe its behavior. The Instron Drop Impact Machine and CEAST Software were utilized to set the experiment variables, including the impact energy, impact velocity, falling height, total mass, and applied current. The parameters set to the Instron Drop Impact Machine during the experiment are shown in Table 2. Figure 2 shows the experimental setup that was conducted in the Automotive Lab at Universiti Pertahanan Nasional Malaysia (UPNM).

Table 2. Parameters of The Drop Impact Test.

Parameters	Input Values
Impact Energy	13.8 J
Impact Velocity	2.24 m/s
Falling Height	256 mm
Total Mass	5.5 kg
Current	0–2 A
Contact area	13.87 cm^2

The experiments were conducted to investigate the behavior of the MRE under different current inputs to the coils. In order to achieve this, varying current values were applied to the coils at 0, 0.5, 1, 1.5, and 2 Amperes. To ensure the accuracy and consistency of the data collected, each experiment was repeated multiple times at each current level. This approach helps to reduce the effects of random errors and improves the reliability of the experimental results. Figure 3 presents the experimental results displaying the force-displacement relationship with respect to the different current inputs. The figure clearly illustrates the changes in stiffness and damping properties of the MRE with varying current inputs. Upon analyzing each force-displacement curve, it was found that the upper slope of the curve represents the response of the MRE undergoing compression, while the lower slope represents the response of the MRE undergoing retraction.

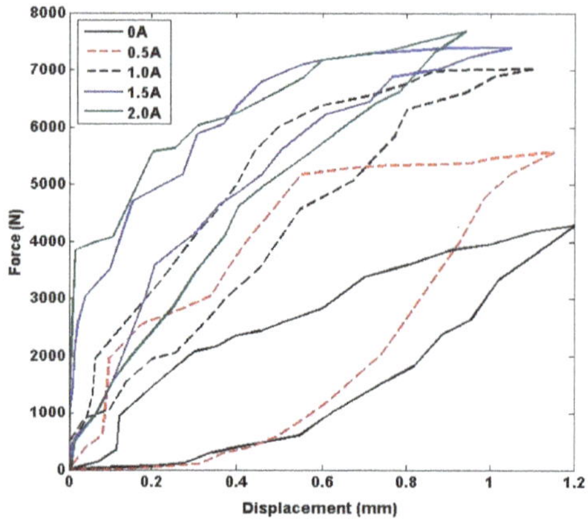

Figure 3. Force-Displacement Characteristics of MREs in Various Currents.

4. Modeling the Hysteresis Characteristics of MREs Using a Multilayer Exponential-Based Preisach Model Optimized with PSO

This section describes the algorithm for a multilayer exponential-based Preisach model that was developed using experimental data on the force-displacement characteristics of MREs under impact loading conditions for input currents injected to the coils at 0, 0.5, 1, 1.5, and 2 Amperes. Additionally, the interpolation method used to obtain the force produced by the MREs for input currents between the specified values is explained. Finally, an optimization tool called Particle Swarm Optimization (PSO) is also discussed.

4.1. Multilayer Exponential-Based Preisach Model

The Preisach model is a mathematical tool used to describe hysteresis, which is a phenomenon in which the output of a system depends not only on the current input but also on its history [22,23]. The model consists of a set of hysterons, which are basic units that represent the behavior of the system. Each hysteron is associated with a particular input/output history and has a unique switching threshold. The model describes the behavior of the system by representing it as a distribution of hysterons, each of which contributes to the overall response of the system [24]. The Preisach model consists of many relay hysterons connected in parallel, given weights, and summed. This can be visualized by a block diagram as shown in Figure 4. Each of these relays has different α and β thresholds and is scaled by μ. By increasing the number of N, the true hysteresis curve can be better approximated [25].

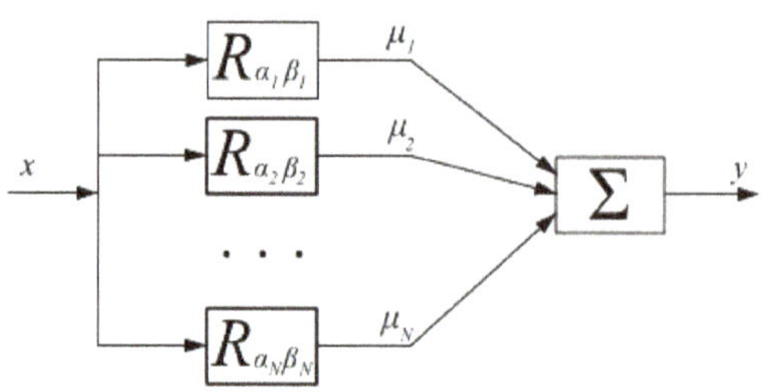

Figure 4. Block Diagram of Preisach Model.

An exponential function is a mathematical function in the form of $f(x) = e^x$. Exponential functions are used to model phenomena that grow or decay at a constant percentage rate over time, such as population growth, radioactive decay, or compound interest. They have a characteristic curve that starts rapidly and then increases or decreases slowly. First, second and n-th hysterons are defined as follows:

$$\mu_1 R_{\alpha_1\beta_1} = a_1 e^{b_1 x}$$

$$\mu_2 R_{\alpha_2\beta_2} = a_2 e^{b_2 x}$$

$$\ldots$$

$$\mu_N R_{\alpha_N\beta_N} = a_N e^{b_N x}$$

The general relationship between input-output of the Preisach model is written as follows:

$$f(x) = a_1 e^{b_1 x} + a_2 e^{b_2 x} + \cdots + a_N e^{b_N x}$$

Referring to Figure 3, there are five hysteresis loops representing the force-displacement characteristics of MREs under impact loading for different amounts of current injected into the coils. The currents include 0, 0.5, 1, 1.5, and 2 Amperes. Each loop has two Preisach model setups representing the upper and lower slopes.

In this proposed model, some controlled parameters are optimized using PSO to accurately represent the hysteresis response of MREs under impact loadings. They are: a_1, $a_2, \ldots, a_N; b_1, b_2, \ldots, b_N; c_1, c_2, \ldots, c_N; d_1, d_2, \ldots, d_N; \ldots; s_1, s_2, \ldots, s_N; t_1, t_2, \ldots, t_N$.

$$f(x)_0^{up} = a_1 e^{b_1 x} + a_2 e^{b_2 x} + \cdots + a_N e^{b_N x}$$

$$f(x)_0^{lo} = c_1 e^{d_1 x} + c_2 e^{d_2 x} + \cdots + c_N e^{d_N x}$$

$$f(x)_{0.5}^{up} = e_1 e^{f_1 x} + e_2 e^{f_2 x} + \cdots + e_N e^{f_N x}$$

$$f(x)_{0.5}^{lo} = g_1 e^{h_1 x} + g_2 e^{h_2 x} + \cdots + g_N e^{h_N x}$$

$$f(x)_1^{up} = i_1 e^{j_1 x} + i_2 e^{j_2 x} + \cdots + i_N e^{j_N x}$$

$$f(x)_1^{lo} = k_1 e^{l_1 x} + k_2 e^{l_2 x} + \cdots + k_N e^{l_N x}$$

$$f(x)_{1.5}^{up} = m_1 e^{n_1 x} + m_2 e^{n_2 x} + \cdots + m_N e^{m_N x}$$

$$f(x)_{1.5}^{lo} = o_1 e^{p_1 x} + o_2 e^{p_2 x} + \cdots + o_N e^{p_N x}$$

$$f(x)_2^{up} = q_1 e^{r_1 x} + q_2 e^{r_2 x} + \cdots + q_N e^{r_N x}$$

$$f(x)_2^{lo} = s_1 e^{t_1 x} + s_2 e^{t_2 x} + \cdots + s_N e^{t_N x}$$

As previously mentioned, the proposed model was developed based on experimental data obtained from the force-displacement measurements for input currents of 0, 0.5, 1, 1.5, and 2 Amperes. To calculate the force generated by the MREs for input currents between 0 A and 0.5 A, an interpolation approach was used based on the known or specified displacement. The interpolation algorithm is detailed in Figure 5, which uses the experimental data for 0 Ampere and 0.5 Ampere as an example. Similar algorithms are used to calculate the force generated by MREs for input currents between 0.5 A and 1 A, between 1 A and 1.5 A, and between 1.5 A and 2 A. The algorithm for calculating force produced by MREs is as follows:

$$F(x)_i^{up} = F(x)_0^{up} + \left(F(x)_{0.5}^{up} - F(x)_0^{up}\right) i$$

$$F(x)_i^{lo} = F(x)_0^{lo} + \left(F(x)_{0.5}^{lo} - F(x)_0^{lo}\right) i$$

where

i: Current injected to the MREs coils (between 0 A to 0.5 A)

$F(x)_i^{up}$: Force produced by MRE at current i during compression

$F(x)_i^{lo}$: Force produced by MRE at current i during retraction
$F(x)_0^{up}$: Force at upper slope of 0 A
$F(x)_0^{lo}$: Force at lower slope of 0 A
$F(x)_{0.5}^{up}$: Force at upper slope of 0.5 A
$F(x)_0^{lo}$: Force at lower slope of 0.5 A

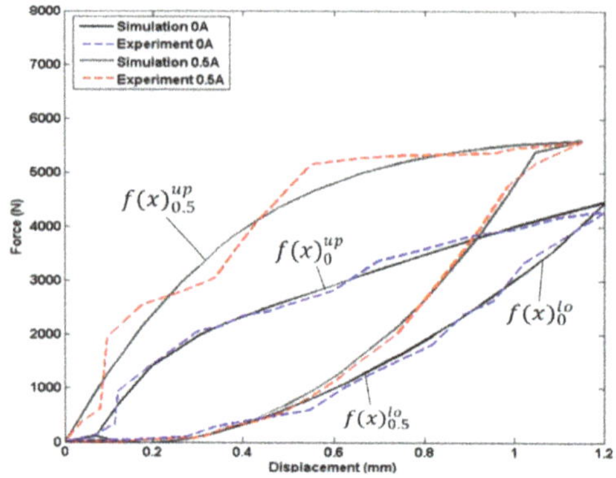

Figure 5. Interpolation of Force-Displacement Characteristics for Input Current Between 0–0.5 Ampere.

4.2. Optimization of Multilayers Sigmoidal Functions Using PSO

PSO is a metaheuristic optimization algorithm that is inspired by the social behavior of bird flocking or fish schooling [26]. In PSO, a set of particles are initialized randomly in the search space and move towards the optimal solution by updating their positions based on their own previous positions and the positions of the best-performing particles in the swarm. The basic theory of PSO can be summarized in the following steps [27]:

1. Initialization: A population of particles is generated randomly in the search space. Each particle represents a potential solution to the optimization problem.
2. Evaluation: The fitness of each particle is evaluated based on the objective function to be optimized.
3. Update of the particle's best position: Each particle keeps track of the best position it has visited so far, denoted as P_{best}. If the fitness of the current position is better than its P_{best}, the particle updates its P_{best}.
4. Update of the swarm's best position: The best position among all the P_{best} positions of the particles is denoted as G_{best}. If the fitness of the current G_{best} position is better than its previous value, the swarm updates its G_{best}.
5. Update of the particle's velocity and position: Each particle updates its velocity and position based on its current velocity, its distance from its P_{best}, and its distance from the G_{best}. The velocity and position updates are given by the following equations:
   Velocity update:

$$V_{(t+1)}^i = w \cdot V_{(t)}^i + c_1 \cdot r_1 \cdot \left(P_{best}^i - X_{(t)}^i\right) + c_2 \cdot r_2 \cdot \left(G_{best} - X_{(t)}^i\right)$$

   Position update:

$$X_{(t+1)}^i = X_{(t)}^i + V_{(t+1)}^i$$

where $V_{(t)}^i$ and $X_{(t)}^i$ are the velocity and position of particle i at time t, w is the inertia weight, c_1 and c_2 are the acceleration coefficients, and r_1 and r_2 are random numbers between 0 and 1.

6. Termination: The algorithm terminates when a stopping criterion is met, such as reaching a maximum number of iterations or a satisfactory fitness level.

7. By iterating through these steps, the particles in the swarm collectively move toward the optimal solution of the optimization problem.

In this study, the parameters of the multilayer exponential-based Preisach model that will be tuned using PSO are $a_1, a_2, \ldots, a_N; b_1, b_2, \ldots, b_N; c_1, c_2, \ldots, c_N; d_1, d_2, \ldots, d_N; \ldots; s_1, s_2, \ldots, s_N; t_1, t_2, \ldots, t_N$. In PSO, there are several optimization parameters that need to be set to ensure the algorithm performs optimally. The choice of parameter values for PSO can depend on the specific problem being solved and the characteristics of the search space. A common approach is to use a parameter tuning method, such as grid search or random search, to find the optimal parameter values. These parameters include:

- Swarm size: 40
- Maximum number of iterations: 100
- Inertia weight: 0.9
- Acceleration coefficients (c_1 and c_2): 1.42

5. Results and Discussions

This section presents a comparison between the simulated responses of the proposed MRE model and the corresponding experimental data. The maximum error of the predicted force will also be analyzed and discussed. Finally, the accuracy of the model will be tested and validated using input currents ranging from 0 to 0.5 A, 0.5 to 1 A, 1 to 1.5 A, and 1.5 to 2 A. Effects of varying the swarm size and the number of iterations on the PSO are also analyzed.

5.1. Comparison between the Simulated Response of Multilayer Exponential-Based Preisach Model with the Experimental Data

Figure 6 compares the simulated model response to experimental data where no current was injected into the coils. The model's response closely matches the experimental data, indicating a high degree of accuracy in the simulation. However, a maximum error of 5% occurred in area A, which may be attributed to limitations in the experimental setup or inaccuracies in the simulation model. Further investigation and refinement of the model could potentially reduce this error and improve the overall accuracy of the simulation.

In Figure 7, the simulated model response is compared to experimental data for an input current of 0.5 A injected into the coils. The simulation results exhibit a high degree of accuracy and closely match the experimental data, but the maximum error occurs in area B with a percentage error of approximately 10%. This discrepancy could be due to limitations in the experimental setup, variations in the material properties, or uncertainties in the simulation model. Despite this limitation, the close agreement between the simulated and experimental data validates the model's capability to predict the system's behavior accurately under different input conditions.

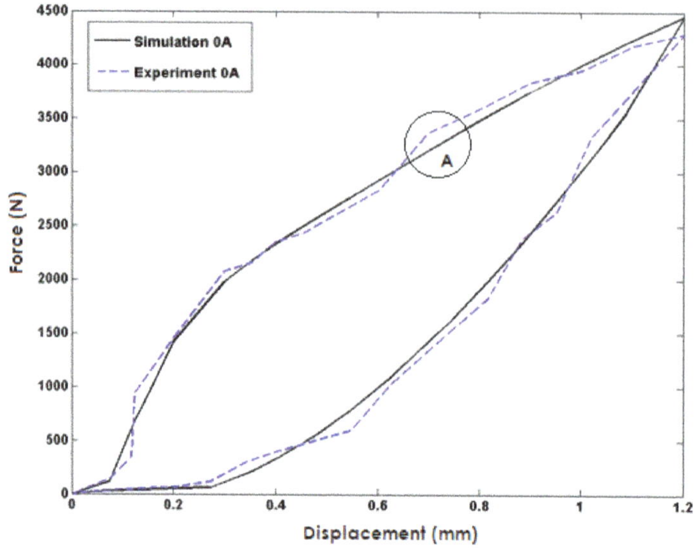

Figure 6. Comparisons with the Model Response and Experimental Data for the Input Current of 0 Ampere.

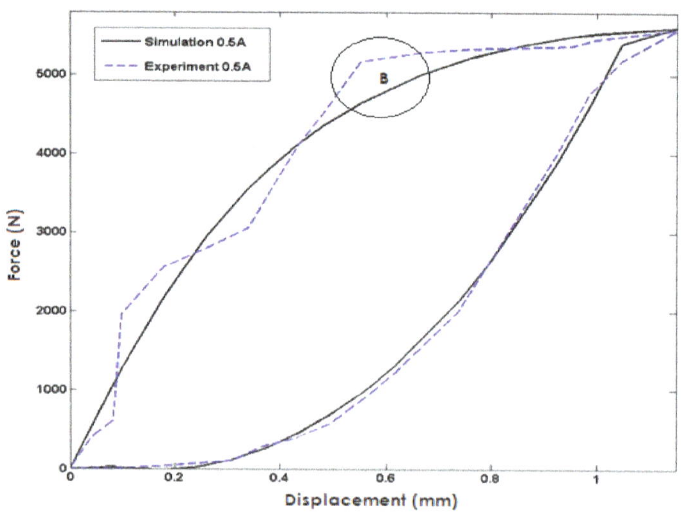

Figure 7. Comparisons with the Model Response and Experimental Data for the Input Current of 0.5 Ampere.

Figure 8 compares the simulated model response to experimental data for an input current of 1 A injected into the coils. The simulation results exhibit a high degree of accuracy and closely match the experimental data, but the maximum error is observed in area C with a percentage error of approximately 4.8%. Again, this discrepancy may be due to experimental limitations, material property variations, or uncertainties in the simulation model. However, the close agreement between the simulated and experimental data provides confidence in the model's ability to accurately predict the system's behavior.

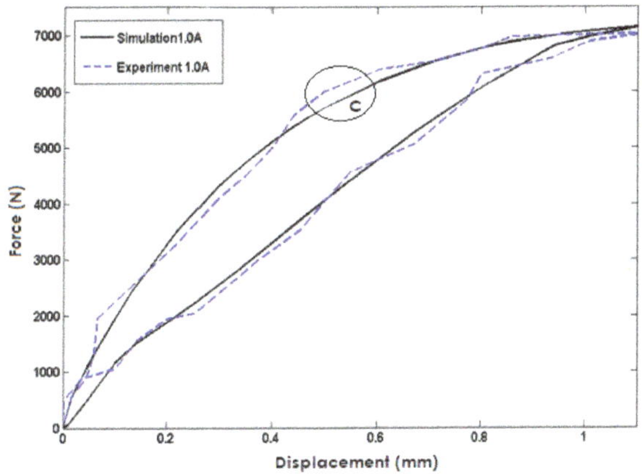

Figure 8. Comparisons with the Model Response and Experimental Data for the Input Current of 1 Ampere.

Figure 9 compares the simulated model response to experimental data for the input current of 1.5 A injected into the coils. The simulation results closely match the experimental data, demonstrating the model's high degree of accuracy in predicting the system's response under this input condition. However, it is worth noting that the maximum error occurs in area D, with a percentage error of approximately 4.7%. The close agreement between the simulated and experimental data validates the model's capability to accurately predict the system's behavior.

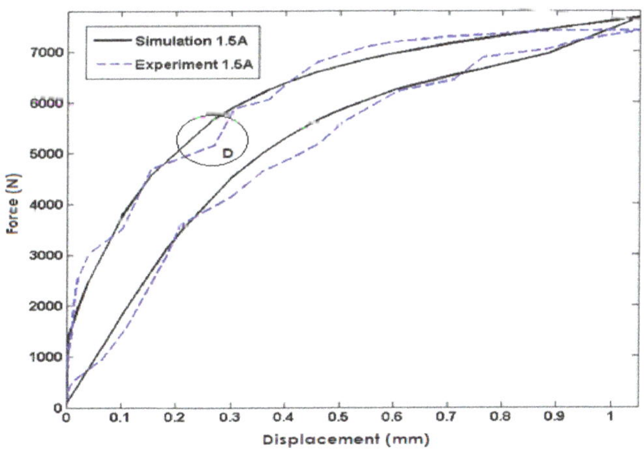

Figure 9. Comparisons with the Model Response and Experimental Data for the Input Current of 1.5 Ampere.

Lastly, in Figure 10, the simulated model response is compared to experimental data for an input current of 2 A injected into the coils. The simulation results closely match the experimental data, indicating that the model accurately predicts the system's response under this input condition. However, the maximum error is observed in area E with a percentage error of approximately 8.1%. Nevertheless, the close agreement between the simulated and experimental data validates the model's capability to predict the system's behavior accurately. The maximum error of the predicted force of the proposed model is summarized in Table 3.

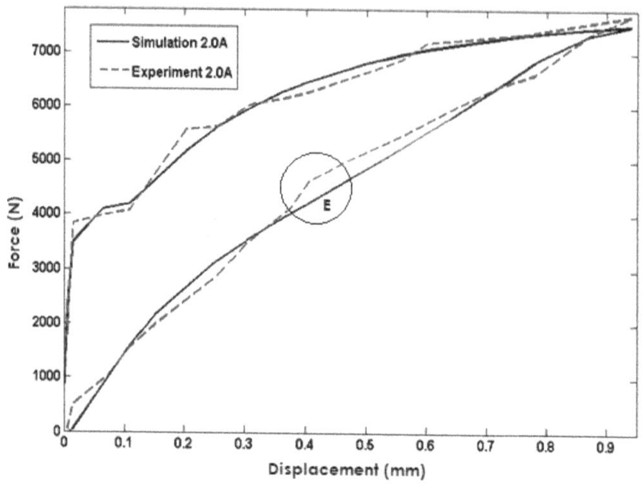

Figure 10. Comparisons with the Model Response and Experimental Data for the Input Current of 2 Ampere.

Table 3. The Maximum Percentage of Error in All Five Cases.

Current	Experimental Data	Simulation Data	Percentage of Error (%)
0 A	3374.804	3206.756	4.98
0.5 A	5171.588	4642.620	10.23
1.0 A	6008.422	5718.536	4.83
1.5 A	7093.411	6762.593	4.66
2.0 A	5572.439	5123.559	8.1

5.2. Validation of the Interpolated Model

Previously, it was mentioned that the force-displacement characteristics of MREs under impact loading for input currents ranging from 0–0.5, 0.5–1, 1–1.5, and 1.5–2 Amperes were predicted using an interpolation algorithm. This algorithm is commonly used to estimate data points within a range of values based on known data points. To validate the accuracy of this interpolation algorithm, experimental works were conducted using drop impact tests. These tests were carried out with the input currents injected into MRE coils of 0.3, 0.7, 1.3, and 1.7 Amperes, and the resulting force-displacement data were recorded. The experimental data obtained from the drop impact tests were then compared to the corresponding model response with the same input current as shown in Figure 11. It can be seen from the figure that the interpolation algorithm is valid. The model responses closely follow the experimental results with an acceptable error. The maximum error of the predicted force of the proposed model in the interpolation regions is summarized in Table 4.

Table 4. The Maximum Percentage of Error for the Interpolated Model.

Current	Experimental Data	Simulation Data	Percentage Error (%)
0.3 A	4321.450	3954.718	8.486
0.7 A	5263.420	5072.986	3.618
1.3 A	6746.820	6518.139	3.389
1.7 A	4982.910	4707.04	5.536

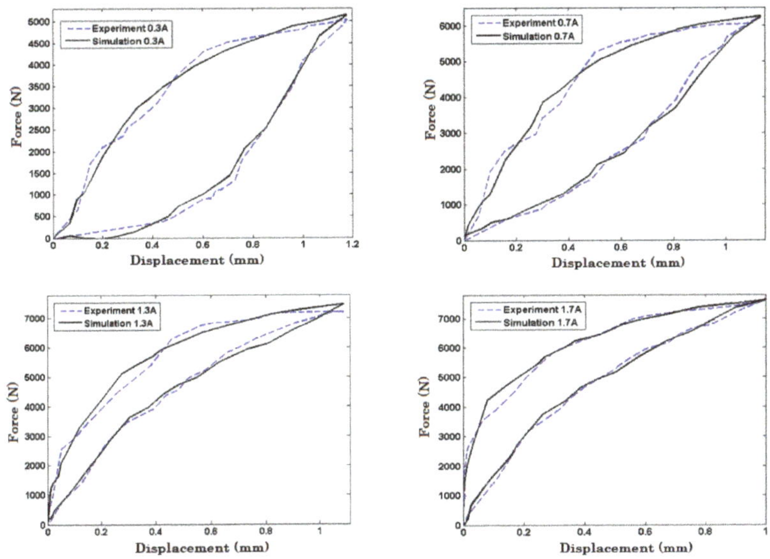

Figure 11. Validation of the Interpolated Model in Various Input Currents.

5.3. Effects of Varying Swarm Size and the Number of Iterations

The number of iterations is an important parameter in PSO, as it determines the length of time the particles are allowed to search for the optimal solution. In general, increasing the number of iterations in PSO can improve the algorithm's ability to find the global optimum, as it allows the particles more time to explore the search space and converge on the best solution. However, at a certain point, additional iterations may not lead to any further improvement in the solution, as the particles may have already converged to a local optimum. On the other hand, decreasing the number of iterations can lead to faster execution times, but at the cost of potentially missing out on better solutions. Therefore, it is important to strike a balance between the number of iterations and the desired level of performance index and execution time. Referring to Figure 12, the number of iterations selected was 100 since it has a fast convergence rate and is able to achieve a performance index as good as the performance index for 120 iterations.

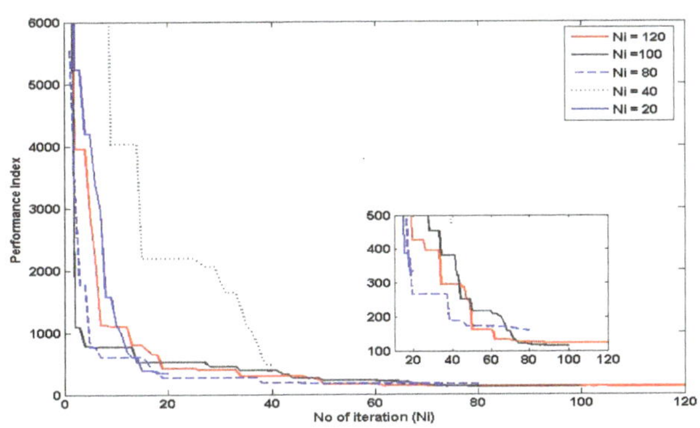

Figure 12. Effects of Varying the Number of Iterations on the Performance of PSO.

The swarm size refers to the number of particles in the population and it determines the diversity and convergence rate of the swarm. Increasing the swarm size can lead to better global exploration, as there are more particles searching the solution space. However, it can also increase the computational cost and reduce the convergence rate, as there are more particles to communicate and update. On the other hand, reducing the swarm size can improve the convergence rate, as there are fewer particles to communicate and update. However, it may also decrease the diversity of the swarm, which can lead to premature convergence and suboptimal solutions. Referring to Figure 13, the swarm size selected in this study was 40 as it shows a fast convergence rate and achieved a performance index as good as a swarm size of 100.

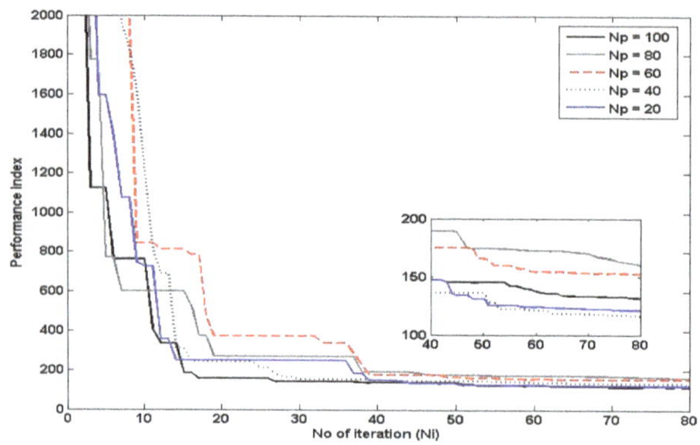

Figure 13. Effects of Varying the Number of Particles on the Performance of PSO.

6. Conclusions

This paper presents a comprehensive study on the hysteresis behavior modeling of magnetorheological elastomers under impact loadings. The study proposes a multilayered exponential-based Preisach model that is enhanced with particle swarm optimization to provide a reliable and accurate framework for capturing the complex hysteresis behavior of the material. The developed model demonstrates excellent performance in capturing the dynamic response of magnetorheological elastomers under various impact-loading scenarios. The results show that the model's response closely matches the experimental data, with a maximum prediction error of 10% or less. The interpolated model's response also shows good agreement with the experimental data, with a maximum percentage error of less than 8.5%. The study also examines the effects of varying the number of iterations and the number of particles on the performance of PSO. Overall, the findings suggest that the proposed model provides a promising approach for accurately predicting the hysteresis behavior of magnetorheological elastomers under impact loadings.

Author Contributions: Conceptualization, K.H. and Z.A.K.; Design/Fabrication, A.H.M.A.; Testing, A.H.M.A. and Z.A.K.; Data Analysis, A.H.M.A. and N.H.A.; Matlab Programming, A.H.M.A. and Z.A.K.; Optimization, A.H.M.A. and N.H.A.; Paper writing, A.H.M.A. and K.H. All authors have read and agreed to the published version of the manuscript.

Funding: This research is funded by the Ministry of Higher Education (MOHE) of Malaysia.

Institutional Review Board Statement: Not applicable.

Data Availability Statement: Not applicable.

Acknowledgments: This research is fully supported by FRGS grant, FRGS/1/2021/TK02/UPNM/02/1 led by K. Hudha. The authors fully acknowledged the Ministry of Higher Education (MOHE) and the National Defence University of Malaysia for the approved funding which makes this important research viable and effective.

Conflicts of Interest: The authors declare that there is no conflict of interest.

Appendix A

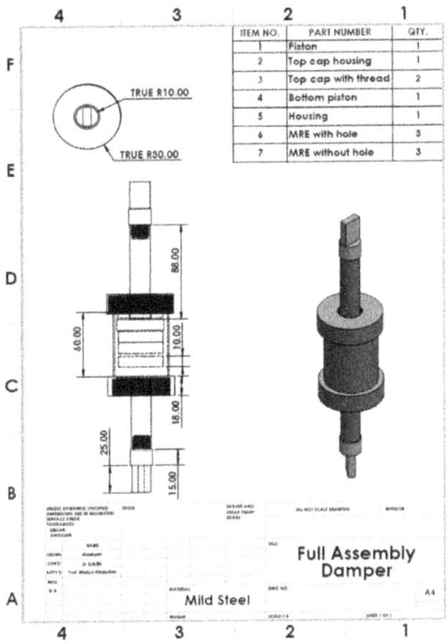

Figure A1. 2D Drawing of the Dual Acting MRE.

References

1. Arslan Hafeez, M.; Usman, M.; Umer, M.A.; Hanif, A. Recent Progress in Isotropic Magnetorheological Elastomers and Their Properties: A Review. *Polymers* **2020**, *12*, 3023. [CrossRef] [PubMed]
2. Li, Y.; Li, J.; Li, W.; Du, H. A state-of-the-art review on magnetorheological elastomer devices. *Smart Mater. Struct.* **2014**, *23*, 12. [CrossRef]
3. Kang, S.; Choi, K.; Nam, J.D. Magnetorheological Elastomers: Fabrication, Characteristics, and Applications. *Materials* **2020**, *13*, 4597. [CrossRef] [PubMed]
4. Masa'id, A.; Lenggana, B.W.; Ubaidillah, U.; Susilo, D.D.; Choi, S.-B. A Review on Vibration Control Strategies Using Magnetorheological Materials Actuators: Application Perspective. *Actuators* **2023**, *12*, 113. [CrossRef]
5. Li, R.P.; Du, C.B.; Guo, F.; Yu, G.J.; Lin, X.G. Performance of Variable Negative Stiffness MRE Vibration Isolation System. *Adv. Mater. Sci. Eng.* **2015**, *9*, 1–8. [CrossRef]
6. Sun, S.; Yang, J.; Yildirim, T.; Ning, D.; Zhu, X.; Du, H.; Zhang, S.; Nakano, M.; Li, W. A magnetorheological elastomer rail damper for wideband attenuation of rail noise and vibration. *J. Intell. Mater. Syst. Struct.* **2020**, *31*, 220–228. [CrossRef]
7. Bira, N.; Dhagat, P.; Davidson, J.R. A Review of Magnetic Elastomers and Their Role in Soft Robotics. *Front. Robot.* **2020**, *7*, 1–9. [CrossRef]
8. Ubaidillah; Mazlan, S.A.; Sutrisno, J.; Zamzuri, H. Potential applications of magnetorheological elastomers. *Appl. Mech. Mater.* **2014**, *663*, 695–699. [CrossRef]
9. Brancati, R.; Di Massa, G.; Pagano, S.; Strano, S. Seismic isolation system for lightweight structures based on MRE devices. *Lect. Notes Eng. Comput. Sci.* **2015**, *2218*, 1313–1318.
10. Skalski, P.; Kalita, K. Implementation of Magnetorheological Elastomers in Transport. *Trans. Inst. Aviat.* **2016**, *245*, 189–198. [CrossRef]
11. Sedlacik, M.; Ronzova, A.; Munteanu, A.; Kutalkova, E.; Drohsler, P.; Moucka, R.; Kracalik, M.; Bilek, O.; Mazlan, S.A. Reprocessed magnetorheological elastomers with reduced carbon footprint and their piezoresistive properties. *Sci. Rep.* **2022**, *12*, 1–12.
12. Jaafar, M.F.; Mustapha, F.; Mustapha, M. Review of current research progress related to magnetorheological elastomer material. *J. Mater. Res. Technol.* **2021**, *15*, 5010–5045. [CrossRef]
13. Pokaad, A.Z.B.; Hudha, K.; Nasir, M.Z.B.M. Simulation and experimental studies on the behaviour of a magnetorheological damper under impact loading. *Int. J. Struct. Eng.* **2011**, *2*, 164–187. [CrossRef]
14. He, H.W.; Wang, Z.H.; Zou, Z.J.; Liu, Y. Nonparametric modeling of ship maneuvering motion based on self-designed fully connected neural network. *Ocean Eng.* **2022**, *251*, 111113. [CrossRef]
15. Taheri, S.M.; Hesamian, G. Non-parametric statistical tests for fuzzy observations: Fuzzy test statistic approach. *Int. J. Fuzzy Log. Intell. Syst.* **2017**, *17*, 145–153. [CrossRef]

16. Thorson, J.T.; Taylor, I.G. A comparison of parametric, semi-parametric, and non-parametric approaches to selectivity in age-structured assessment models. *Fish* **2014**, *158*, 74–83. [CrossRef]
17. Wang, T.; Zhi-Wen, Z. A new type of nonlinear hysteretic model for magnetorheological elastomer and its application. *Mater. Lett.* **2021**, *301*, 130176. [CrossRef]
18. Shahar, S.F.M.; Mazlan, S.A.; Johari, N.; Johari, M.A.F.; Aziz, S.A.A.; Khairi, M.H.A.; Nordin, N.A.; Hapipi, N.M. Mechanical Properties and Microstructural Behavior of Uniaxial Tensile-Loaded Anisotropic Magnetorheological Elastomer. *Actuators* **2022**, *11*, 306. [CrossRef]
19. Brancati, R.; Di Massa, G.; Di Vaio, M.; Pagano, S.; Santini, S. *Experimental Investigation on Magneto-Rheological Elastomers*; Springer International Publishing: Berlin/Heidelberg, Germany, 2019; Volume 68.
20. Zhang, B.; Cao, Y.; Qie, J. Study on the dynamic mechanical properties of magnetorheological elastomer (MRE). *J. Intell. Mater. Syst. Struct.* **2022**, *33*, 1115–1125. [CrossRef]
21. Hudha, K.; Sobri, N.S.; Sumasundram, K.; Haniffah, N.A.; Kadir, Z.A.; Rahmat, M.S. Investigation on the Effect of the Ferrous Particles Size on the Impact Absorption Capability of Magnetorheological Elastomer. In Proceedings of the 2022 IEEE 18th International Colloquium on Signal Processing & Applications, Selangor, Malaysia, 12 May 2022; pp. 299–303.
22. Kuczmann, M.M. Dynamic Preisach hysteresis model. *J. Adv. Res. Phys.* **2010**, *1*, 1–5.
23. Natale, C.; Velardi, F.; Visone, C. Identification and compensation of Preisach hysteresis models for magnetostrictive actuators. *Phys. B Condens. Matter.* **2001**, *306*, 161–165. [CrossRef]
24. Willerich, S.; Herzog, H.G. A continuous vector preisach model based on vectorial relay operators. *IEEE Trans. Magn.* **2020**, *56*, 1–4. [CrossRef]
25. Nguyen, P.B.; Choi, S.B.; Song, B.K. Development of a novel diagonal-weighted Preisach model for rate-independent hysteresis. *Proc. Inst. Mech. Eng. Part C J. Mech. Eng. Sci.* **2017**, *231*, 961–976. [CrossRef]
26. Marini, F.; Walczak, B. Particle swarm optimization (PSO). A tutorial. *Chemom. Intell. Lab. Syst.* **2015**, *149*, 153–165. [CrossRef]
27. Chan, C.L.; Chen, C.L. A cautious PSO with conditional random. *Expert Syst. Appl.* **2015**, *42*, 4120–4125. [CrossRef]

polymers

MDPI

Article

Predicting the Performance of Functional Materials Composed of Polymeric Multicomponent Systems Using Artificial Intelligence—Formulations of Cleansing Foams as an Example

Masugu Hamaguchi [1,2,*], Hideki Miwake [3], Ryoichi Nakatake [3] and Noriyoshi Arai [1]

[1] Department of Mechanical Engineering, Keio University, 3-14-1 Hiyoshi, Kohoku-ku, Yokohama 223-8522, Kanagawa, Japan; arai@mech.keio.ac.jp
[2] Kirin Central Research Institute, Kirin Holdings, 26-1, Muraoka-Higashi 2-Chome, Fujisawa 251-8555, Kanagawa, Japan
[3] Research Institute, Fancl Corporation, 12-13 Kamishinano, Totsuka-ku, Yokohama 244-0806, Kanagawa, Japan
* Correspondence: hamaguchi@keio.jp; Tel.: +81-80-1123-3846

Abstract: Cleansing foam is a common multicomponent polymeric functional material. It contains ingredients in innumerable combinations, which makes formulation optimization challenging. In this study, we used artificial intelligence (AI) with machine learning to develop a cleansing capability prediction system that considers the effects of self-assembled structures and chemical properties of ingredients. Over 500 cleansing foam samples were prepared and tested. Molecular descriptors and Hansen solubility index were used to estimate the cleansing capabilities of each formulation set. We used five machine-learning models to predict the cleansing capability. In addition, we employed an in silico formulation by generating virtual formulations and predicting their cleansing capabilities using an established AI model. The achieved accuracy was $R^2 = 0.770$. Our observations revealed that mixtures of cosmetic ingredients exhibit complex interactions, resulting in nonlinear behavior, which adds to the complexity of predicting cleansing performance. Nevertheless, accurate chemical property descriptors, along with the aid of in silico formulations, enabled the identification of potential ingredients. We anticipate that our system will efficiently predict the chemical properties of polymer-containing blends.

Keywords: QSPR; AI; machine learning; cleansing capability; super-multicomponent system

1. Introduction

Quantitative structure–property relationships (QSPRs) are used in several fields, such as environmental chemistry [1], drug design [2], and materials science [3], demonstrating their versatility and ability in scientific research and industrial applications. The QSPR is a statistical and mathematical method for expressing the relationship between chemical structure and physical properties, which enables the rapid prediction of physical properties. This expectation is particularly high in materials science, where recent advances in synthetic technology have enabled the creation of nearly an infinite variety of polymers. The materials produced by this synthesis require highly desirable physical properties. In recent years, several materials have been improved by adding different mixtures. For example, adding nanoparticles or nanofillers to a material to tailor a nanocomposite improves its thermal response and ionic conductivity [4–8]. Several attempts have been made to predict the physical and chemical properties of such complex systems using QSPRs. In the context of the prediction of physical or chemical properties of mixtures, reports predicting the flash point [9], diffusion coefficient [10], boiling point [11], refractive indices [12], or toxicity [13] of binary mixtures are available. Surfactants are representative functional polymers, and studies related to surfactants were performed that predicted the properties such as critical micelle concentration, cloud point, and the hydrophilic–lipophilic balance of binary mixtures [14,15]. From the perspective of multicomponent mixtures comprising three or more substances, examples of predicting properties such as vapor–liquid critical volume are available [16].

In several reports, molecular descriptors were acquired, and mixtures were characterized using methods such as weighted averages. Subsequently, predictive models are often constructed using linear prediction models (or their derivatives), machine learning, or deep-learning approaches. However, in the case of multicomponent systems, particularly with large-molecular-weight materials, challenges still exist for QSPRs regarding prediction accuracy [17], and reports using predictive models for estimating mixture effects are scarce. The lack of extensive research on predicting the properties of multicomponent mixtures, particularly those involving surfactants, and applying these predictive models to infer mixture effects and facilitate product design underscores the novelty and significance of this study.

The cleansing foam used in cosmetics is a typical polymeric multicomponent system, and polymers, such as polyethylene glycol or polyglycerin, are often used as ingredients. It is used to wash excess sebum and dirt from the skin, and a makeup remover is used to remove makeup cosmetics. In recent years, a growing need has been observed for highly functional cleansing foams because people want to use only one product for cleansing foam and makeup removal for reasons of time (shortening procedures) and ecology (saving water and reducing chemical emissions into the environment) [18]. Solvent-based cleansing agents, such as makeup remover oils, are highly soluble in makeup products, which contain oil and pigments, resulting in excellent removability. However, solvent-based cleansing agents are associated with problems such as high environmental impact, high material costs, and a feeling of residual oiliness after rinsing [19]. By contrast, surfactant-based cleansing agents such as cleansing foams have excellent rinsing properties but weak oil removability because they are primarily water-based. In this study, the latter approach was used to improve the cleansing performance of the foams.

Cleansing foams are composed of numerous components. They contain several types of surfactants, polyols, pH adjusters, and water, making optimizing the formulations difficult because an infinite number of ingredient combinations are possible. Therefore, artificial intelligence (AI) using machine learning has been introduced into formulation design to construct a cleansing capability prediction system that considers the effects of surfactant self-assembly and chemical characteristics of ingredients.

The focus of this study is twofold. First, it aims to extend our understanding of property prediction for multicomponent mixtures. Second, it aims to determine high-performance mixing conditions using predictive models. To achieve these goals, we employed various machine-learning methods. In addition, in silico simulations have been introduced to assist human formulators in achieving desirable products during product development.

2. Materials and Methods

2.1. Evaluation of Cleansing Capability

Cleaning foams consisting of 537 samples of ionic surfactants, amphoteric surfactants, nonionic surfactants, polyols, a pH adjuster, and water were prepared by thorough mixing and stirring. Examples of ingredients and formulations are listed in Tables 1 and 2, respectively. Each sample comprised ~20% ionic surfactants, 10% nonionic surfactants, 10% polyols, 1% citric acid, and 60% water by weight. To study the prepared samples, a waterproof eyeliner pencil was placed on a piece of white artificial leather that was dried for 30 min. Then, 0.1 mL of the corresponding cleansing foam sample was added to the dried eyeliner, rubbed 30 times, rinsed, and dried. A schematic of all the procedures is shown in Figure 1.

Table 1. Examples of ingredients used in the formulations prepared in this study.

Category	The Number of Ingredients	Ingredient Examples
Anionic surfactant	8	Potassium cocoyl glutamate Potassium cocoyl glycinate
Amphoteric surfactant	4	Lauramidopropyl hydroxysultaine Sodium cocoamphoacetate
Nonionic surfactant	24	Eicosaglycerol hexacaprylate Decaglycerol isostearate PEG-20 glyceryl triisostearate
Polyol	33	PPG-9 diglyceryl ether Cyclohexylglycerin Glycerin
pH adjuster	1	Citric acid
Base	1	Water

Table 2. Example of cleansing foam samples in this study.

Category	Material Name						
Anionic surfactant	Potassium cocoyl glutamate	6	7	13	8	8	7
Amphoteric surfactant	Lauramidopropyl Hydroxysultaine	5	9	4	9	7	7
Nonionic surfactants	Eicosaglycerol hexacaprylate	10	10			10	
	Decaglycerol isostearate			10			
	PEG-20 glyceryl triisostearate				10		10
Polyols	PPG-9 diglyceryl ether		11		11		
	Cyclohexylglycerin	9		14			
	Glycerin					11	13
pH adjuster	Citric acid	0.8	0.8	0.8	0.8	0.8	0.8
Base	Water	69.2	62.2	58.2	61.2	63.2	62.2

The cleansing capability was evaluated using the eyeliner pencil residual ratio, which was calculated using the color differences as follows:

$$\text{Cleansing capability } (\%) = \frac{\sqrt{\left(L_1^* - L_2^*\right)^2 + \left(a_1^* - a_2^*\right)^2 + \left(b_1^* - b_2^*\right)^2}}{\sqrt{\left(L_1^* - L_0^*\right)^2 + \left(a_1^* - a_0^*\right)^2 + \left(b_1^* - b_0^*\right)^2}} * 100 \quad (1)$$

where L^* indicates lightness, and a^* and b^* indicate chromaticity. (L^*, a^*, b^*) represents the color space value measured using a colorimeter (CM-2600d, Konica Minolta, Inc., Tokyo, Japan). (L^*_0, a^*_0, b^*_0), (L^*_1, a^*_1, b^*_1), and (L^*_2, a^*_2, b^*_2) represent the color space values of the white artificial leather before applying the eyeliner pencil, after applying it, and after cleaning it, respectively [20].

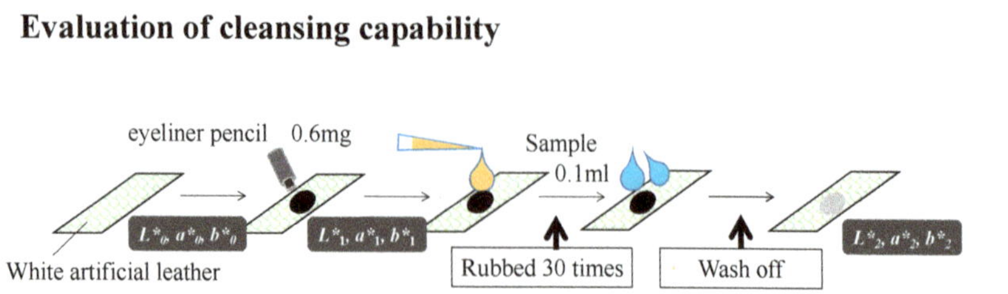

Figure 1. Schematic of the evaluation test to determine the cleansing capability of the prepared samples, Equation (1).

2.2. Modeling of AI

2.2.1. Data Processing

We trained the AI on the prescribing data and modeled them using descriptors and Hansen dissolution parameters to incorporate chemical information. Figure 2 shows the data-processing flowchart.

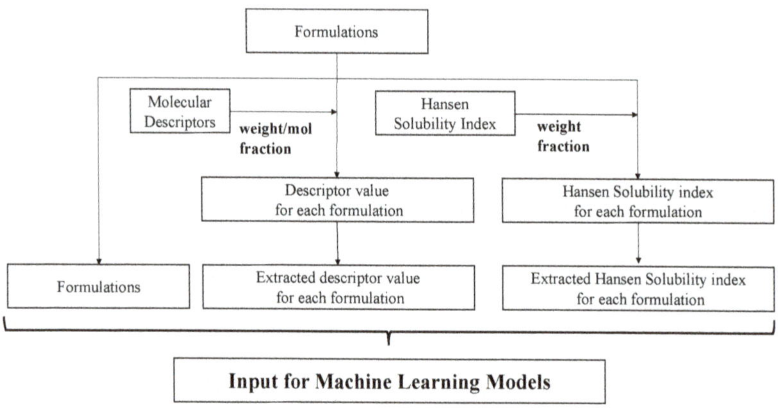

Figure 2. Data-processing flow.

2.2.2. Molecular Descriptors

A molecular descriptor is a numerical molecular property extracted from a chemical structure. Each type of molecular descriptor is related to a specific type of interaction between chemical groups in a particular molecule. Descriptors are used to predict the chemical properties of not only single chemicals but also chemical mixtures. In addition, descriptors have been applied to predict the critical micelle concentration (CMC) of gemini surfactants [21,22]. Therefore, we extracted information from ingredients and predicted the cleansing capabilities of the prepared formulations using molecular descriptors. The structural formula of each ingredient was determined using ChemDraw and converted into a Simplified Molecular Input Line Entry System (SMILES). Regarding the specification of the degree of polymerization, we adopted the representative degree of polymerization for each raw material. Descriptor values were then calculated from the SMILES of each ingredient using the chemoinformatic tools rdkit [23] and PaDEL-descriptor [24]. Entries with infinite or only one value were removed, and a *k*-NN imputer was applied to predict missing values. The weighted average of each ingredient was then calculated using the molar or weight fraction to estimate the descriptor values of the mixture ingredients.

2.2.3. Hansen Solubility Parameters

We applied Hansen solubility parameters (HSPs) to predict the cleansing capability. The HSPs were developed by Hansen to predict the ability of a material to dissolve in another material, forming a solution. The HSP distance between the solute and solvent is generally calculated to estimate whether a solute dissolves in a solvent. Some studies have used the HSPs to predict the properties of surfactants [25,26]. In this study, instead of the solute and solvent, we calculated the distance between each sample and obtained the cleansing samples with the highest cleansing capability. We adopted this procedure because the solute, the eyeliner in this study, was made of several ingredients, and it was difficult to identify its structural formula. The HSP distance was defined as $\{4^*(dD_1\text{-}dD_2)^2 + (dP_1\text{-}dP_2)^2 + (dH_1\text{-}dH_2)^2\}^{0.5}$, where dD_1, dP_1, and dH_1 are the values of each sample—the average calculated based on each component's proportion by wight in the mixture—and dD_2, dP_2 and dH_2 are the average values of the three highest cleansing capabilities in our samples. The HSPs would better estimate the effects of the interactions between the ingredients in a formulation than the descriptor method, in which measuring the nonlinear effect of the ingredient interactions is challenging. The HSP values were calculated using the Hansen Solubility Parameters in Practice (HSPiP) software of version 5.0.09. Because some HSPs cannot be calculated using HSPiP for molecules with high molecular weights, missing HSP values were imputed using a *k*-NN imputer for the descriptor calculation.

2.2.4. Modeling and Feature Selection

The number of explanatory variables was >1000 when descriptors and HSPs were used. Therefore, we applied machine learning to obtain laws to predict cleansing performance based on these numerous features. Three types of machine-learning algorithms are available: supervised learning, unsupervised learning, and reinforcement learning. Supervised learning (regression) was chosen for this study to predict the results within a continuous output. The input dataset is described in Sections 2.1 and 2.2, and the output is the cleansing capability. In this study, we aimed to capture the inherent behavior of surfactants in cleansing forms, anticipating their utilization in more generic applications, such as predicting properties other than cleansing capability. Therefore, we did not focus on developing prediction models specialized for cleansing capability; instead, we adopted representative machine-learning models. We adopted two decision-tree-based models (random forest and extra tree regressors), two linear-based models (lasso and partial least squares), and one support-vector-machine-based model (support vector regressor). The hyperparameters are listed in Table 3. Each model exhibits unique characteristics: decision-tree-based models help capture nonlinear relationships in the data, and linear models are particularly suitable when a linear relationship is assumed between variables. Support-vector-machine-based models are suitable for predicting high-dimensional data. We aimed to develop versatile models by employing these diverse methodologies and uncover new potential for understanding surfactant behavior in cleansing forms. The hyperparameters were optimized using a grid-search method. All explanatory features were standardized with a mean of zero and standard deviation of one. Because numerous features cause noise in the modeling, we adopted the Boruta method [27] to reduce the noise from unimportant features.

2.2.5. Modeling Evaluation

Herein, we employed a machine-learning model and optimized its hyperparameters using grid-search cross-validation, which is a popular method for hyperparameter tuning that works systematically through multiple combinations of parameter tunings. The scoring metric used to evaluate the performance of the model was the coefficient of determination, denoted as R^2, which represents the proportion of variance for a dependent variable that is explained by the independent variables. The R^2 values were calculated as follows:

$$R^2 = 1 - \frac{\sum_1^n (y_i - \hat{y}_i)^2}{\sum_1^n (y_i - \overline{y})^2}, \quad (2)$$

where $\hat{y}_i$, y_i, and $\overline{y}$ represent the predicted, actual, and mean values of the actual output, respectively. The dataset comprised 537 samples, and ten-fold cross-validation was applied to calculate the accuracy. For the computations, 90% of the data were allocated as training data, and the remaining 10% as test data. This computation was conducted ten times, ensuring that all data were used as test data at some point. The average value of the ten R^2 scores of the foldout data was accepted as the model performance. The modeling was executed five times with different random seeds, which were applied to the modeling of tree models and cross-validation split, and the averaged values were calculated as a result of accuracy.

Table 3. Hyperparameter set for modeling.

Model Name	Hyperparameter	Values
Extra tree regressor	The number of trees	10, 50, 100, 500, 1000, 2000
	The number of features to consider when looking for the best split	(the number of features)$^{0.5}$, $\log_2$(the number of features), the number of features
	The maximum depth of the tree	10, 20, 30
Random forest regressor	The number of trees	10, 50, 100, 500, 1000, 2000
	The number of features to consider when looking for the best split	sqrt, log2, none
	The maximum depth of the tree	10, 20, 30
Support vector regressor	C	0.1, 1, 10
	Epsilon	0.01, 0.1, 1
	Kernel	Linear, poly, rbf
Partial least squares regressor	The number of components	1 to 20 (integer)
Lasso regressor	Alpha	0.0001, 0.001, 0.01, 0.1, 10

2.3. *In Silico Formulation*

To evaluate whether the AI models could support human formulators, formulations were virtually created with a computer using the rules described below. We call this procedure the 'in silico formulation'.

- All ingredients were assigned to one of six categories (the same categories described in Table 1): anionic surfactants, amphoteric surfactants, nonionic surfactants, polyols, a pH adjuster (only citric acid), and a base (only water).
- To compare the predicted and actual cleansing capabilities, the selection of anionic and amphoteric surfactants was restricted to one type: the anionic surfactant was restricted to potassium cocoyl glutamate, and amphoteric surfactant was restricted to lauramidopropyl hydroxysultaine.
- Only one ingredient was selected from each category; for example, two nonionic surfactants could not be selected for one formulation.
- The addition rates of each ingredient, except for the pH adjuster (citric acid) and water, were randomized for each category within the predefined ranges described in Table 4. The addition rate of citric acid was fixed with the value of 0.8 weight%, and the addition rate of water was calculated such that the sum of all the ingredients was 100%.
- In the procedure, 10^5 formulations were made, which were predicted with the best model described in Section 2.2.4.

Table 4. Condition of in silico formulation.

Category	Material Name	Randomize the Addition Rate	Minimum Weight% *	Maximum Weight% *
Anionic surfactant	Potassium cocoyl glutamate	Yes	3	20
Amphoteric surfactant	Lauramidopropyl hydroxysultaine	Yes	3	20
Nonionic surfactant	Eicosaglycerol hexacaplyrate PPG-20 glyceryl triisosterarate etc. (24 kinds in total)	Yes	0	10
Polyols	PPG-9 diglyceryl ether glycerin etc. (33 kinds in total)	Yes	0	30
pH adjuster	Citric acid	No	0.8% (fixed ratio)	
Base	Water	No	100-Σ (other material amount)	

* Water content was excluded in the weight% expression.

To validate the predictions made by the in silico formulation, the actual cleansing capabilities of some formulations were measured experimentally (the formulations of the measured samples are shown in the Section 3).

3. Results

3.1. Evaluation of AI Modeling

An AI model was established to predict the cleansing capability. The prediction accuracy of each model is listed in Table 5. The best prediction accuracy was obtained with the R^2 value of 0.770. The prediction accuracy increased significantly with the use of the descriptors. The results of the best model, extra tree regressor, using molecular descriptors, Hansen solubility index, and feature extraction are shown in Figure 3. An R^2 value of 0.770 translates to 15% when converted to a root-mean-square error (RMSE). We believe this accuracy level is sufficient for screening purposes, such as opting not to conduct low-predictive cleansing capability experiments before engaging in experiments using actual substances. Such preliminary filtering enables a more efficient allocation of resources to experiments with higher probabilities of success, thereby optimizing the overall research process.

Table 5. Prediction accuracy of each model.

	Mol/Weight Fraction	Descriptors	Hansen Solubility Index	Feature Extraction	Extra Tree Regressor	Random Forest Regressor	SVR	Lasso	PLS
1	Weight	Not used	Not used	Not used	0.661	0.652	0.488	0.526	0.503
2	Mol	Not used	Not used	Not used	0.668	0.643	0.469	0.503	0.472
3	Weight	Used	Not used	Not used	0.751	0.715	0.567	0.565	0.541
4	Mol	Used	Not used	Not used	0.753	0.751	0.509	0.526	0.519
5	Weight	Not used	Used	Not used	0.725	0.657	0.320	0.582	0.535
6	Mol	Not used	Used	Not used	0.719	0.654	0.305	0.556	0.505
7	Weight	Used	Used	Not used	0.748	0.711	0.568	0.582	0.543
8	Mol	Used	Used	Not used	0.751	0.749	0.524	0.537	0.526
9	Weight	Used	Used	Used	0.770	0.738	0.671	0.564	0.532
10	Mol	Used	Used	Used	0.768	0.753	0.620	0.523	0.504

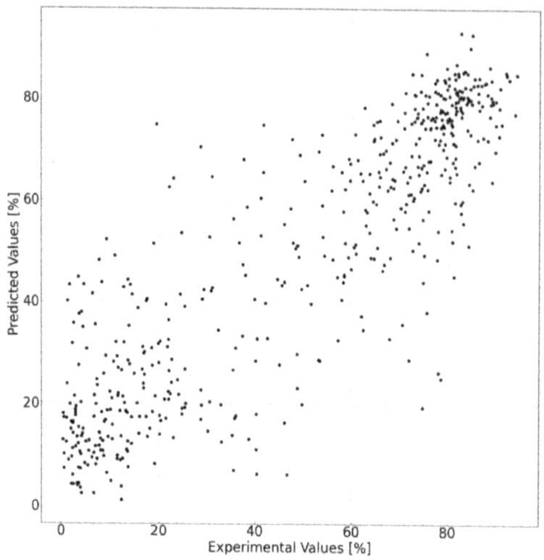

Figure 3. Experimental vs. predicted values of the cleansing capabilities of the cleansing foam formulations with the best model (10-fold cross validation was executed and all foldout data are shown in this plot).

3.2. *In Silico Formulation and Actual Cleansing Capabilities*

The cleansing capabilities of the in silico formulations are shown in Figures 4 and 5. A box with light and dark gray color in these figures indicates the middle 50% of the data (that is, the middle two quartiles of the data distribution), and horizontal bars display all points within 1.5 times the interquartile range (that is, all points within 1.5 times the width of the adjoining box), or all points at the maximum or minimum extent of the data. As shown in Figure 4, eicosaglycerol hexacaprylate exhibited the highest cleansing capability at both the median value (middle horizontal line in each box) and best value (top horizontal line).

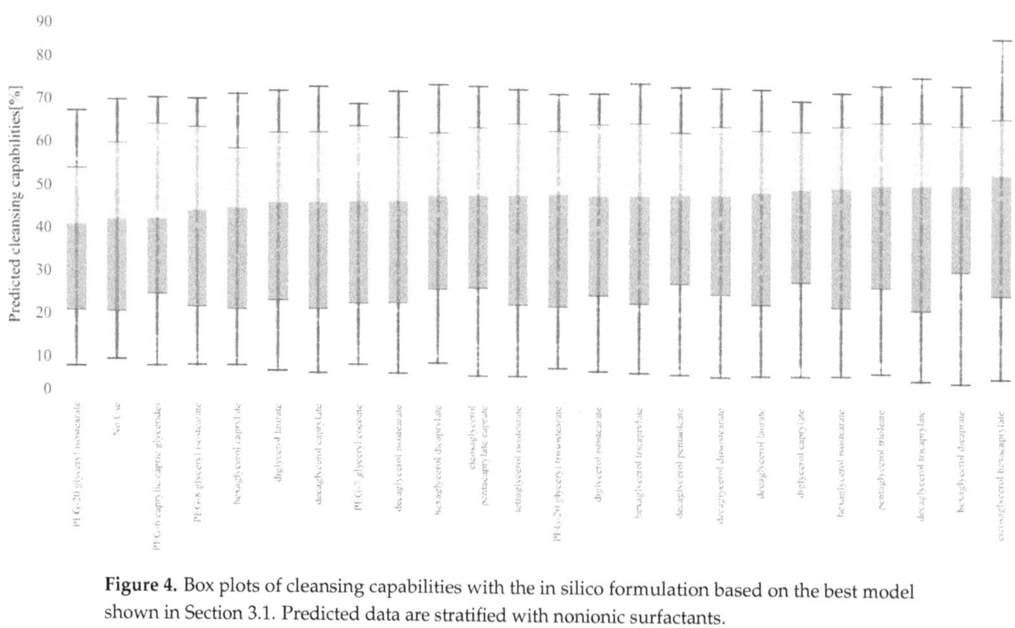

Figure 4. Box plots of cleansing capabilities with the in silico formulation based on the best model shown in Section 3.1. Predicted data are stratified with nonionic surfactants.

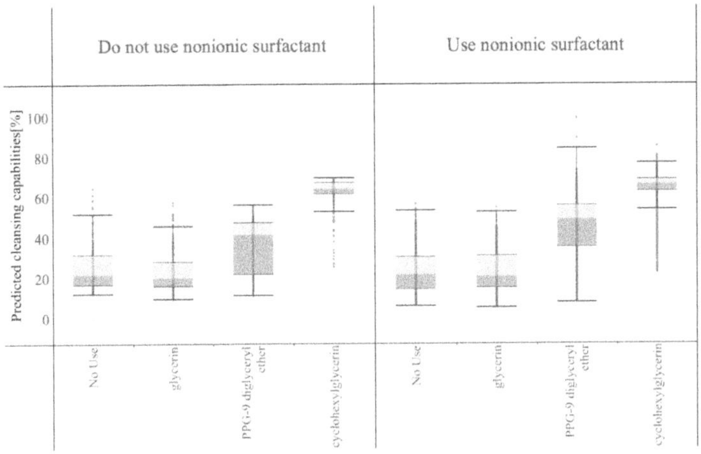

Figure 5. Box plots of predicted cleansing capabilities with the in silico formulation based on the best model shown in Section 3.1. Predicted data are stratified with nonionic surfactants and polyols.

In Figure 5, the formulation data were stratified into two categories: those not using nonionic surfactants and those using nonionic surfactants. Next, each category was stratified into subcategories based on the polyols to estimate their interactions with nonionic surfactants and polyols. The addition of nonionic surfactants increased the cleansing capabilities, and hydrophobic PPG-9 diglyceryl ether or cyclohexylglycerin, which have lower inorganic and organic balance (IOB) values, boosted the cleansing capability more than glycerin.

Several formulations were selected to validate the predicted data obtained from the in silico formulation, and their cleansing capabilities were measured. The formulations and results are shown in Table 6 and Figure 6. Nonionic surfactants eicosaglycerol hexacaprylate and cyclohexylglycerin/PPG-9 diglyceryl ether, (A) and (B), showed the highest cleansing capabilities among the actual formulations. Formulations with other nonionic surfactants and cyclohexylglycerin/PPG-9, (C) and (D), showed lower cleansing capabilities. Formulations with glycerin, (E) and (F), showed much lower cleansing capabilities, regardless of the type of nonionic surfactant. These tendencies correspond to the results shown in Figures 4 and 5.

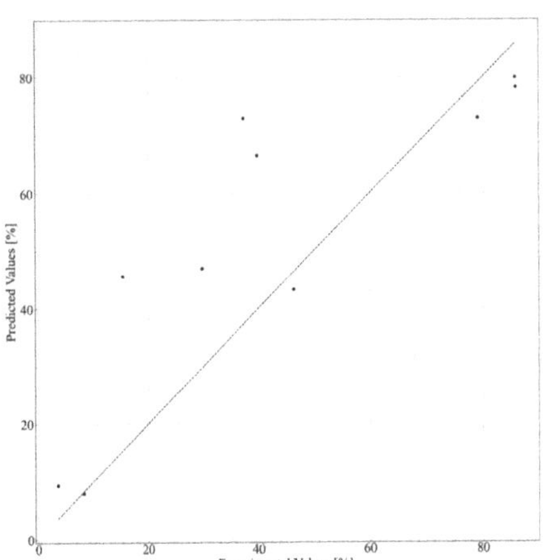

Figure 6. Experimental vs. predicted values of the cleansing capabilities for in silico formulations.

Table 6. Formulations for comparison of predicted and actual cleansing capabilities.

Category	Material Name	(A)	(B)	(C)	(D)	(E)	(F)
Anionic surfactant	Potassium cocoyl glutamate	6	7	13	8	8	7
Amphoteric surfactant	Lauramidopropyl Hydroxysultaine	5	9	4	9	7	7
Nonionic surfactants	Eicosaglycerol hexacaprylate	10	10			10	
	Decaglycerol isostearate			10			
	PEG-20 glyceryl triisostearate				10		10
Polyols	PPG-9 diglyceryl ether		11		11		
	Cyclohexylglycerin	9		14			
	Glycerin					11	13
pH adjuster	Citric acid	0.8	0.8	0.8	0.8	0.8	0.8
Base	Water	69.2	62.2	58.2	61.2	63.2	62.2
Cleansing capability test	Prediction/%	81.2	78.3	71.9	42.1	7.4	14.4
	Actual/%	85.8	85.9	39.8	46.3	8.3	3.6

4. Discussion

The use of descriptors increased the prediction accuracy of all the models, indicating that the chemical properties expressed as molecular descriptors successfully enabled the prediction of cleansing capabilities. HSPs improved the prediction accuracy for several models, but the accuracy was insufficient for models with descriptors, indicating that descriptors were more informative than HSPs.

Furthermore, weight% was suitable for linear-based models for the weighted average calculation, whereas no difference was observed for tree-based models. The mol% of the weighted average is potentially more accurate based on stoichiometry. However, because water constitutes > 97 mol% on average in the formulations owing to the high molecular weights of the surfactants, the influence of water was more dominant in the mol% calculation. Linear-based models were more affected by this influence than tree-based models. To predict cleansing capability, nonlinear behavior should also be considered owing to the interactions between surfactants and water molecules and their self-assembly. Tree-based models are typically more suitable for nonlinear predictions; therefore, their prediction accuracies are higher than those of linear-based models. The most accurate method is the elastic tree regressor. Because this method is based on decision tree models, its accuracy will decline when the data intended for inference fall into extrapolation regions relative to the training data. In such cases, additional experiments must be conducted to augment data and retrain the model. Although we employed descriptors to enhance the generalization capability of the model, it was estimated that the accuracy of predicting the cleansing capability of samples made with ingredients not present in the training data would be lower than the accuracy calculated in this study.

The raw ingredients used included polymer-based components such as PEG-20 glyceryl triisostearate; however, the prediction accuracy was maintained. In addition, as shown in Figure 4 and Table 6, the formulations using PEG-20 glyceryl triisostearate exhibited a lower cleansing performance than those using shorter-chain raw materials. This suggests that the length of the polymer may not be a key factor influencing the cleansing performance; instead, it is likely that the higher-order structure between the ingredients plays a more significant role in cleansing.

The in silico formulation helped us understand the effect not only of each material on cleansing capabilities but also of combinations of materials with the consequence of molecular interactions. The in silico formulation assisted in developing formulations with higher cleansing capabilities.

In this study, applying prediction and in silico formulation methods, we identified a cleansing foam formulation consisting of eicosaglycerol hexacaprylate and cyclohexylglycerin/PPG-9 that exhibited a high cleansing capability of >85% for the removal of waterproof eyeliners.

5. Conclusions

Using AI with machine learning, we built QSPR models that incorporated super-multicomponent ingredients, including polymers, to estimate the effects of surfactant self-assembly and chemical characteristics of the ingredients. An accuracy of R^2 = 0.770 (RMSE = 15%) was obtained to predict the cleansing performance, which was sufficient for ingredient screening. Nonlinear behavior, i.e., interactions among cosmetic ingredients in formulations, makes it more difficult for formulators to predict their performance. However, a high accuracy was obtained by incorporating chemical characteristics with descriptors. Based on the molecular structure of the ingredients and surfactant self-assembly, this AI prediction model showed higher accuracy than conventional approaches, such as multiple linear regression. Using in silico formulations, formulators can obtain information on which ingredients should be selected to achieve the highest cleansing capabilities. These findings suggest that cleansing performance is not merely dependent on the polymer length of the ingredients but also on the higher-order structures resulting from the interactions between the ingredients. This prediction model and in silico formulation significantly reduce the effort required for cosmetic development. In summary, we achieved the following results:

- A QSPR model was constructed for super-multicomponent ingredients, including polymers, achieving an accuracy of R^2 = 0.770 (RMSE = 15%), sufficient for product development screening.
- Using an in silico formulation, we predicted the optimal combination of the ingredients.
- The application of these technologies reduces the developmental effort and optimizes the overall development process.

Herein, we demonstrated the applicability of QSPRs to multicomponent mixtures, focusing specifically on cleansing foams. Based on the insights provided by QSPRs, we successfully identified an optimal combination of ingredients suitable for product development. We believe this methodology has high generalizability, facilitating the discovery of ideal combinations with minimal experimentation in various fields, not limited to cosmetics but also in drug development or other material designs. Therefore, the QSPR-based approach can emerge as a potent tool, yielding significant benefits in these industries.

Author Contributions: Conceptualization, M.H. and N.A.; methodology, M.H. and H.M.; software, M.H.; validation, H.M., R.N. and N.A.; formal analysis, M.H.; investigation, M.H. and H.M.; resources, H.M.; data curation, H.M.; writing—original draft preparation, M.H.; writing—review and editing, H.M., R.N. and N.A.; visualization, M.H.; supervision, N.A.; project administration, N.A.; funding acquisition, M.H. and H.M. All authors have read and agreed to the published version of the manuscript.

Funding: This research received no external funding.

Institutional Review Board Statement: Not applicable.

Data Availability Statement: The data presented in this study are available on request from the corresponding author. However, some data are not publicly available because the database comprising individual formulation data is utilized for product development.

Conflicts of Interest: The authors declare no conflict of interest.

References

1. Wu, J.; Wang, S.; Zhou, L.; Ji, X.; Dai, Y.; Dang, Y.; Kraft, M. Deep-learning architecture in QSPR modeling for the prediction of energy conversion efficiency of solar cells. *Ind. Eng. Chem. Res.* **2020**, *59*, 18991–19000. [CrossRef]
2. Hayashi, Y.; Marumo, Y.; Takahashi, T.; Nakano, Y.; Kosugi, A.; Kumada, S.; Hirai, D.; Takayama, K.; Onuki, Y. In silico predictions of tablet density using a quantitative structure-property relationship model. *Int. J. Pharm.* **2019**, *558*, 351–356. [CrossRef] [PubMed]
3. Mallakpour, S.; Hatami, M.; Golmohammadi, H. Prediction of inherent viscosity for polymers containing natural amino acids from the theoretical derived molecular descriptors. *Polymer* **2010**, *51*, 3568–3574. [CrossRef]
4. Gordillo, M.A.; Benavides, P.A.; Ma, K.; Saha, S. Transforming an insulating metal–organic framework (MOF) into semiconducting MOF/gold nanoparticle (AuNP) and MOF/polymer/AuNP composites to gain electrical conductivity. *ACS Appl. Mater. Interfaces* **2022**, *5*, 13912–13920. [CrossRef]
5. Xiu, X.; Mao, W.; Gong, J.; Liu, H.; Shao, Y.; Sun, L.; Wang, H.; Wang, C. Enhanced electrochemical performance of PEO-based composite polymer electrolyte with single-ion conducting polymer grafted SiO_2 nanoparticles. *Polymers* **2023**, *15*, 394.
6. Tekell, M.C.; Nikolakakou, G.; Glynos, E.; Kumar, S.K. Ionic conductivity and mechanical reinforcement of well-dispersed polymer nanocomposite electrolytes. *ACS Appl. Mater. Interfaces* **2023**, *15*, 30756–30768. [CrossRef]
7. Ortac, B.; Mutlu, S.; Baskan, T.; Yilmaz, S.S.; Yilmax, A.H.; Erol, B. Thermal conductivity and phase-change properties of boron nitride–lead oxide nanoparticle-doped polymer nanocomposites. *Polymers* **2023**, *15*, 2326. [CrossRef]
8. Yu, C.; Gong, X.; Wang, M.; Li, L.; Ren, S. Hyper-cross-linked nanoparticle reinforced composite polymer electrolytes with enhanced ionic conductivity and thermal stability for lithium-ion batteries. *ACS Appl. Mater. Interfaces* **2023**, *5*, 1509–1519. [CrossRef]
9. Gaudin, T.; Rotureau, P.; Fayet, G. Mixture descriptors toward the development of quantitative structure–property relationship models for the flash points of organic mixtures. *Ind. Eng. Chem. Res.* **2015**, *54*, 6596–6604. [CrossRef]
10. Abbasi, A.; Eslamloueyan, R. Determination of binary diffusion coefficients of hydrocarbon mixtures using MLP and ANFIS networks based on QSPR method. *Chemom. Intellig. Lab. Syst.* **2014**, *132*, 39–51. [CrossRef]
11. Faramarzi, Z.; Abbasitabar, F.; Zare-Shahabadi, V.; Jahromi, H.J. Novel mixture descriptors for the development of quantitative structure−property relationship models for the boiling points of binary azeotropic mixtures. *J. Mol. Liq.* **2019**, *296*, 111854. [CrossRef]
12. Khan, P.M.; Rasulev, B.; Roy, K. QSPR modeling of the refractive index for diverse polymers using 2D descriptors. *ACS Omega* **2018**, *3*, 13374–13386. [CrossRef] [PubMed]
13. Chatterjee, M.; Roy, K. Prediction of aquatic toxicity of chemical mixtures by the QSAR approach using 2D structural descriptors. *J. Hazard. Mater.* **2021**, *408*, 124936. [CrossRef]
14. Baghban, A.; Sasanipour, J.; Sarafbidabad, M.; Piri, A.; Razavi, R. On the Prediction of Critical Micelle Concentration for Sugar-Based Non-Ionic Surfactants. *Chem. Phys. Lipids* **2018**, *214*, 46–57. [CrossRef]
15. Cheng, K.C.; Khoo, Z.S.; Lo, N.W.; Tan, W.J.; Chemmangattuvalappil, N.G. Design and Performance Optimisation of Detergent Product Containing Binary Mixture of Anionic-Nonionic Surfactants. *Heliyon* **2020**, *6*, e03861. [CrossRef] [PubMed]
16. Sobati, M.A.; Abooali, D.; Maghbooli, B.; Najafi, H. A New Structure-Based Model for Estimation of True Critical Volume of Multi-Component Mixtures. *Chemom. Intellig. Lab. Syst.* **2016**, *155*, 109–119. [CrossRef]
17. Muratov, E.N.; Varlamova, E.V.; Artemenko, A.G.; Polishchuk, P.G.; Kuz'min, V.E. Existing and developing approaches for QSAR analysis of mixtures. *Mol. Inform.* **2012**, *31*, 202–221. [CrossRef]
18. Watanabe, K.; Sakurai, N.; Meno, T.; Yasuda, C.; Takahashi, S.; Hori, A.; Tsuchiya, K.; Sakai, K. Novel spontaneous cleansing feature of foam–Hybrid bicontinuous-microemulsion-type foamy makeup remover. *J. Soc. Cosmet. Chem. Jpn.* **2021**, *55*, 19–27. [CrossRef]
19. Watanabe, K.; Masuda, M.; Nakamura, K.; Inaba, T.; Noda, A.; Yanagida, T.; Yanaki, T. A new makeup remover prepared with a system comprising dual continuous channels (bicontinuous phase) of silicone oil and water. *IFSCC Mag.* **2004**, *7*, 310–318. [CrossRef]
20. Iwanaga, T.; Uchida, K.; Takeuchi, N.; Abe, Y. Development of oil-type make-up remover prepared with polyglycerol fatty acid esters. *J. Soc. Cosmet. Chem. Jpn.* **2005**, *39*, 186–194. [CrossRef]
21. Jiao, L.; Wang, Y.; Qu, L.; Xue, Z.; Ge, Y.; Liu, H.; Lei, B.; Gao, Q.; Li, M. Hologram QSAR Study on the Critical Micelle Concentration of Gemini Surfactants. *Colloids Surf. A Physicochem. Eng. Asp.* **2020**, *586*, 124226. [CrossRef]
22. Absalan, G.; Hemmateenejad, B.; Soleimani, M.; Akhond, M.; Miri, R. Quantitative Structure–micellization Relationship Study of Gemini Surfactants Using Genetic-PLS and Genetic-MLR. *QSAR Comb. Sci.* **2004**, *23*, 416–425. [CrossRef]
23. RDKit: Open-Source Cheminformatics Software. Available online: https://www.knime.com/rdkit (accessed on 5 September 2023).
24. Yap, C.W. PaDEL-descriptor: An open source software to calculate molecular descriptors and fingerprints. *J. Comput. Chem.* **2011**, *32*, 1466–1474. [CrossRef] [PubMed]
25. Faasen, D.P.; Jarray, A.; Zandvliet, H.J.; Kooij, E.S.; Kwiecinski, W. Hansen solubility parameters obtained via molecular dynamics simulations as a route to predict siloxane surfactant adsorption. *J. Colloid Interface Sci.* **2020**, *575*, 326–336. [CrossRef] [PubMed]

26. Afzal, O.; Alshammari, H.A.; Altamimi, M.A.; Hussain, A.; Almohaywi, B.; Altamimi, A.S. Hansen solubility parameters and green nanocarrier based removal of trimethoprim from contaminated aqueous solution. *J. Mol. Liq.* **2022**, *361*, 119657. [CrossRef]
27. Kursa, M.B.; Jankowski, A.; Rudnicki, W.R. Boruta—A system for feature selection. *Fundam. Inform.* **2010**, *101*, 271–285. [CrossRef]

polymers

MDPI

Article

Estimation and Prediction of the Polymers' Physical Characteristics Using the Machine Learning Models

Ivan Pavlovich Malashin [1,*], Vadim Sergeevich Tynchenko [1,2,3,*], Vladimir Aleksandrovich Nelyub [1], Aleksei Sergeevich Borodulin [1] and Andrei Pavlovich Gantimurov [1]

1 Artificial Intelligence Technology Scientific and Education Center, Bauman Moscow State Technical University, 105005 Moscow, Russia; vladimir.nelub@emtc.ru (V.A.N.); alexey.borodulin@emtc.ru (A.S.B.); agantimurov@emtc.ru (A.P.G.)
2 Information-Control Systems Department, Institute of Computer Science and Telecommunications, Reshetnev Siberian State University of Science and Technology, 660037 Krasnoyarsk, Russia
3 Department of Technological Machines and Equipment of Oil and Gas Complex, School of Petroleum and Natural Gas Engineering, Siberian Federal University, 660041 Krasnoyarsk, Russia
* Correspondence: ivan.p.malashin@gmail.com (I.P.M.); vadimond@mail.ru (V.S.T.); Tel.: +7-926-875-7128 (I.P.M.)

Abstract: This article investigates the utility of machine learning (ML) methods for predicting and analyzing the diverse physical characteristics of polymers. Leveraging a rich dataset of polymers' characteristics, the study encompasses an extensive range of polymer properties, spanning compressive and tensile strength to thermal and electrical behaviors. Using various regression methods like Ensemble, Tree-based, Regularization, and Distance-based, the research undergoes thorough evaluation using the most common quality metrics. As a result of a series of experimental studies on the selection of effective model parameters, those that provide a high-quality solution to the stated problem were found. The best results were achieved by Random Forest with the highest R^2 scores of 0.71, 0.73, and 0.88 for glass transition, thermal decomposition, and melting temperatures, respectively. The outcomes are intricately compared, providing valuable insights into the efficiency of distinct ML approaches in predicting polymer properties. Unknown values for each characteristic were predicted, and a method validation was performed by training on the predicted values, comparing the results with the specified variance values of each characteristic. The research not only advances our comprehension of polymer physics but also contributes to informed model selection and optimization for materials science applications.

Keywords: physical characteristics analysis; machine learning; polymers; predictive analytics; random forest; properties

1. Introduction

The article explores the application of ML techniques in predicting and analyzing the physical characteristics of polymers. Harnessing the power of ML algorithms, the study delves into diverse polymer properties, ranging from compressive and tensile strength to thermal and electrical behavior. The prediction of physical characteristics in polymers is of paramount importance, spanning various industrial and scientific applications. This predictive capability not only enhances our fundamental understanding of polymer behavior [1] but also catalyzes advancements in materials science [2], manufacturing processes [3], and product development [4]. Let us describe the following examples of the polymers' properties prediction needs:

- **Material Design and Engineering**. Precise predictions of properties such as tensile strength, elasticity, and thermal conductivity empower material scientists in designing polymers with tailored attributes [5]. This facilitates the creation of innovative materials for specific applications, ranging from lightweight composites in aerospace engineering [6] to durable polymers in medical devices [7].

- **Process Optimization.** Understanding and predicting physical characteristics play a crucial role in optimizing manufacturing processes. For instance, predicting melt viscosity in polymer processing aids [8] in controlling the extrusion process, ensuring the production of consistent and high-quality polymer products [9].
- **Quality Control in Polymer Manufacturing**. The ability to predict physical characteristics is instrumental in quality control within polymer manufacturing [10]. Predictive models can assist in identifying deviations in real-time, enabling timely adjustments in the production process to maintain desired material properties.
- **Environmental Impact Assessment**. Predicting properties is essential in determining their biodegradability and recyclability [11]. It contributes to the assessment of a polymer's environmental impact. This knowledge is particularly relevant in the development of sustainable materials, aligning with the growing emphasis on eco-friendly practices.
- **Pharmaceutical and Medical Applications**. In the field of pharmaceuticals, predicting characteristics can help to determine drug release rates from polymer matrices [12]. It is vital for designing controlled drug delivery systems. Similarly, in medical applications, predicting the mechanical properties of biocompatible polymers is crucial for developing implants and medical devices.

The research employs a variety of regression models, including Lasso Regression [13], Elastic Net [14], Decision Tree Regressor [15], Bagging Regressor [16], AdaBoost Regressor [17], XGBoost Regressor [18], Support Vector Regressor [19], Gradient Boosting Regressor [20], Linear Regression [21], and Random Forest Regressor [22].

Lasso Regression shines in feature selection by inducing sparsity through the regularization of some coefficients to zero [23]. While promoting model simplicity, it does come with the caveat of potentially discarding relevant features and displaying sensitivity to outliers.

Linear Regression, known for its simplicity and interpretability, is suitable for capturing linear relationships [24]. However, its assumption of linearity may limit its performance with intricate, non-linear data. On the other hand, Polynomial Regression, offering flexibility to capture non-linear relationships, is susceptible to overfitting, particularly with higher-degree polynomials.

Support Vector Regression (SVR), effective in high-dimensional spaces and robust to outliers, demands careful selection of kernel and parameters due to its computational intensity [25]. Decision Tree Regression, with its capability to handle non-linearity and interactions, is visually interpretable but prone to overfitting and sensitive to small variations in data.

Random Forest Regression, an ensemble of decision trees, mitigates overfitting but introduces complexity and challenges in interpretation [26].

Gradient Boosting Regression, known for its high predictive accuracy by correcting errors of previous models sequentially, is susceptible to overfitting and requires meticulous hyperparameter tuning [27].

Elastic Net combines the strengths of Lasso and Ridge Regression, offering a balance between feature selection and regularization. However, navigating the optimal mix of L1 and L2 penalties poses a challenge [28].

Decision Tree Regressor excels in capturing non-linear relationships and intricate interactions within the data. Its visual interpretability is a notable asset, but caution is warranted as decision trees are susceptible to overfitting, particularly with complex data [29].

Bagging Regressor, an ensemble technique, mitigates overfitting by aggregating the predictions of multiple decision trees. While enhancing model robustness, it introduces complexity and may be less interpretable [30].

AdaBoost Regressor focuses on sequentially improving model performance by emphasizing misclassified instances. It tends to be less prone to overfitting but is sensitive to noisy data [31].

Gradient Boosting Regressor iteratively builds models, correcting the errors of previous ones [32]. It boasts high predictive accuracy but demands careful parameter tuning to avoid overfitting.

XGBoost Regressor, an extension of Gradient Boosting, excels in predictive accuracy and handles missing data effectively [33]. However, it necessitates careful tuning of hyperparameters and can be computationally intensive.

When generating input for models predicting various physical characteristics of polymers, a diverse set of features such as melting temperature, density and others, and processing conditions are meticulously considered. The inclusion of these multifaceted attributes ensures a comprehensive representation of the intricate relationships governing the polymers' behavior, enhancing the models' predictive capabilities.

Each model undergoes rigorous assessment using metrics such as Mean Squared Error [34], R-squared [35], Root Mean Squared Error [36], Normalized Mean Squared Error [37], Mean Absolute Error [38], and Mean Percentage Error [39]. Due to the varying dimensions of the characteristics and the unequal number of non-zero values for each characteristic, it did not make sense to consider Mean Squared Error (MSE) and Mean Absolute Error (MAE). Since Normalized Mean Squared Error (NMSE) is expressed as $1 - R^2$, only the coefficient of determination (R^2) and Mean Percentage Error (MPE) were considered as objective metrics. The outcomes are then compared and contrasted, shedding light on the effectiveness of different ML approaches for predicting polymer properties.

The findings not only contribute to advancing the understanding of polymer physics but also offer valuable insights into the selection and optimization of ML models for materials science applications. This research is a significant step towards leveraging ML to enhance our comprehension of complex material behaviors, paving the way for more efficient and accurate predictions in polymer science.

2. Materials and Methods

2.1. Dataset Preparation

The original dataset contained information on 66,981 different characteristics [40] of polymer materials, representing 18,311 unique polymers with 99 unique physical characteristics, each characterized by varying quantities of known physical attributes [41]. Among these characteristics is crucial information in the form of Simplified Molecular Input Line Entry System (SMILES) strings.

In Figures 1 and 2, the vertical bars represent the count of non-null values for each characteristic across the dataset. The index corresponds to the names of the characteristics, and the vertical axis indicates the count of non-null values. For understanding the completeness of the dataset the numerical annotations on top of each bar provided.

Tables A1 and A2 provide an overview of key characteristics, including counts, means, standard deviations, minimum and maximum values, medians, and units, offering a comprehensive understanding of the dataset under consideration.

The SMILES strings in the dataset adds a significant dimension to the information available for each polymer material [42]. SMILES provides a standardized and human-readable representation of the chemical structure of molecules. This chemical notation system not only facilitates the accurate identification of distinct polymers but also opens avenues for exploring the relationship between molecular structure and physical characteristics.

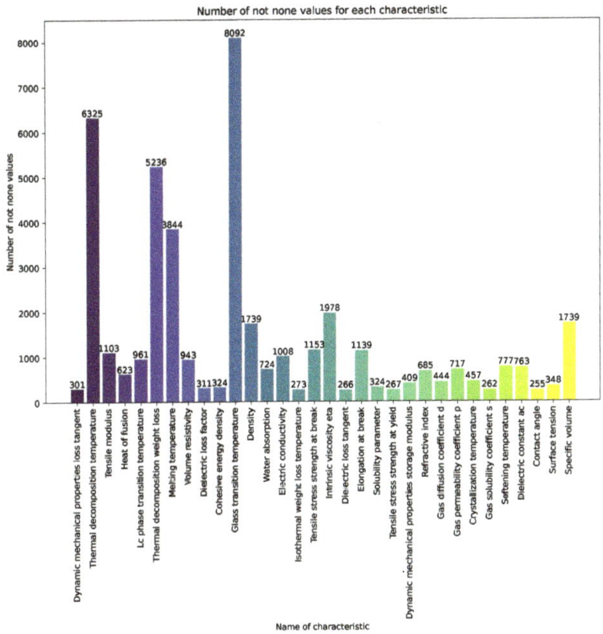

Figure 1. Count of non-null values for each characteristic across the dataset for a count exceeding 250.

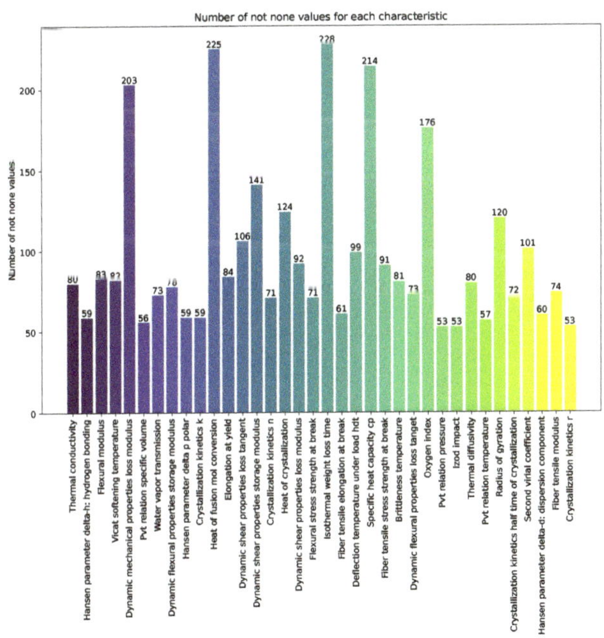

Figure 2. Count of non-null values for each characteristic across the dataset for a count ranging from 50 to 250.

The representation of the dataset transformation process is shown in Figure 3.

Figure 3. Schematic representation of the dataset transformation process. The original dataset undergoes a series of transformations to create datasets of varying dimensions, each tailored for training on individual physical characteristics. The vectorization involves encoding the SMILES notations into 1024 binary features, facilitating machine learning model training on diverse molecular attributes. Then Prediction Imputation has been used to estimate unknown values for each characteristic. Subsequently, a validation of the method was conducted by training on the predicted values and comparing the outcomes with the specified variance values associated with each characteristic.

For each polymer, there was information on the median value of the physical characteristic and the possible variance, although often information about the variance was missing. None of the polymers had complete information on all characteristics.

To initiate the machine learning process, the original dataset underwent a structural transformation. Each row now represents the following structure: the first column contains the material's name, the second contains the corresponding SMILES string, the third indicates the number of known characteristics for that material, and the fourth lists the names of these characteristics. The subsequent 98 columns contain the median values of all characteristics, and another 98 columns contain the range values for each of these characteristics. This new data structure provides convenience for further analysis and the application of machine learning methods.

The process of vectorizing SMILES into a binary feature vector using RDKit Python library is a crucial step in the analysis of polymer materials [43]. SMILES serves as a string representation of chemical compound structures, and its vectorization is a key stage for applying machine learning methods. To achieve this transformation, a technique is utilized that assigns a unique binary code to each SMILES character. The resulting binary vectors, with a length of 1024, constitute a set of bits reflecting the chemical structure of compounds. This process provides an efficient representation of information about the molecular structure, making it accessible for analysis and processing by machine learning algorithms. Through the vectorization of SMILES, unique numerical representations are created, serving as a valuable tool in addressing tasks related to predicting the physical characteristics of polymers.

2.2. Model Training for Predicting the Physical Characteristics of Polymer

In the process of preparing the dataset for predicting the physical characteristics of polymers, multiple transformations were applied to create an optimal data structure. The original dataset, comprising 66,981 unique characteristics of various polymer materials, included information about median values and dispersion. However, this information was often incomplete. To enhance the efficiency of machine learning model training, it was decided to iteratively create new datasets, each consisting of 1024 columns for representing SMILES and an additional column for each physical characteristic containing non-empty values.

Subsequently, each of these created datasets was split into training and testing sets at an 80% to 20% ratio, respectively. In the training phase, diverse machine learning regression models, including but not limited to KNeighborsRegressor, Lasso, Elastic Net, Decision Tree, Bagging, AdaBoost, XGBoost, SVR, Gradient Boosting, Linear Regression, and Random Forest, were utilized to optimize the prediction of physical characteristics in polymer materials. Model performance was evaluated using metrics like MSE (Mean Squared Error), RMSE (Root Mean Squared Error), NMSE (Normalized Mean Squared Error), MAE (Mean Absolute Error), MPE (Mean Percentage Error), R^2. Additionally, a custom metric was introduced, accounting for the difference between predicted and true values, considering a predefined non-zero dispersion value. The obtained evaluation results enable more effective utilization of trained models for predicting the physical characteristics of polymer materials.

Hyperparameter optimization has been conducted for each model to maximize its predictive capability. Techniques such as grid search, random search to systematically explore the hyperparameter space and identify configurations that yield improved model performance [44].

Subsequently, all the obtained metrics for each feature with post-training on every model were saved in separate files. Following this, a graph analytical processing of these files was conducted to determine the optimal machine learning models for each characteristic.

2.3. *Using Prediction Method for Imputation of Missing Values of Polymer Physical 98 Characteristics*

In contemporary polymer research, extensive datasets of physical characteristics are often analyzed, providing valuable information about material properties. However, the data collection process introduces the challenge of missing values, creating a hurdle in accurately reconstructing the complete dataset. This study introduces a novel approach to address this issue, based on the Prediction Imputation method.

The Prediction Imputation method [45] is a way to fill missing values in data by utilizing machine learning models. In this research, we applied this method to predict missing values for each polymer's physical characteristic, with the number of missing values varying for each characteristic.

The process involved selecting a suitable machine learning regression model, training it on known data, and then using the trained model to predict values where they were missing. The evaluation of the method included comparing predicted values with real ones, where available.

This innovative approach to handling missing data opens new perspectives for accurate analysis of polymer physical characteristics, improving data recovery and providing more reliable research results.

The analysis of obtained metrics identified optimal regressors for each characteristic, forming a diverse set of best machine learning models. Each applied model was saved using the joblib library for subsequent use.

Subsequently, in accordance with information about the best models, missing values for each characteristic were predicted using the corresponding optimal regressor. These predicted values were merged with the known values, creating a dataset where all characteristics were filled according to the best models used.

Thus, this approach not only efficiently utilizes predictive models for recovering missing data but also allows adapting model selection for each specific characteristic, ensuring more accurate investigation of polymer physical properties.

2.4. *Examination of Our Approach*

To assess the quality of predicted characteristic values, the same series of experiments were conducted to evaluate the consistency between predicted and actual data. For each of the 66 characteristics (for three out of 68 characteristics for which the number of non-zero values was initially greater than 50, the model could not be saved), where the initially known values exceeded 50, an 11-fold experiment was performed.

The specificity of the experiment involved using only predicted values as the training set, while the test set consisted of actually known characteristic values. This approach allowed for evaluating the accuracy of predictive models, considering real data, and provided more reliable indicators than using random or other sample separation methods.

Consistency assessment was conducted using the variance metric. The results of these experiments provide information about the degree of alignment between predicted values and actual data for each regression model, as well as a comprehensive picture across all characteristics.

An important implication of these experiments is the possibility of selecting the most effective models for each specific characteristic, ultimately enhancing the accuracy and reliability of predicting polymer physical property values. The obtained assessments can be utilized to choose optimal regressors for further research in materials science and polymer science.

2.5. *Categories of Characteristics*

The dataset encompasses a diverse array of physical characteristics, each contributing valuable insights into the multifaceted nature of polymer materials. These characteristics are systematically categorized to capture the wide-ranging aspects of a material's behavior. Compression characteristics and tensile property delve into the material's response to forces, providing crucial information about its strength and deformability. Creep characteristics illuminate the material's behavior under sustained loads over time, offering insights into long-term structural integrity. Dilute solution property and rheological property focus on the material's behavior in solution and its flow properties, respectively, aiding in applications like polymer processing.

The dataset also includes categories such as electric property, shedding light on the material's conductivity and dielectric properties. Flexural property and shear property offer a nuanced understanding of the material's response to bending and shearing forces, respectively. Hardness quantifies the material's resistance to indentation or scratching, while impact strength gauges its ability to absorb sudden impacts. Optical property provides insights into light interaction, and heat characteristics and thermal property delve into the material's response to temperature changes, including its thermal conductivity and expansion.

Heat resistance and combustion characterize the material's performance under elevated temperatures, contributing to applications where heat stability is crucial. Other physical property and physicochemical property serve as comprehensive categories that encompass a broad spectrum of diverse properties, ensuring a holistic examination. This systematic categorization enhances the dataset's utility, facilitating targeted exploration and modeling of specific polymer traits for various industrial applications. Figure 4 illustrates the comprehensive spectrum of physical polymer characteristics explored in this study.

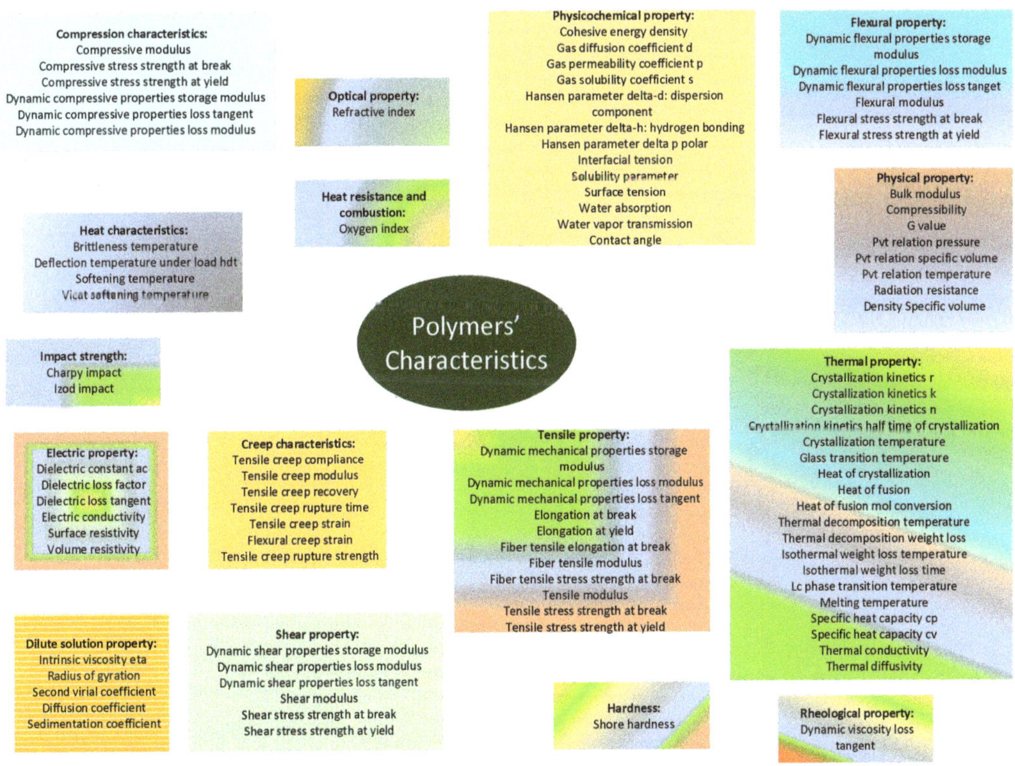

Figure 4. Graph depicting all possible characteristics of physical polymers featured in this study. The characteristics are grouped according to their categories, providing a systematic overview of various aspects of the physical properties of polymer materials.

The characteristics are systematically arranged based on their respective categories, offering a structured representation of diverse aspects of polymer material properties. The description of each physical characteristic presented in the dataset is provided in Appendix B.

3. Results

Experimental conditions involved transforming SMILES representations into binary features and training models individually on each characteristic using non-empty values. The experiment utilized an Intel(R) Core(TM) i7-7700 CPU @ 3.60 GHz for computational tasks [46].

In Figure 5, R^2 scores are illustrated for 68 characteristics, each of which has more than 50 non-zero values in the original dataset. Thirty-one (31) characteristics exhibit R^2 values within the range of 0.5 to 1.

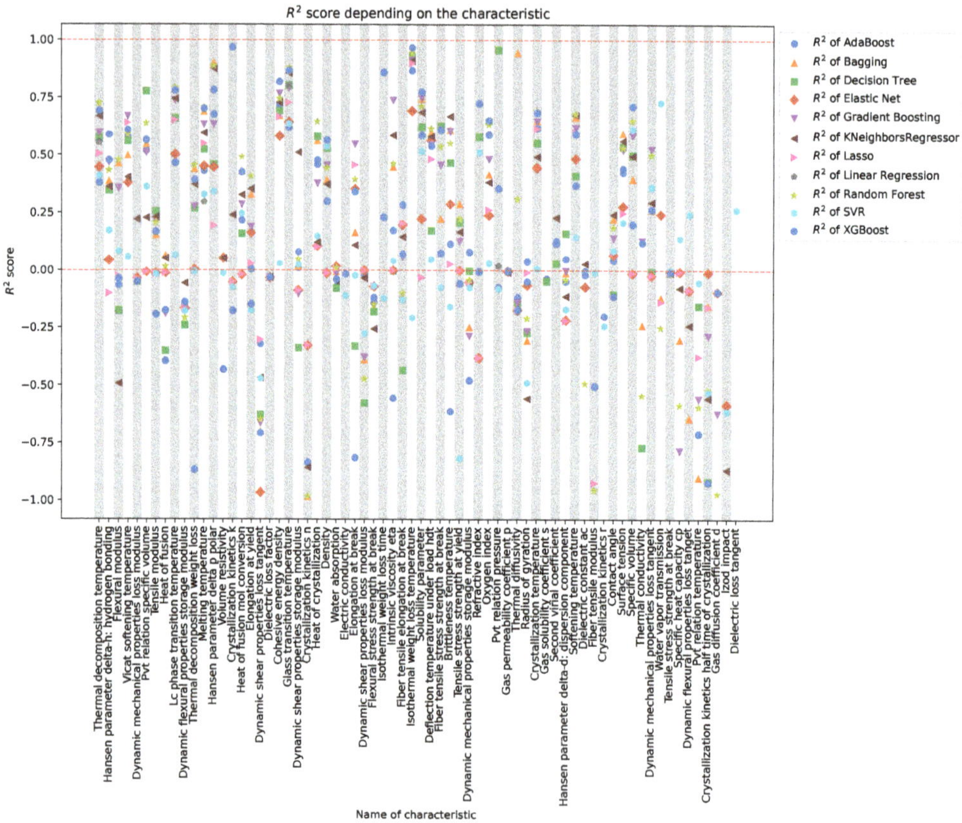

Figure 5. Array of R^2 Scores, each point is a testament to the mastery of machine learning models in deciphering the intricacies of physical traits.

The alignment of optimal metric values across all regression models for each characteristic highlights a consistent pattern. This alignment emphasizes the robust performance of machine learning models in predicting physical characteristics of polymers, particularly for the identified subset of characteristics. The coherence in results across various models underscores the reliability and effectiveness of the chosen models in capturing the underlying patterns in the dataset.

Optimal regression models and metrics for physical characteristics are shown in Table 1. The table presents the most effective regression models and associated metrics for predicting various physical characteristics of polymers. Each row corresponds to a specific characteristic, showcasing the selected regression model, the maximum R^2 score achieved, and the corresponding Normalized Mean Squared Error (NMSE). The models were carefully evaluated, and the results offer insights into the predictive performance for different characteristics in the polymer dataset.

Table 1. Optimal regression models and R^2 best scores for physical characteristics.

Characteristic	Data Size [1]	Best Regressor	Max R^2	MPE
Glass transition temperature	8092	Random Forest	0.88	1.23
Thermal decomposition temperature	6325	Random Forest	0.73	2.25
Melting temperature	3844	Random Forest	0.71	1.05
Intrinsic viscosity ETA	1978	Gradient Boosting	0.74	
Specific volume	1739	XGBoost	0.71	2.75
Density	1739	XGBoost	0.56	0.5
Elongation at break	1139	Gradient Boosting	0.55	
LC phase transition temperature	961	Random Forest	0.79	3.02
Softening temperature	777	Random Forest	0.68	20.73
Refractive index	685	XGBoost	0.73	0.91
Crystallization temperature	457	Random Forest	0.69	6.3
Surface tension	348	Bagging	0.59	0.06
Solubility parameter	324	XGBoost	0.77	0.04
Cohesive energy density	324	XGBoost	0.82	0.96
Dynamic mechanical properties loss tangent	301	Gradient Boosting	0.52	
Isothermal weight loss temperature	273	XGBoost	0.97	0.13
Isothermal weight loss time	228	XGBoost	0.86	
Oxygen index	176	XGBoost	0.65	12.24
Dynamic shear properties storage modulus	141	KNeighborsRegressor	0.51	
Heat of crystallization	124	Random Forest	0.65	
Deflection temperature under load HDT	99	Random Forest	0.61	4.0
Fiber tensile stress strength at break	91	Decision Tree	0.63	1.1
Vicat softening temperature	82	Gradient Boosting	0.67	0.45
Brittleness temperature	81	KNeighborsRegressor	0.67	1.2
Thermal diffusivity	80	Bagging	0.94	4.13
Water vapor transmission	73	SVR	0.73	
Hansen parameter delta p polar	59	Bagging	0.9	0.45
Hansen parameter delta-h: hydrogen bonding	59	AdaBoost	0.59	2.56
Crystallization kinetics k	59	XGBoost	0.97	
PVT relation specific volume	56	Decision Tree	0.78	0.01
PVT relation pressure	53	Decision Tree	0.96	

[1] The reported data size corresponds to the count of non-none values for each median parameter for each characteristic.

Figure 6 depicts a graph of variance metric values for all initially known characteristics of polymers. Different characteristics are marked on the x-axis, while the y-axis displays the values of the variance metric, measuring the degree of correspondence between predicted and actual values. Each stripe on the graph corresponds to one of the 66 characteristics, where the initial number of known values exceeds 50.

The presented graph allowing for a visual assessment of the predictive models' effectiveness for each specific characteristic. Stripes rising above indicate high accuracy in predictions, while those descending below may suggest some disparities between predicted and real values.

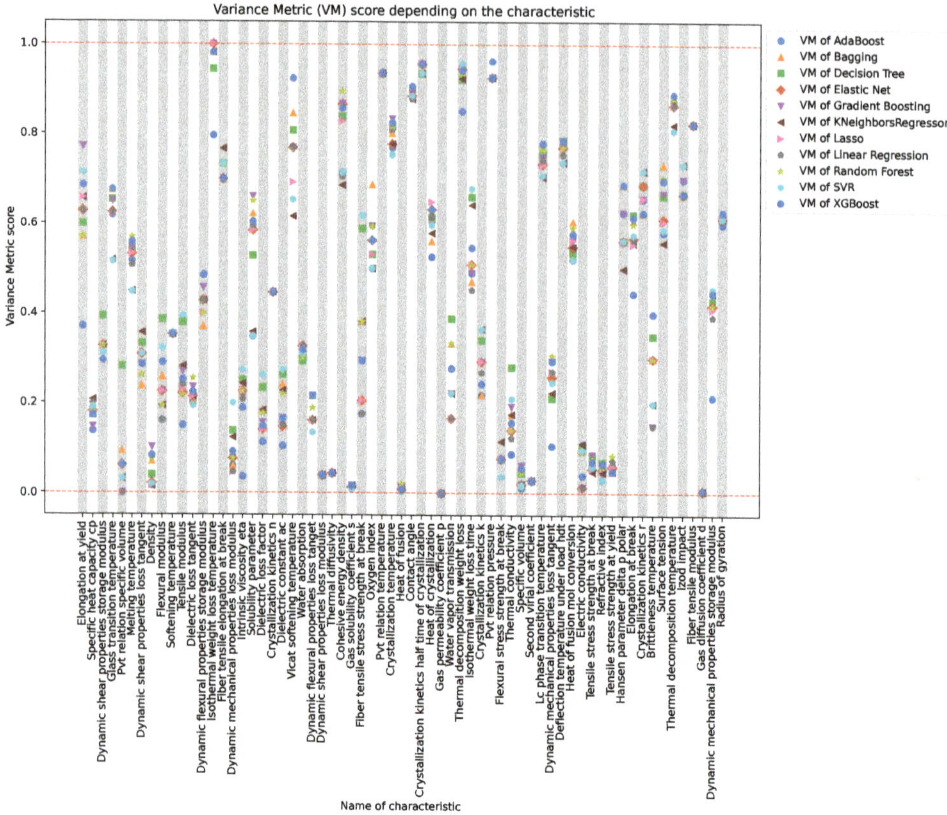

Figure 6. Array of VM Scores, each point is a testament to the mastery of machine learning models in deciphering the intricacies of physical traits.

Table 2 presents optimal variance metrics obtained by training on predicted values and subsequently predicting known values for validation, demonstrating the model's performance on the dataset.

Table 2. Optimal regression models and best variance metrics (VM) scores for physical characteristics.

Characteristic	Data Size [1]	Best Regressor	Max VM
Isothermal weight loss temperature	219	Elastic Net	1.0
PVT relation pressure	26	AdaBoost	0.96
Thermal decomposition weight loss	3567	SVR	0.96
Crystallization kinetics half time of crystallization	26	AdaBoost	0.96
PVT relation temperature	26	AdaBoost	0.94
Vicat softening temperature	56	Random Forest	0.92
Contact angle	116	Random Forest	0.91
Cohesive energy density	219	Random Forest	0.9
Thermal decomposition temperature	2968	XGBoost	0.89
Crystallization temperature	331	Gradient Boosting	0.83
Fiber tensile modulus	40	AdaBoost	0.82

Table 2. *Cont.*

Characteristic	Data Size [1]	Best Regressor	Max VM
Deflection temperature under load HDT	38	Bagging	0.79
LC phase transition temperature	430	XGBoost	0.78
Elongation at yield	49	Gradient Boosting	0.77
Fiber tensile elongation at break	31	KNeighborsRegressor	0.77
Izod impact	23	KNeighborsRegressor	0.73
Surface tension	176	Bagging	0.73
Crystallization kinetics r	21	KNeighborsRegressor	0.72
Oxygen index	144	Bagging	0.69
Hansen parameter delta p polar	43	Bagging	0.69
Isothermal weight loss time	175	SVR	0.68
Glass transition temperature	6278	Random Forest	0.68
Solubility parameter	218	Gradient Boosting	0.66
Heat of crystallization	67	Lasso	0.65
Radius of gyration	45	Bagging	0.63
Elongation at break	854	Decision Tree	0.62
Fiber tensile stress strength at break	57	SVR	0.62
Heat of fusion mol conversion	154	Bagging	0.61
Melting temperature	2182	Random Forest	0.57

[1] The reported data size corresponds to the count of non-null values for each variance parameter for each characteristic.

4. Discussion

In recent years, there has been substantial interest in accelerating materials design and discovery, spurred by initiatives like the Materials Genome Initiative and Integrated Computational Materials Engineering [47]. This perspective aims to outline general problems, information science methods, and outstanding challenges in the field of materials informatics [48]. For example, ref. [49] introduces Polymer Genome [50], a web-based machine-learning capability for near-instantaneous predictions of polymer properties. Study [51] explores computational alternatives, Group Interaction Modeling (GIM), and Machine Learning (ML), for predicting thermal and mechanical properties of polymers. The paper [52] addresses challenges in utilizing machine learning for polymer discovery, focusing on accurately representing complex, multi-scale structures. Ref. [53] introduces a promising CGCNN framework that directly learns material properties from crystal structures, offering a universal and interpretable representation. The paper [54] addresses the ongoing debate in molecular property prediction by comparing two prominent classes of models—neural networks applied to computed molecular fingerprints or expert-crafted descriptors and graph convolutional neural networks. The authors of [55] introduce a data-driven framework for predicting work functions of complex compounds, showcasing the effectiveness of a random forest model in achieving high accuracy. The paper [56] contributes significantly to the chemistry machine learning field by showcasing the potential of machine learning methods in predicting bulk properties of molecules, specifically crystalline density.

In this study, we observed distinct patterns in the performance of regression models across various characteristics, particularly focusing on the glass transition temperature, thermal decomposition temperature, and melting temperature, which exhibited substantial counts of non-null values (3844, 6325, and 8092, respectively).

This study presents an evaluation of various regression methods. The study does not merely apply ML algorithms but goes further, conducting experimental studies to select the best model for each physical characteristic. This meticulous approach showcases a commitment to refining models for improved predictive accuracy, providing valuable insights into the strengths and weaknesses of different ML approaches.

For characteristics with a higher number of non-null values, we noticed a proportional increase in the R^2 score as the data size expanded. Specifically, the R^2 scores for glass transition temperature, thermal decomposition temperature, and melting temperature were 0.71, 0.73, and 0.88, respectively. Random Forest emerged as the optimal regression model for these characteristics, showcasing its ability to handle larger datasets and capture complex relationships. These scores serve as a baseline for future work and highlight the inherent strengths and limitations of each regressor in its default configuration.

On the other hand, for characteristics with a data size ranging from 176 to 2000, XGBoost and Gradient Boosting demonstrated superior performance, outshining other regression models. These findings suggest that these boosting algorithms excel in capturing intricate patterns within datasets of a moderate size.

Interestingly, for characteristics with lower data sizes, ranging from 59 to 141, a mix of regression models, including Decision Tree, Bagging, KNeighborsRegressor, AdaBoost, and SVR, displayed competitive performance. The diversity in optimal models for these characteristics implies that the choice of the most suitable regression algorithm may depend on the specific characteristics of the dataset, and a one-size-fits-all approach might not be appropriate.

Several factors could contribute to these observations. Firstly, the complexity of the relationship between molecular features (captured by SMILES strings) and physical characteristics may vary across different characteristics, influencing the model's performance. Additionally, the nature of the dataset, including the distribution of non-null values and the diversity of polymer structures, might impact the effectiveness of certain regression models. Further investigations into the specific molecular features contributing to the predictive power of each model and a deeper understanding of the underlying chemical processes could provide valuable insights into the observed patterns.

The results of the analysis of the variance metric presented in Table 2 yielded insightful observations regarding the performance of predictive models based solely on predicted values derived from optimal regressors. Surprisingly, a significant alignment was observed between the majority of characteristics and the models, resulting in notably high variance metric scores. This consistency suggests a robust predictive capability of the chosen regressors across various physical properties of polymers.

Several factors contribute to the success of the models, while also shedding light on potential pitfalls. Characteristics exhibiting exceptionally high values, such as volume resistivity, might present challenges in prediction due to their intrinsic variability or non-linear dependencies on other factors. Additionally, features with substantial data dispersion or limited data points may introduce uncertainties, influencing the precision of predictions.

The remarkable congruence between characteristics and models implies a certain universality in the efficacy of the selected regressors. The results underscore the adaptability of these models across diverse physical attributes of polymers, enhancing their utility in materials science research.

However, it is essential to acknowledge that the success of predictive modeling is contingent on the nature of the characteristic being predicted. While the variance metric serves as a comprehensive metric, its applicability can be context-dependent. High metric scores indicate successful prediction, but the interpretation should consider the specific challenges associated with each characteristic.

Depending on the size of the dataset, different models are considered the best fit for various physical properties of polymers. This variation in model suitability can be attributed to the complex and heterogeneous nature of polymer systems [57]. Large datasets may facilitate the application of more complex models, such as ensemble methods or deep learning, to capture intricate relationships [58], while smaller datasets may benefit from simpler models to avoid overfitting [59].

The reasons for the different natures of models include the distribution of data [60], the presence of non-linearities and interactions in physical properties [61], and the dimensionality of the feature space. In high-dimensional feature spaces, models like Lasso Regression or Elastic Net may be preferred for feature selection and regularization, while simpler models like linear regression may suffice for fewer features.

Similar phenomena are observed in other systems such as colloids [62], proteins [63], and nucleic acids [64]. The optimal choice of models for predicting physical descriptors varies based on the nature of the system and the characteristics of the data.

- **Colloids:** different models may be suitable for predicting properties such as particle size, shape, and stability, considering the diverse interactions and conditions influencing colloidal systems [65].
- **Proteins:** the structure and function of proteins may require distinct modeling approaches. For example, machine learning models like Random Forests may be effective for predicting protein-ligand binding affinities [66], while simpler models may suffice for secondary structure prediction [67].
- **Nucleic Acids:** the unique properties of nucleic acids, such as DNA or RNA, may demand different models for predicting structural features [68], interaction energies, or other physical descriptors based on the specific characteristics of the dataset.

The discussions highlight both the achievements and challenges encountered in using prediction imputation for estimating missing values in polymer physical characteristics. The positive alignment of characteristics and models indicates the promising potential of this approach, opening avenues for further refinement and application in the field of materials science and polymer research.

5. Conclusions

In conclusion, this study aimed to predict missing values for various physical characteristics of polymers using machine learning techniques. The predictive models, including Random Forest, Gradient Boosting, and XGBoost, demonstrated strong performance, with the Random Forest model achieving the highest R^2 scores of 0.71, 0.73, and 0.88 for glass transition temperature, thermal decomposition temperature, and melting temperature, respectively. The validation process involved predicting unknown values, showcasing the reliability of the models.

The best-performing model, Random Forest, displayed promising results in handling the complexity of polymer characteristics. Future research could explore refining the models further, considering additional feature engineering, and expanding the dataset to enhance predictive accuracy. Additionally, investigating the transferability of the models to different polymer datasets could contribute to the broader applicability of the developed predictive framework. Some directions for further research are:

1. **Feature Engineering and Selection**: explore advanced feature engineering techniques and refine feature selection methods to identify the most influential characteristics. Investigate the impact of incorporating domain-specific knowledge to enhance the models' ability to capture subtle nuances in polymer behavior.
2. **Model Optimization**: this includes experimenting with different ensemble methods, regularization techniques, and model architectures to achieve a more robust and accurate predictive framework.
3. **Dataset Expansion**: consider augmenting the dataset by incorporating data from diverse polymer sources. A larger and more diverse dataset could provide a comprehensive understanding of polymer characteristics, enabling models to generalize better across different types of polymers.
4. **Cross-Dataset Validation**: evaluate the transferability of the developed models by validating them on external polymer datasets. Assessing the models' performance on different datasets will provide insights into their robustness and applicability across various polymer compositions and properties.
5. **Incorporating Temporal Aspects**: if applicable, consider incorporating temporal aspects into the models to capture any time-dependent trends or changes in polymer characteristics. This could involve analyzing how polymers evolve over time under different conditions.

6. **Interpretability and Explainability**: enhance the interpretability of the models to provide clearer insights into the features driving predictions. This could involve employing techniques such as SHAP (SHapley Additive exPlanations) values to explain the contribution of each feature to the model's output.
7. **Uncertainty Quantification**: integrate methods for uncertainty quantification to provide more reliable predictions and confidence intervals. This is particularly important in applications where understanding the uncertainty associated with predictions is crucial for decision-making.
8. **Collaboration with Domain Experts**: foster collaboration between data scientists and domain experts in polymer science to gain deeper insights into the underlying physics and chemistry. Leveraging domain knowledge can lead to the development of more informed models and a better understanding of the relationships between polymer characteristics.

By addressing these avenues, future research endeavors can contribute to the continuous improvement and applicability of machine learning models in predicting and understanding the complex characteristics of polymers.

Author Contributions: Conceptualization, I.P.M., V.S.T. and A.S.B.; Data curation, I.P.M., A.S.B. and A.P.G.; Formal analysis, I.P.M., V.S.T., V.A.N. and A.P.G.; Funding acquisition, V.S.T. and V.A.N.; Investigation, I.P.M. and V.A.N.; Methodology, V.S.T. and V.A.N.; Project administration, V.S.T. and A.S.B.; Resources, V.A.N.; Software, I.P.M. and A.P.G.; Supervision, V.S.T. and V.A.N.; Validation, I.P.M., V.S.T., A.S.B. and A.P.G.; Visualization, I.P.M.; Writing—original draft, I.P.M., V.S.T., V.A.N., A.S.B. and A.P.G.; Writing—review & editing, I.P.M., V.S.T., V.A.N., A.S.B. and A.P.G. All authors have read and agreed to the published version of the manuscript.

Funding: The studies were carried out within the program of the Russian Federation of strategic academic leadership "Priority-2030", aimed at supporting the development programs of educational institutions of higher education, and the scientific project PRIOR/SN/NU/23/SP5/5 "Development of a polymer classification model based on molecular descriptors using machine learning."

Institutional Review Board Statement: Not applicable.

Data Availability Statement: All code, datasets, and images referenced in this article are publicly available in the following GitHub repository: https://github.com/catauggie/polymersML (accessed on 24 November 2023). Researchers are encouraged to refer to this repository for access to the complete set of resources used in the study.

Conflicts of Interest: The authors declare no conflicts of interest.

Appendix A. Data Description

Table A1. Summary of physical characteristics (more than 300 values).

Characteristic	Count	Mean	Std	Min	Max	50%	Unit
Dynamic mechanical properties loss tangent	301	0.56	1.12	0.0	11.6	0.14	
Thermal decomposition temperature	6325	401.0	112.87	18.0	1000.0	403.0	C
Tensile modulus	1103	3.69	13.03	0.0	202.0	2.1	GPa
Heat of fusion	623	0.01	0.01	0.0	0.12	0.01	kcal/g
LC phase transition temperature	961	191.76	95.37	−90.0	528.0	175.0	C
Thermal decomposition weight loss	5236	10.25	13.09	0.0	100.0	5.0	%
Melting temperature	3844	194.93	108.24	−54.0	580.0	186.35	C
Volume resistivity	943	1.15×10^{16}	1.21×10^{17}	0.0	3.1×10^{18}	45×10^{8}	ohm·cm
Dielectric loss factor	311	775.51	13608.97	0.0	240000.0	0.1	
Cohesive energy density	324	112.21	60.3	0.0	626.0	96.0	cal/cm^3
Glass transition temperature	8092	145.18	110.69	−123.0	495.0	138.0	C
Density	1739	1.24	0.2	0.23	3.03	1.23	g/cm^3
Water absorption	724	10.95	48.98	0.0	1065.0	2.5	wt%

Table A1. *Cont.*

Characteristic	Count	Mean	Std	Min	Max	50%	Unit
Electric conductivity	1008	1.96×10^6	61.73×10^6	0.0	19.6×10^9	0.0	1/(ohm·cm)
Elongation at break	1139	51.98	157.26	0.26	3000.0	10.1	%
Tensile stress strength at break	1153	0.19	2.14	0.0	64.02	0.08	GPa
Intrinsic viscosity ETA	1978	1.43	12.4	0.0	495.0	0.52	dl/g
Solubility parameter	324	21.08	5.0	0.0	51.2	20.0	$(J/cm^3)^{1/2}$
Dynamic mechanical properties storage modulus	409	2.28	4.58	0.0	64.6	1.3	GPa
Refractive index	685	1.65	0.86	0.49	23.0	1.6	
Gas diffusion coefficient d	444	0.0	0.0	0.0	0.0	0.0	cm^2/s
Gas permeability coefficient p	717	0.0	0.0	0.0	0.0	0.0	cm^3(STP)cm/(cm^2·s·Pa)
Crystallization temperature	457	138.4	105.61	−120.0	496.0	124.0	C
Softening temperature	777	176.31	103.88	−185.0	800.0	173.0	C
Dielectric constant AC	763	22.51	403.72	0.12	11,002.15	3.26	
Surface tension	348	30.95	13.08	5.75	72.5	31.13	mN/m
Specific volume	1739	0.83	0.15	0.33	4.3	0.81	cm^3/g
Dielectric loss tangent	266.0	0.74	4.8	−0.03	55.0	0.02	
Isothermal weight loss temperature	273.0	389.35	165.13	100.0	900.0	350.0	C
Tensile stress strength at yield	267.0	0.07	0.05	0.0	0.4	0.06	GPa
Contact angle	255.0	73.96	19.85	15.0	158.9	76.0	degree
Gas solubility coefficient s	262.0	0.01	0.06	0.0	0.69	0.0	cm^3(STP)/(cm^3·Pa)

Table A2. Summary of Physical Characteristics (More than 50 up to 250 Values).

Characteristic	Count	Mean	Std	Min	Max	50%	Unit
Thermal conductivity	80	0.81	2.95	0.01	23.0	0.22	W/(m·K)
Hansen parameter delta−h: hydrogen bonding	59	8.03	3.5	0.0	16.0	7.4	$(J/cm^3)^{1/2}$
Flexural modulus	83	8.27	21.18	0.04	108.0	2.61	GPa
Vicat softening temperature	82	137.08	59.47	29.7	380.0	133.0	C
Dynamic mechanical properties loss modulus	203	2.47	22.2	0.0	260.0	0.1	GPa
PVT relation specific volume	56	0.85	0.17	0.4	1.17	0.85	cm^3/g
Water vapor transmission	73	0.82	2.38	0.0	15.0	0.01	g·mil/(cm^2·24 h)
Dynamic flexural properties storage modulus	78	1.71	4.36	0.0	37.0	0.79	GPa
Hansen parameter delta p polar	59	7.11	4.84	1.1	19.5	6.1	$(J/cm^3)^{1/2}$
Crystallization kinetics k	59	0.66	2.25	0.0	15.07	0.01	
Heat of fusion mol conversion	225	3.99	3.33	0.0	21.0	3.32	kcal/mol
Elongation at yield	84	22.45	50.18	0.08	334.0	8.3	%
Dynamic shear properties loss tangent	106	1.88	14.6	0.0	150.0	0.07	
Dynamic shear properties storage modulus	141	0.43	0.67	0.0	3.65	0.03	GPa
Crystallization kinetics n	71	2.59	0.72	0.59	4.15	2.6	
Heat of crystallization	124.0	10.39	9.95	0.29	49.3	8.3	cal/g
Dynamic shear properties loss modulus	92	0.05	0.11	0.0	0.7	0.0	GPa
Flexural stress strength at break	71	0.15	0.29	0.0	1.84	0.09	GPa
Isothermal weight loss time	228	86.9	333.57	0.18	2500.0	13.8	h
Fiber tensile elongation at break	61	39.65	48.71	2.25	242.34	21.0	%
Deflection temperature under load HDT	99	189.38	87.48	45.0	417.0	197.0	C
Specific heat capacity CP	214	0.38	0.25	0.0	2.52	0.35	cal/(g·C)
Fiber tensile stress strength at break	91	50.8	329.54	0.17	3090.0	3.6	g/denier
Brittleness temperature	81	−22.15	35.05	−80.0	90.0	−26.0	C
Dynamic flexural properties loss tanget	73	0.59	0.71	0.0	3.02	0.17	
Oxygen index	176	35.85	14.05	4.5	95.0	34.0	%
PVT relation pressure	53	74.91	135.54	0.0	598.8	35.0	MPa
Izod impact	53	161.43	350.27	0.02	1990.0	40.0	kJ/m
Thermal diffusivity	80	0.0	0.0	0.0	0.0	0.0	m^2/s

Table A2. *Cont.*

Characteristic	Count	Mean	Std	Min	Max	50%	Unit
PVT relation temperature	57	216.95	523.81	4.0	3822.0	87.5	C
Radius of gyration	120	33.15	36.39	0.5	264.35	21.72	nm
Crystallization kinetics half time of crystallization	72	2389.53	6064.79	11.1	35,400.0	289.5	s
Second virial coefficient	101	0.15	1.07	−0.0	8.95	0.0	$cm^3 \cdot mol/g^2$
Hansen parameter delta−d: dispersion component	60	16.54	4.09	0.0	21.5	17.53	$(J/cm^3)^{1/2}$
Fiber tensile modulus	74	90.59	156.26	3.86	847.0	43.5	g/denier
Crystallization kinetics r	53	1986.31	10,850.65	0.02	79,175.0	97.0	nm/s

Appendix B. Physical Characteristics

Appendix B.1. Physical Properties

1. **Bulk Modulus:** measures a material's resistance to volume change under pressure. It is crucial for understanding how a material responds to changes in pressure [69].
2. **Compressibility:** describes the degree to which a material can be compressed. It is the reciprocal of bulk modulus and helps assess a material's response to external pressure [70].
3. **G Value:** represents the ratio of the strain energy stored in a material to the kinetic energy. It provides insights into a material's elastic behavior under deformation [71].
4. **PVT Relation Pressure:** describes the relationship between pressure and specific volume in a material. It is essential for understanding the material's response to changes in pressure and volume [72].
5. **PVT Relation Specific Volume:** defines the correlation between specific volume and pressure in a material. It is crucial for analyzing the material's behavior under varying pressure conditions [73].
6. **PVT Relation Temperature:** illustrates the relationship between temperature and specific volume in a material. It is essential for studying how temperature influences the material's volume properties [74].
7. **Radiation Resistance:** measures a material's ability to withstand the effects of ionizing radiation. This property is vital for materials used in radiation-exposed environments [75].
8. **Density:** represents the mass of a material per unit volume. Density is a fundamental property that influences various material characteristics [76].
9. **Specific Volume:** describes the volume occupied by a unit mass of a material. It is the reciprocal of density and provides insights into material compactness [77].

Appendix B.2. Compression Characteristics

1. **Compressive Modulus:** measures the material's resistance to compression. Essential in the construction of structural elements made of polymers [78].
2. **Compressive Stress Strength at Break:** determines the maximum pressure a polymer can withstand before breaking. Important for assessing the resilience of polymer structures to mechanical forces [79].
3. **Compressive Stress Strength at Yield:** measures the strength of a polymer under pressure before plastic deformation begins. Important for the preliminary evaluation of the material's structural reliability [80].
4. **Dynamic Compressive Properties Storage Modulus:** characterizes the material's ability to store energy under dynamic loading. Important for materials subjected to cyclic loads, such as in damping materials [81].
5. **Dynamic Compressive Properties Loss Tangent:** reflects the fraction of energy loss due to dynamic deformation. Important in the development of materials with effective damping properties [82].

6. **Dynamic Compressive Properties Loss Modulus:** determines the energy loss during dynamic deformation. Important for materials designed for sound absorption or vibration reduction [83].

Appendix B.3. Creep Characteristics

1. **Tensile Creep Compliance:** determines the polymer's ability to undergo deformation under constant tensile load. This is crucial for assessing the long-term stability of polymer materials under constant force or load [84].
2. **Tensile Creep Modulus:** measures the elasticity of the polymer when deformed under constant force. This parameter is useful in designing materials for applications where resistance to constant mechanical loads is important [85].
3. **Tensile Creep Recovery:** evaluates the polymer's ability to return to its original shape after deformation under tensile loading. This is important, for example, for materials used in springs or elastic elements [86].
4. **Tensile Creep Rupture Time:** specifies the period during which the polymer undergoes deformation before rupture under tensile loading. This is an important characteristic for assessing the material's resistance to long-term mechanical loads [87].
5. **Tensile Creep Strain:** measures the level of deformation a polymer can undergo under constant tensile force. This is important for understanding the material's behavior under constant load and can be used in the design of structural elements [88].
6. **Flexural Creep Strain:** evaluates the deformation of the polymer under constant load during bending. This characteristic is important, for example, when using polymer materials in structures subjected to constant bending forces [89].
7. **Tensile Creep Rupture Strength:** determines the maximum load a polymer can withstand before rupture under constant tensile force. This is a crucial parameter for assessing the durability and resilience of polymer materials under constant mechanical loads [90].

Appendix B.4. Dilute Solution Property

1. **Intrinsic Viscosity (η):** measures the polymer's resistance to flow in a dilute solution, providing insights into its molecular size and structure. Intrinsic viscosity is crucial for understanding the polymer's solubility and processing behavior [91].
2. **Radius of Gyration:** defines the average distance of polymer segments from the center of mass, indicating the spatial extent of the polymer chain in solution. This property is significant in studying polymer conformations [92].
3. **Second Virial Coefficient:** describes the non-ideality of polymer solutions, providing information about the intermolecular interactions and solute-solvent interactions. This coefficient influences the solution behavior and phase separation [93].
4. **Diffusion Coefficient:** represents the rate at which polymer molecules spread through the solution, influencing mass transport and the polymer's ability to interact with its surroundings [94].
5. **Sedimentation Coefficient:** measures the rate at which polymer particles settle under the influence of gravity in a centrifugal field, providing information about particle size and shape in solution [95].

Appendix B.5. Electric Property

1. **Dielectric Constant (AC):** reflects the material's ability to store electrical energy in an alternating current (ac) field. The dielectric constant influences the capacitance of electronic components [96].
2. **Dielectric Loss Factor:** measures the efficiency with which a dielectric material converts electrical energy into heat. This property is crucial in applications where minimal energy loss is desired [97].

3. **Dielectric Loss Tangent:** describes the ratio of the dielectric loss factor to the dielectric constant, providing insights into the material's efficiency in handling electrical energy [98].
4. **Electric Conductivity:** represents the ability of a material to conduct electric current. This property is essential in various electronic and electrical applications [99].
5. **Surface Resistivity:** defines the electrical resistance across the surface of a material, influencing its performance in applications where surface conductivity is critical [100].
6. **Volume Resistivity:** measures the electrical resistance through the volume of a material, providing information about its overall resistance to electric current flow [101].

Appendix B.6. Flexural Property

1. **Dynamic Flexural Properties Storage Modulus:** characterizes the material's ability to store energy under dynamic flexural (bending) loading conditions. Important for materials subjected to cyclic loads [102].
2. **Dynamic Flexural Properties Loss Modulus:** determines the energy dissipation capacity of the material during dynamic flexural deformation. Relevant for applications requiring effective damping [103].
3. **Dynamic Flexural Properties Loss Tangent:** reflects the ratio of the loss modulus to the storage modulus in dynamic flexural deformation, providing insights into the material's damping behavior [104].
4. **Flexural Modulus:** measures the material's stiffness and resistance to bending deformation. Crucial in designing structural components where flexural strength is essential [105].
5. **Flexural Stress Strength at Break:** indicates the maximum stress a material can withstand before fracturing under bending stress. Important for evaluating the material's structural integrity [106].
6. **Flexural Stress Strength at Yield:** measures the material's stress resistance under bending before exhibiting plastic deformation. Important for assessing structural reliability under flexural loads [107].

Appendix B.7. Hardness

1. **Shore Hardness:** measures the resistance of the material to indentation or penetration. Shore hardness is a valuable indicator of a material's overall hardness and durability [108].

Appendix B.8. Heat Characteristics

1. **Brittleness Temperature:** indicates the temperature at which a material transitions from a flexible to a brittle state, providing insight into its low-temperature performance [109].
2. **Deflection Temperature under Load (HDT):** represents the temperature at which a standard test bar experiences a specified deformation under a specific load. HDT is crucial for understanding a material's ability to withstand elevated temperatures while supporting a load [110].
3. **Softening Temperature:** defines the temperature range at which a material starts to soften, losing its rigidity. Softening temperature is essential for assessing a material's behavior under heat [111].
4. **Vicat Softening Temperature:** determines the temperature at which a needle penetrates a material under a specified load. Vicat softening temperature provides insights into the heat resistance and stability of a material [112].

Appendix B.9. Heat Resistance and Combustion

1. **Oxygen Index:** measures the minimum concentration of oxygen in a mixture with an inert gas that supports the combustion of a material. This parameter is crucial for evaluating a material's fire resistance and combustion characteristics [113].

Appendix B.10. Impact Strength

1. **Charpy Impact:** assesses a material's resistance to sudden impact by measuring the amount of energy absorbed during fracture. Charpy impact testing is widely used to evaluate the toughness of materials [114].
2. **Izod Impact:** similar to Charpy impact testing, Izod impact testing measures a material's resistance to impact. It assesses the energy required to break a notched specimen under a sudden impact [115].

Appendix B.11. Optical Property

1. **Refractive Index:** determines the degree to which light is refracted or bent as it passes through a material. Refractive index is essential for understanding optical transparency and performance in various applications [116].

Appendix B.12. Physicochemical Property

1. **Cohesive Energy Density:** represents the energy required to separate unit volumes of material. It is a measure of the cohesive forces within a substance [117].
2. **Gas Diffusion Coefficient (D):** describes the rate at which gas molecules diffuse through a substance. It is crucial for understanding gas transport properties [118].
3. **Gas Permeability Coefficient (P):** measures a material's ability to allow gas permeation. It is essential for applications where gas barrier properties are significant [119].
4. **Gas Solubility Coefficient (S):** represents the capacity of a material to dissolve gases. This property is vital for understanding gas absorption in polymers [120].
5. **Hansen Parameter $\delta - d$: Dispersion Component:** describes the dispersion forces within a material. It is part of the Hansen solubility parameters, which characterize solute-solvent interactions [121].
6. **Hansen Parameter $\delta - h$: Hydrogen Bonding:** represents the hydrogen bonding contribution to the Hansen solubility parameters. It provides insights into materials' compatibility with various solvents [122].
7. **Hansen Parameter $\delta - p$: Polar:** describes the polar forces within a material. It is another component of the Hansen solubility parameters [123].
8. **Interfacial Tension:** measures the energy required to increase the surface area between two phases. It is crucial for understanding interactions at material interfaces [124].
9. **Solubility Parameter:** represents the overall solubility characteristics of a substance. It is a combination of the Hansen parameters and is used to predict material compatibility [125].
10. **Surface Tension:** describes the force acting on the surface of a liquid that tends to minimize the area. Surface tension is vital for understanding wetting and adhesion [126].
11. **Water Absorption:** measures the ability of a material to absorb water. It is essential for assessing the material's response to humid environments [127].
12. **Water Vapor Transmission:** describes the rate at which water vapor permeates through a material. It is crucial for applications requiring water vapor barrier properties [128].
13. **Contact Angle:** represents the angle formed between a liquid droplet and a solid surface. It provides insights into the wettability of a material [129].

Appendix B.13. Rheological Property

1. **Dynamic Viscosity Loss Tangent:** describes the ratio of the loss modulus to the storage modulus in the context of dynamic viscosity. It provides insights into the energy dissipation behavior of the material under dynamic conditions [130].

Appendix B.14. Shear Property

1. **Dynamic Shear Properties Storage Modulus:** represents the ability of a material to store elastic energy under shear stress in dynamic conditions [131].

2. **Dynamic Shear Properties Loss Modulus:** describes the portion of energy that a material loses as heat under shear stress in dynamic conditions [132].
3. **Dynamic Shear Properties Loss Tangent:** represents the ratio of the loss modulus to the storage modulus in the context of dynamic shear properties. It provides insights into the material's response to shear forces [133].
4. **Shear Modulus:** measures a material's resistance to deformation under shear stress. It is crucial for understanding a material's shear behavior [134].
5. **Shear Stress Strength at Break:** represents the maximum shear stress a material can withstand before experiencing failure. It is an essential parameter for evaluating the material's shear strength [135].
6. **Shear Stress Strength at Yield:** measures the shear stress a material can withstand before undergoing plastic deformation. This parameter is crucial for assessing the material's yield strength under shear forces [136].

Appendix B.15. Tensile Property

1. **Dynamic Mechanical Properties Storage Modulus:** represents the material's ability to store elastic energy under dynamic tensile conditions [137].
2. **Dynamic Mechanical Properties Loss Modulus:** describes the portion of energy that a material loses as heat under dynamic tensile conditions [138].
3. **Dynamic Mechanical Properties Loss Tangent:** represents the ratio of the loss modulus to the storage modulus in the context of dynamic tensile properties. It provides insights into the material's response to dynamic tensile forces [139].
4. **Elongation at Break:** measures the extent to which a material can stretch before experiencing rupture. It is a crucial parameter for evaluating the material's ductility [140].
5. **Elongation at Yield:** measures the material's deformation before it starts yielding under tensile stress. This parameter provides insights into the material's yield behavior under tension [141].
6. **Fiber Tensile Elongation at Break:** describes the elongation capability of fiber materials before experiencing rupture under tensile stress [142].
7. **Fiber Tensile Modulus:** represents the stiffness of a fiber material under tensile stress. It is a critical parameter for assessing the material's tensile rigidity [143].
8. **Fiber Tensile Stress Strength at Break:** represents the maximum tensile stress a fiber material can withstand before undergoing rupture [144].
9. **Tensile Modulus:** measures the material's resistance to deformation under tensile stress. It is crucial for understanding the material's tensile behavior [145].
10. **Tensile Stress Strength at Break:** represents the maximum tensile stress a material can withstand before experiencing failure [146].
11. **Tensile Stress Strength at Yield:** measures the tensile stress a material can withstand before undergoing plastic deformation. This parameter is crucial for assessing the material's yield strength under tensile forces [147].

Appendix B.16. Thermal Property

1. **Crystallization Kinetics r:** characterizes the crystallization kinetics of a material, representing the rate of crystallization [148].
2. **Crystallization Kinetics k:** represents a parameter in the crystallization kinetics equation, providing insights into the crystallization process [149].
3. **Crystallization Kinetics n:** another parameter in the crystallization kinetics equation, influencing the rate of crystallization [150].
4. **Crystallization Kinetics Half Time of Crystallization:** describes the time required for half of the crystallization process to occur [151].
5. **Crystallization Temperature:** represents the temperature at which a material undergoes crystallization [152].
6. **Glass Transition Temperature:** indicates the temperature at which an amorphous material transitions from a rigid to a rubbery state [153].

7. **Heat of Crystallization:** represents the amount of heat released or absorbed during the crystallization process [154].
8. **Heat of Fusion:** describes the heat energy required to change a substance from a solid to a liquid state at a constant temperature [155].
9. **Heat of Fusion Mol Conversion:** provides insights into the heat energy required for the conversion of a mole of substance from solid to liquid state [156].
10. **Thermal Decomposition Temperature:** represents the temperature at which a material starts to decompose thermally [157].
11. **Thermal Decomposition Weight Loss:** describes the weight loss associated with the thermal decomposition of a material [158].
12. **Isothermal Weight Loss Temperature:** represents the temperature maintained during a process where a material experiences weight loss [159].
13. **Isothermal Weight Loss Time:** describes the duration of time during which a material undergoes weight loss under isothermal conditions [160].
14. **LC Phase Transition Temperature:** represents the temperature at which a phase transition occurs in the liquid crystalline state [161].
15. **Melting Temperature:** indicates the temperature at which a material transitions from a solid to a liquid state [162].
16. **Specific Heat Capacity C_p:** describes the amount of heat energy required to raise the temperature of a unit mass of a material by one degree Celsius at constant pressure [163].
17. **Specific Heat Capacity C_v:** similar to C_p but at constant volume, representing the heat energy required to raise the temperature at constant volume [164].
18. **Thermal Conductivity:** describes the ability of a material to conduct heat [165].
19. **Thermal Diffusivity:** represents the ability of a material to conduct heat relative to its ability to store heat. It is the ratio of thermal conductivity to volumetric heat capacity [166].

References

1. Bates, F.S. Polymer-polymer phase behavior. *Science* **1991**, *251*, 898–905. [CrossRef]
2. Jenkins, A.D. *Polymer Science: A Materials Science Handbook*; Elsevier: Amsterdam, The Netherlands, 2013.
3. Ligon, S.C.; Liska, R.; Stampfl, J.; Gurr, M.; Mulhaupt, R. Polymers for 3D printing and customized additive manufacturing. *Chem. Rev.* **2017**, *117*, 10212–10290. [CrossRef]
4. Aidoo, R.P.; Depypere, F.; Afoakwa, E.O.; Dewettinck, K. Industrial manufacture of sugar-free chocolates–Applicability of alternative sweeteners and carbohydrate polymers as raw materials in product development. *Trends Food Sci. Technol.* **2013**, *32*, 84–96. [CrossRef]
5. Li, V.C. Tailoring ECC for special attributes: A review. *Int. J. Concr. Struct. Mater.* **2012**, *6*, 135–144. [CrossRef]
6. Kesarwani, S. Polymer composites in aviation sector. *Int. J. Eng. Res* **2017**, *6*, 10. [CrossRef]
7. Jenkins, M.; Stamboulis, A. *Durability and Reliability of Medical Polymers*; Elsevier: Amsterdam, The Netherlands, 2012.
8. Hong, Y.; Cooper-White, J.; Mackay, M.; Hawker, C.; Malmström, E.; Rehnberg, N. A novel processing aid for polymer extrusion: Rheology and processing of polyethylene and hyperbranched polymer blends. *J. Rheol.* **1999**, *43*, 781–793. [CrossRef]
9. Ohshima, M.; Tanigaki, M. Quality control of polymer production processes. *J. Process Control* **2000**, *10*, 135–148. [CrossRef]
10. Stevenson, S.; Vaisey-Genser, M.; Eskin, N. Quality control in the use of deep frying oils. *J. Am. Oil Chem. Soc.* **1984**, *61*, 1102–1108. [CrossRef]
11. Del Nobile, M.A.; Conte, A.; Buonocore, G.G.; Incoronato, A.; Massaro, A.; Panza, O. Active packaging by extrusion processing of recyclable and biodegradable polymers. *J. Food Eng.* **2009**, *93*, 1–6. [CrossRef]
12. Borgquist, P.; Körner, A.; Piculell, L.; Larsson, A.; Axelsson, A. A model for the drug release from a polymer matrix tablet—effects of swelling and dissolution. *J. Control. Release* **2006**, *113*, 216–225. [CrossRef]
13. Ranstam, J.; Cook, J. LASSO regression. *J. Br. Surg.* **2018**, *105*, 1348–1348. [CrossRef]
14. De Mol, C.; De Vito, E.; Rosasco, L. Elastic-net regularization in learning theory. *J. Complex.* **2009**, *25*, 201–230. [CrossRef]
15. Xu, M.; Watanachaturaporn, P.; Varshney, P.K.; Arora, M.K. Decision tree regression for soft classification of remote sensing data. *Remote Sens. Environ.* **2005**, *97*, 322–336. [CrossRef]
16. Breiman, L. Bagging predictors. *Mach. Learn.* **1996**, *24*, 123–140. [CrossRef]
17. Solomatine, D.P.; Shrestha, D.L. AdaBoost. RT: A boosting algorithm for regression problems. In Proceedings of the 2004 IEEE International Joint Conference on Neural Networks, Budapest, Hungary, 25–29 July 2004; Volume 2, pp. 1163–1168.
18. Zhang, X.; Yan, C.; Gao, C.; Malin, B.A.; Chen, Y. Predicting missing values in medical data via XGBoost regression. *J. Healthc. Inform. Res.* **2020**, *4*, 383–394. [CrossRef]
19. Awad, M.; Khanna, R.; Awad, M.; Khanna, R. Support vector regression. *Efficient Learning Machines: Theories, Concepts, and Applications for Engineers and System Designers*; Springer: Berlin, Germany, 2015; pp. 67–80.

20. Prettenhofer, P.; Louppe, G. Gradient boosted regression trees in scikit-learn. In Proceedings of the PyData 2014, London, UK, 21–23 February 2014.
21. Weisberg, S. *Applied Linear Regression*; John Wiley & Sons: Hoboken, NJ, USA, 2005; Volume 528.
22. Liu, Y.; Wang, Y.; Zhang, J. New machine learning algorithm: Random forest. In Proceedings of the Information Computing and Applications: Third International Conference, ICICA 2012, Chengde, China, 14–16 September 2012; pp. 246–252.
23. Li, F.; Yang, Y.; Xing, E. From lasso regression to feature vector machine. *Adv. Neural Inf. Process. Syst.* **2005**, *18*, 18.
24. James, G.M.; Wang, J.; Zhu, J. Functional linear regression that's interpretable. *Ann. Statist.* **2009**, *37*, 2083–2108. [CrossRef]
25. Mohammed Rashid, A.; Midi, H.; Dhhan, W.; Arasan, J. Detection of outliers in high-dimensional data using nu-support vector regression. *J. Appl. Stat.* **2022**, *49*, 2550–2569. [CrossRef]
26. Segal, M.R. Machine learning benchmarks and random forest regression. *J. Data Anal. Inf. Process.* **2004**, *8*, 4.
27. Koyamparambath, A.; Adibi, N.; Szablewski, C.; Adibi, S.A.; Sonnemann, G. Implementing artificial intelligence techniques to predict environmental impacts: Case of construction products. *Sustainability* **2022**, *14*, 3699. [CrossRef]
28. Sancar, N.; Onakpojeruo, E.P.; Inan, D.; Ozsahin, D.U. Adaptive Elastic Net Based on Modified PSO for Variable Selection in Cox Model with High-dimensional Data: A Comprehensive Simulation Study. *IEEE Access* **2023**, *11*, 127302–127316. [CrossRef]
29. Paez, A.; López, F.; Ruiz, M.; Camacho, M. Inducing non-orthogonal and non-linear decision boundaries in decision trees via interactive basis functions. *Expert Syst. Appl.* **2019**, *122*, 183–206. [CrossRef]
30. Florez-Lopez, R.; Ramon-Jeronimo, J.M. Enhancing accuracy and interpretability of ensemble strategies in credit risk assessment. A correlated-adjusted decision forest proposal. *Expert Syst. Appl.* **2015**, *42*, 5737–5753. [CrossRef]
31. Cao, J.; Kwong, S.; Wang, R. A noise-detection based AdaBoost algorithm for mislabeled data. *Pattern Recognit.* **2012**, *45*, 4451–4465. [CrossRef]
32. Otchere, D.A.; Ganat, T.O.A.; Ojero, J.O.; Tackie-Otoo, B.N.; Taki, M.Y. Application of gradient boosting regression model for the evaluation of feature selection techniques in improving reservoir characterisation predictions. *J. Pet. Sci. Eng.* **2022**, *208*, 109244. [CrossRef]
33. Ahmed, A.; Song, W.; Zhang, Y.; Haque, M.A.; Liu, X. Hybrid BO-XGBoost and BO-RF Models for the Strength Prediction of Self-Compacting Mortars with Parametric Analysis. *Materials* **2023**, *16*, 4366. [CrossRef]
34. Wang, Z.; Bovik, A.C. Mean squared error: Love it or leave it? A new look at signal fidelity measures. *IEEE Signal Process. Mag.* **2009**, *26*, 98–117. [CrossRef]
35. Miles, J. R-squared, adjusted R-squared. In *Encyclopedia of Statistics in Behavioral Science*; Wiley: Hoboken, NJ, USA, 2005.
36. Chai, T.; Draxler, R.R. Root mean square error (RMSE) or mean absolute error (MAE)? Arguments against avoiding RMSE in the literature. *Geosci. Model Dev.* **2014**, *7*, 1247–1250. [CrossRef]
37. Händel, P. Understanding normalized mean squared error in power amplifier linearization. *IEEE Microw. Wirel. Components Lett.* **2018**, *28*, 1047–1049. [CrossRef]
38. Willmott, C.J.; Matsuura, K. Advantages of the mean absolute error (MAE) over the root mean square error (RMSE) in assessing average model performance. *Clim. Res.* **2005**, *30*, 79–82. [CrossRef]
39. Jiang, Y. Estimation of monthly mean daily diffuse radiation in China. *Appl. Energy* **2009**, *86*, 1458–1464. [CrossRef]
40. Polymer Database (PoLyInfo). Available online: https://polymer.nims.go.jp/ (accessed on 18 October 2023).
41. Otsuka, S.; Kuwajima, I.; Hosoya, J.; Xu, Y.; Yamazaki, M. PoLyInfo: Polymer database for polymeric materials design. In Proceedings of the 2011 International Conference on Emerging Intelligent Data and Web Technologies, Tirana, Albania, 7–9 September 2011; pp. 22–29.
42. Weininger, D. SMILES, a chemical language and information system. 1. Introduction to methodology and encoding rules. *J. Chem. Inf. Comput. Sci.* **1988**, *28*, 31–36. [CrossRef]
43. Landrum, G. RDKit: A software suite for cheminformatics, computational chemistry, and predictive modeling. *Greg Landrum* **2013**, *8*, 31.
44. Yang, L.; Shami, A. On hyperparameter optimization of machine learning algorithms: Theory and practice. *Neurocomputing* **2020**, *415*, 295–316. [CrossRef]
45. Moons, K.G.; Donders, R.A.; Stijnen, T.; Harrell Jr, F.E. Using the outcome for imputation of missing predictor values was preferred. *J. Clin. Epidemiol.* **2006**, *59*, 1092–1101. [CrossRef] [PubMed]
46. Charles, J.; Jassi, P.; Ananth, N.S.; Sadat, A.; Fedorova, A. Evaluation of the intel® core™ i7 turbo boost feature. In Proceedings of the 2009 IEEE International Symposium on Workload Characterization (IISWC), Austin, TX, USA, 4–6 October 2009; pp. 188–197.
47. Lookman, T.; Alexander, F.J.; Rajan, K. *Information Science for Materials Discovery and Design*; Springer: Berlin, Germany, 2016; Volume 225.
48. Mannodi-Kanakkithodi, A.; Chandrasekaran, A.; Kim, C.; Huan, T.D.; Pilania, G.; Botu, V.; Ramprasad, R. Scoping the polymer genome: A roadmap for rational polymer dielectrics design and beyond. *Mater. Today* **2018**, *21*, 785–796. [CrossRef]
49. Doan Tran, H.; Kim, C.; Chen, L.; Chandrasekaran, A.; Batra, R.; Venkatram, S.; Kamal, D.; Lightstone, J.P.; Gurnani, R.; Shetty, P.; et al. Machine-learning predictions of polymer properties with Polymer Genome. *J. Appl. Phys.* **2020**, *128*, 10. [CrossRef]
50. Kim, C.; Chandrasekaran, A.; Huan, T.D.; Das, D.; Ramprasad, R. Polymer genome: A data-powered polymer informatics platform for property predictions. *J. Phys. Chem. C* **2018**, *122*, 17575–17585. [CrossRef]
51. Kazemi-Khasragh, E.; Blázquez, J.P.F.; Gómez, D.G.; González, C.; Haranczyk, M. Facilitating polymer property prediction with machine learning and group interaction modelling methods. *Int. J. Solids Struct.* **2024**, *286*, 112547. [CrossRef]

52. Antoniuk, E.R.; Li, P.; Kailkhura, B.; Hiszpanski, A.M. Representing Polymers as Periodic Graphs with Learned Descriptors for Accurate Polymer Property Predictions. *J. Chem. Inf. Model.* **2022**, *62*, 5435–5445. [CrossRef]
53. Xie, T.; Grossman, J.C. Crystal graph convolutional neural networks for an accurate and interpretable prediction of material properties. *Phys. Rev. Lett.* **2018**, *120*, 145301. [CrossRef]
54. Yang, K.; Swanson, K.; Jin, W.; Coley, C.; Eiden, P.; Gao, H.; Guzman-Perez, A.; Hopper, T.; Kelley, B.; Mathea, M.; et al. Analyzing learned molecular representations for property prediction. *J. Chem. Inf. Model.* **2019**, *59*, 3370–3388. [CrossRef] [PubMed]
55. Schindler, P.; Antoniuk, E.R.; Cheon, G.; Zhu, Y.; Reed, E.J. Discovery of materials with extreme work functions by high-throughput density functional theory and machine learning. *arXiv* **2020**, arXiv:2011.10905.
56. Nguyen, P.; Loveland, D.; Kim, J.T.; Karande, P.; Hiszpanski, A.M.; Han, T.Y.J. Predicting energetics materials' crystalline density from chemical structure by machine learning. *J. Chem. Inf. Model.* **2021**, *61*, 2147–2158. [CrossRef] [PubMed]
57. Leblanc, J.L. Rubber–filler interactions and rheological properties in filled compounds. *Prog. Polym. Sci.* **2002**, *27*, 627–687. [CrossRef]
58. Ganaie, M.A.; Hu, M.; Malik, A.; Tanveer, M.; Suganthan, P. Ensemble deep learning: A review. *Eng. Appl. Artif. Intell.* **2022**, *115*, 105151. [CrossRef]
59. Ying, X. An overview of overfitting and its solutions. *J. Phys. Conf. Ser.* **2019**, *1168*, 022022. [CrossRef]
60. Wu, X.; Zhu, X.; Wu, G.Q.; Ding, W. Data mining with big data. *IEEE Trans. Knowl. Data Eng.* **2013**, *26*, 97–107.
61. Rodríguez-Caballero, E.; Cantón, Y.; Lazaro, R.; Solé-Benet, A. Cross-scale interactions between surface components and rainfall properties. Non-linearities in the hydrological and erosive behavior of semiarid catchments. *J. Hydrol.* **2014**, *517*, 815–825. [CrossRef]
62. Molnar, I.L.; Johnson, W.P.; Gerhard, J.I.; Willson, C.S.; O'carroll, D.M. Predicting colloid transport through saturated porous media: A critical review. *Water Resour. Res.* **2015**, *51*, 6804–6845. [CrossRef]
63. Chen, C.W.; Lin, M.H.; Liao, C.C.; Chang, H.P.; Chu, Y.W. iStable 2.0: Predicting protein thermal stability changes by integrating various characteristic modules. *Comput. Struct. Biotechnol. J.* **2020**, *18*, 622–630. [CrossRef] [PubMed]
64. Sim, A.Y.; Minary, P.; Levitt, M. Modeling nucleic acids. *Curr. Opin. Struct. Biol.* **2012**, *22*, 273–278. [CrossRef] [PubMed]
65. Moore, T.L.; Rodriguez-Lorenzo, L.; Hirsch, V.; Balog, S.; Urban, D.; Jud, C.; Rothen-Rutishauser, B.; Lattuada, M.; Petri-Fink, A. Nanoparticle colloidal stability in cell culture media and impact on cellular interactions. *Chem. Soc. Rev.* **2015**, *44*, 6287–6305. [CrossRef] [PubMed]
66. Ballester, P.J.; Mitchell, J.B. A machine learning approach to predicting protein–ligand binding affinity with applications to molecular docking. *Bioinformatics* **2010**, *26*, 1169–1175. [CrossRef] [PubMed]
67. Garnier, J.; Osguthorpe, D.J.; Robson, B. Analysis of the accuracy and implications of simple methods for predicting the secondary structure of globular proteins. *J. Mol. Biol.* **1978**, *120*, 97–120. [CrossRef] [PubMed]
68. Sun, L.Z.; Zhang, D.; Chen, S.J. Theory and modeling of RNA structure and interactions with metal ions and small molecules. *Annu. Rev. Biophys.* **2017**, *46*, 227–246. [CrossRef]
69. Mott, P.H.; Dorgan, J.R.; Roland, C. The bulk modulus and Poisson's ratio of "incompressible" materials. *J. Sound Vib.* **2008**, *312*, 572–575. [CrossRef]
70. Ito, T. Compressibility of the polymer crystal. *Polymer* **1982**, *23*, 1412–1434. [CrossRef]
71. Favier, V.; Chanzy, H.; Cavaillé, J. Polymer nanocomposites reinforced by cellulose whiskers. *Macromolecules* **1995**, *28*, 6365–6367. [CrossRef]
72. Rodgers, P.A. Pressure–volume–temperature relationships for polymeric liquids: A review of equations of state and their characteristic parameters for 56 polymers. *J. Appl. Polym. Sci.* **1993**, *48*, 1061–1080. [CrossRef]
73. Goyanes, S.; Salgueiro, W.; Somoza, A.; Ramos, J.; Mondragon, I. Direct relationships between volume variations at macro and nanoscale in epoxy systems. PALS/PVT measurements. *Polymer* **2004**, *45*, 6691–6697. [CrossRef]
74. Kowalska, B. Processing aspects of pvT relationship. *Polimery* **2006**, *51*, 862–865. [CrossRef]
75. Nambiar, S.; Yeow, J.T. Polymer-composite materials for radiation protection. *ACS Appl. Mater. Interfaces* **2012**, *4*, 5717–5726. [CrossRef] [PubMed]
76. Robertson, R.E. Polymer order and polymer density. *J. Phys. Chem.* **1965**, *69*, 1575–1578. [CrossRef]
77. Fox, T.; Loshaek, S. Influence of molecular weight and degree of crosslinking on the specific volume and glass temperature of polymers. *J. Polym. Sci.* **1955**, *15*, 371–390. [CrossRef]
78. Wongpa, J.; Kiattikomol, K.; Jaturapitakkul, C.; Chindaprasirt, P. Compressive strength, modulus of elasticity, and water permeability of inorganic polymer concrete. *Mater. Des.* **2010**, *31*, 4748–4754. [CrossRef]
79. Martínez-Vázquez, F.J.; Perera, F.H.; Miranda, P.; Pajares, A.; Guiberteau, F. Improving the compressive strength of bioceramic robocast scaffolds by polymer infiltration. *Acta Biomater.* **2010**, *6*, 4361–4368. [CrossRef]
80. Raghava, R.; Caddell, R.M.; Yeh, G.S. The macroscopic yield behaviour of polymers. *J. Mater. Sci.* **1973**, *8*, 225–232. [CrossRef]
81. Zeltmann, S.E.; Prakash, K.A.; Doddamani, M.; Gupta, N. Prediction of modulus at various strain rates from dynamic mechanical analysis data for polymer matrix composites. *Compos. Part B: Eng.* **2017**, *120*, 27–34. [CrossRef]
82. Fan, J.; Weerheijm, J.; Sluys, L. Dynamic compressive mechanical response of a soft polymer material. *Mater. Des.* **2015**, *79*, 73–85. [CrossRef]
83. Liu, G.J.; Bai, E.L.; Xu, J.Y.; Yang, N.; Wang, T.j. Dynamic compressive mechanical properties of carbon fiber-reinforced polymer concrete with different polymer-cement ratios at high strain rates. *Constr. Build. Mater.* **2020**, *261*, 119995. [CrossRef]

84. Plaseied, A.; Fatemi, A. Tensile creep and deformation modeling of vinyl ester polymer and its nanocomposite. *J. Reinf. Plast. Compos.* **2009**, *28*, 1775–1788. [CrossRef]
85. Raghavan, J.; Meshii, M. Creep of polymer composites. *Compos. Sci. Technol.* **1998**, *57*, 1673–1688. [CrossRef]
86. Wilding, M.; Ward, I.M. Tensile creep and recovery in ultra-high modulus linear polyethylenes. *Polymer* **1978**, *19*, 969–976. [CrossRef]
87. Trantina, G.G. Creep analysis of polymer structures. *Polym. Eng. Sci.* **1986**, *26*, 776–780. [CrossRef]
88. Zhang, Z.; Yang, J.L.; Friedrich, K. Creep resistant polymeric nanocomposites. *Polymer* **2004**, *45*, 3481–3485. [CrossRef]
89. Yang, Z.; Wang, H.; Ma, X.; Shang, F.; Ma, Y.; Shao, Z.; Hou, D. Flexural creep tests and long-term mechanical behavior of fiber-reinforced polymeric composite tubes. *Compos. Struct.* **2018**, *193*, 154–164. [CrossRef]
90. Spathis, G.; Kontou, E. Creep failure time prediction of polymers and polymer composites. *Compos. Sci. Technol.* **2012**, *72*, 959–964. [CrossRef]
91. Pamies, R.; Hernández Cifre, J.G.; del Carmen López Martínez, M.; García de la Torre, J. Determination of intrinsic viscosities of macromolecules and nanoparticles. Comparison of single-point and dilution procedures. *Colloid Polym. Sci.* **2008**, *286*, 1223–1231. [CrossRef]
92. Fixman, M. Radius of gyration of polymer chains. *J. Chem. Phys.* **1962**, *36*, 306–310. [CrossRef]
93. Orofino, T.A.; Flory, P. Relationship of the second virial coefficient to polymer chain dimensions and interaction parameters. *J. Chem. Phys.* **1957**, *26*, 1067–1076. [CrossRef]
94. Duda, J.; Vrentas, J.; Ju, S.; Liu, H. Prediction of diffusion coefficients for polymer-solvent systems. *AIChE J.* **1982**, *28*, 279–285. [CrossRef]
95. Closs, W.; Jennings, B.; Jerrard, H. Sedimentation velocity of polymer solutions—I. Concentration dependence of the sedimentation coefficient. *Eur. Polym. J.* **1968**, *4*, 639–649. [CrossRef]
96. Zuo, F.; Angelopoulos, M.; MacDiarmid, A.; Epstein, A.J. AC conductivity of emeraldine polymer. *Phys. Rev. B* **1989**, *39*, 3570. [CrossRef] [PubMed]
97. Zhu, L. Exploring strategies for high dielectric constant and low loss polymer dielectrics. *J. Phys. Chem. Lett.* **2014**, *5*, 3677–3687. [CrossRef] [PubMed]
98. Subodh, G.; Deepu, V.; Mohanan, P.; Sebastian, M. Dielectric response of high permittivity polymer ceramic composite with low loss tangent. *Appl. Phys. Lett.* **2009**, *95*, 062903. [CrossRef]
99. Radzuan, N.A.M.; Sulong, A.B.; Sahari, J. A review of electrical conductivity models for conductive polymer composite. *Int. J. Hydrog. Energy* **2017**, *42*, 9262–9273. [CrossRef]
100. Lekpittaya, P.; Yanumet, N.; Grady, B.P.; O'Rear, E.A. Resistivity of conductive polymer–coated fabric. *J. Appl. Polym. Sci.* **2004**, *92*, 2629–2636. [CrossRef]
101. Weber, M.; Kamal, M.R. Estimation of the volume resistivity of electrically conductive composites. *Polym. Compos.* **1997**, *18*, 711–725. [CrossRef]
102. Zhang, Z.; Wang, P.; Wu, J. Dynamic mechanical properties of EVA polymer-modified cement paste at early age. *Phys. Procedia* **2012**, *25*, 305–310. [CrossRef]
103. Kimoto, M. Flexural properties and dynamic mechanical properties of glass fibre-epoxy composites. *J. Mater. Sci.* **1990**, *25*, 3327–3332. [CrossRef]
104. Hiremath, V.; Shukla, D. Effect of particle morphology on viscoelastic and flexural properties of epoxy–alumina polymer nanocomposites. *Plast. Rubber Compos.* **2016**, *45*, 199–206. [CrossRef]
105. Fu, S.Y.; Hu, X.; Yue, C.Y. The flexural modulus of misaligned short-fiber-reinforced polymers. *Compos. Sci. Technol.* **1999**, *59*, 1533–1542. [CrossRef]
106. Goracci, C.; Cadenaro, M.; Fontanive, L.; Giangrosso, G.; Juloski, J.; Vichi, A.; Ferrari, M. Polymerization efficiency and flexural strength of low-stress restorative composites. *Dent. Mater.* **2014**, *30*, 688–694. [CrossRef] [PubMed]
107. Bae, J.M.; Kim, K.N.; Hattori, M.; Hasegawa, K.; Yoshinari, M.; Kawada, E.; Oda, Y. The flexural properties of fiber-reinforced composite with light-polymerized polymer matrix. *Int. J. Prosthodont.* **2001**, *14*, 33–39.
108. Liao, Z.; Hossain, M.; Yao, X. Ecoflex polymer of different Shore hardnesses: Experimental investigations and constitutive modelling. *Mech. Mater.* **2020**, *144*, 103366. [CrossRef]
109. Brostow, W.; Lobland, H.E.H.; Khoja, S. Brittleness and toughness of polymers and other materials. *Mater. Lett.* **2015**, *159*, 478–480. [CrossRef]
110. Takemori, M.T. Towards an understanding of the heat distortion temperature of thermoplastics. *Polym. Eng. Sci.* **1979**, *19*, 1104–1109. [CrossRef]
111. Van Breemen, L.C.; Engels, T.A.; Klompen, E.T.; Senden, D.J.; Govaert, L.E. Rate-and temperature-dependent strain softening in solid polymers. *J. Polym. Sci. Part B: Polym. Phys.* **2012**, *50*, 1757–1771. [CrossRef]
112. Aouachria, K.; Belhaneche-Bensemra, N. Miscibility of PVC/PMMA blends by vicat softening temperature, viscometry, DSC and FTIR analysis. *Polym. Test.* **2006**, *25*, 1101–1108. [CrossRef]
113. Kambour, R.; Klopfer, H.; Smith, S. Limiting oxygen indices of silicone block polymer. *J. Appl. Polym. Sci.* **1981**, *26*, 847–859. [CrossRef]
114. Nishi, Y.; Inoue, K.; Salvia, M. Improvement of Charpy impact of carbon fiber reinforced polymer by low energy sheet electron beam irradiation. *Mater. Trans.* **2006**, *47*, 2846–2851. [CrossRef]

115. Patterson, A.E.; Pereira, T.R.; Allison, J.T.; Messimer, S.L. IZOD impact properties of full-density fused deposition modeling polymer materials with respect to raster angle and print orientation. *Proc. Inst. Mech. Eng. Part C: J. Mech. Eng. Sci.* **2021**, *235*, 1891–1908. [CrossRef]
116. Liu, J.g.; Ueda, M. High refractive index polymers: Fundamental research and practical applications. *J. Mater. Chem.* **2009**, *19*, 8907–8919. [CrossRef]
117. Bristow, G.; Watson, W. Cohesive energy densities of polymers. Part 1.—Cohesive energy densities of rubbers by swelling measurements. *Trans. Faraday Soc.* **1958**, *54*, 1731–1741. [CrossRef]
118. Tanaka, K.; Kawai, T.; Kita, H.; Okamoto, K.I.; Ito, Y. Correlation between gas diffusion coefficient and positron annihilation lifetime in polymers with rigid polymer chains. *Macromolecules* **2000**, *33*, 5513–5517. [CrossRef]
119. Stern, S.; Fang, S.M.; Frisch, H. Effect of pressure on gas permeability coefficients. A new application of "free volume" theory. *J. Polym. Sci. Part A-2: Polym. Phys.* **1972**, *10*, 201–219. [CrossRef]
120. Michaels, A.S.; Bixler, H.J. Solubility of gases in polyethylene. *J. Polym. Sci.* **1961**, *50*, 393–412. [CrossRef]
121. Liu, C.; Corradini, M.; Rogers, M. Self-assembly of 12-hydroxystearic acid molecular gels in mixed solvent systems rationalized using Hansen solubility parameters. *Colloid Polym. Sci.* **2015**, *293*, 975–983. [CrossRef]
122. Sobodacha, C.J.; Lynch, T.J.; Durham, D.L.; Paradis, V.R. Solvents in novolak synthesis. In Proceedings of the Advances in Resist Technology and Processing X. SPIE, San Jose, CA, USA, 1–2 March 1993 , Volume 1925; pp. 582–592.
123. De La Peña-Gil, A.; Toro-Vazquez, J.F.; Rogers, M.A. Simplifying Hansen solubility parameters for complex edible fats and oils. *Food Biophys.* **2016**, *11*, 283–291. [CrossRef]
124. Wu, S. Calculation of interfacial tension in polymer systems. *J. Polym. Sci. Part Polym. Symp.* **1971**, *34*, 19–30. [CrossRef]
125. Hansen, C.M. The three dimensional solubility parameter. *Dan. Tech. Cph.* **1967**, *14*.
126. Roe, R.J. Surface tension of polymer liquids. *J. Phys. Chem.* **1968**, *72*, 2013–2017. [CrossRef]
127. Baschek, G.; Hartwig, G.; Zahradnik, F. Effect of water absorption in polymers at low and high temperatures. *Polymer* **1999**, *40*, 3433–3441. [CrossRef]
128. Tock, R.W. Permeabilities and water vapor transmission rates for commercial polymer films. *Adv. Polym. Technol. J. Polym. Process. Inst.* **1983**, *3*, 223–231. [CrossRef]
129. Yasuda, T.; Okuno, T.; Yasuda, H. Contact angle of water on polymer surfaces. *Langmuir* **1994**, *10*, 2435–2439. [CrossRef]
130. Ballou, J.; Smith, J. Dynamic measurements of polymer physical properties. *J. Appl. Phys.* **1949**, *20*, 493–502. [CrossRef]
131. Tam, K.; Tiu, C. Steady and dynamic shear properties of aqueous polymer solutions. *J. Rheol.* **1989**, *33*, 257–280. [CrossRef]
132. Saba, N.; Jawaid, M.; Alothman, O.Y.; Paridah, M. A review on dynamic mechanical properties of natural fibre reinforced polymer composites. *Constr. Build. Mater.* **2016**, *106*, 149–159. [CrossRef]
133. Kovacs, A.; Stratton, R.A.; Ferry, J.D. Dynamic mechanical properties of polyvinyl acetate in shear in the glass transition temperature range. *J. Phys. Chem.* **1963**, *67*, 152–161. [CrossRef]
134. Gittes, F.; MacKintosh, F. Dynamic shear modulus of a semiflexible polymer network. *Phys. Rev. E* **1998**, *58*, R1241. [CrossRef]
135. Chua, P.; Piggott, M. The glass fibre-polymer interface: II—Work of fracture and shear stresses. *Compos. Sci. Technol.* **1985**, *22*, 107–119. [CrossRef]
136. Mohammed, A.; Mahmood, W.; Ghafor, K. Shear stress limit, rheological properties and compressive strength of cement-based grout modified with polymers. *J. Build. Pathol. Rehabil.* **2020**, *5*, 1–17. [CrossRef]
137. Wielage, B.; Lampke, T.; Utschick, H.; Soergel, F. Processing of natural-fibre reinforced polymers and the resulting dynamic–mechanical properties. *J. Mater. Process. Technol.* **2003**, *139*, 140–146. [CrossRef]
138. Lewis, T.; Nielsen, L. Dynamic mechanical properties of particulate-filled composites. *J. Appl. Polym. Sci.* **1970**, *14*, 1449–1471. [CrossRef]
139. Wada, Y.; Kasahara, T. Relation between impact strength and dynamic mechanical properties of plastics. *J. Appl. Polym. Sci.* **1967**, *11*, 1661–1665. [CrossRef]
140. Palomba, D.; Vazquez, G.E.; Díaz, M.F. Prediction of elongation at break for linear polymers. *Chemom. Intell. Lab. Syst.* **2014**, *139*, 121–131. [CrossRef]
141. Ward, I.M. The yield behaviour of polymers. *J. Mater. Sci.* **1971**, *6*, 1397–1417. [CrossRef]
142. Rahman, R.; Putra, S.Z.F.S. Tensile properties of natural and synthetic fiber-reinforced polymer composites. *Mechanical and Physical Testing of Biocomposites, Fibre-Reinforced Composites and Hybrid Composites*; Elsevier: Amsterdam, The Netherlands, 2019; pp. 81–102.
143. Yu, T.; Zhang, Z.; Song, S.; Bai, Y.; Wu, D. Tensile and flexural behaviors of additively manufactured continuous carbon fiber-reinforced polymer composites. *Compos. Struct.* **2019**, *225*, 111147. [CrossRef]
144. Tan, E.; Ng, S.; Lim, C. Tensile testing of a single ultrafine polymeric fiber. *Biomaterials* **2005**, *26*, 1453–1456. [CrossRef]
145. Ji, X.L.; Jing, J.K.; Jiang, W.; Jiang, B.Z. Tensile modulus of polymer nanocomposites. *Polym. Eng. Sci.* **2002**, *42*, 983–993. [CrossRef]
146. Smith, P.; Lemstra, P.J.; Pijpers, J.P. Tensile strength of highly oriented polyethylene. II. Effect of molecular weight distribution. *J. Polym. Sci. Polym. Phys. Ed.* **1982**, *20*, 2229–2241. [CrossRef]
147. Argon, A.; Cohen, R. Toughenability of polymers. *Polymer* **2003**, *44*, 6013–6032. [CrossRef]
148. Patel, R.M.; Spruiell, J.E. Crystallization kinetics during polymer processing—Analysis of available approaches for process modeling. *Polym. Eng. Sci.* **1991**, *31*, 730–738. [CrossRef]

149. Mandelkern, L.; Quinn, F.A., Junior; Flory, P.J. Crystallization kinetics in high polymers. I. Bulk polymers. *J. Appl. Phys.* **1954**, *25*, 830–839. [CrossRef]
150. Gedde, U.W.; Hedenqvist, M.S.; Gedde, U.W.; Hedenqvist, M.S. Crystallization kinetics. *Fundamental Polymer Science*; Springer: Berlin, Germany, 2019; pp. 327–386.
151. Jenkins, M.; Harrison, K. The effect of molecular weight on the crystallization kinetics of polycaprolactone. *Polym. Adv. Technol.* **2006**, *17*, 474–478. [CrossRef]
152. Keller, A.; Machin, M. Oriented crystallization in polymers. *J. Macromol. Sci. Part B: Phys.* **1967**, *1*, 41–91. [CrossRef]
153. Meyer, J. Glass transition temperature as a guide to selection of polymers suitable for PTC materials. *Polym. Eng. Sci.* **1973**, *13*, 462–468. [CrossRef]
154. Strobl, G. Colloquium: Laws controlling crystallization and melting in bulk polymers. *Rev. Mod. Phys.* **2009**, *81*, 1287. [CrossRef]
155. Kirshenbaum, I. Entropy and heat of fusion of polymers. *J. Polym. Sci. Part A: Gen. Pap.* **1965**, *3*, 1869–1875. [CrossRef]
156. Flory, P.J. Thermodynamics of crystallization in high polymers. IV. A theory of crystalline states and fusion in polymers, copolymers, and their mixtures with diluents. *J. Chem. Phys.* **1949**, *17*, 223–240. [CrossRef]
157. Beyler, C.L.; Hirschler, M.M. Thermal decomposition of polymers. *SFPE Handb. Fire Prot. Eng.* **2002**, *2*, 111–131.
158. Chrissafis, K.; Bikiaris, D. Can nanoparticles really enhance thermal stability of polymers? Part I: An overview on thermal decomposition of addition polymers. *Thermochim. Acta* **2011**, *523*, 1–24. [CrossRef]
159. Bowles, K.J.; Jayne, D.; Leonhardt, T.A. Isothermal Aging Effects on PMR-15 Resin; Technical Report; NASA: Washington, DC, USA, 1992.
160. Abate, L.; Blanco, I.; Motta, O.; Pollicino, A.; Recca, A. The isothermal degradation of some polyetherketones: A comparative kinetic study between long-term and short-term experiments. *Polym. Degrad. Stab.* **2002**, *75*, 465–471. [CrossRef]
161. Percec, V.; Keller, A. A thermodynamic interpretation of polymer molecular weight effect on the phase transitions of main-chain and side-chain liquid-crystal polymers. *Macromolecules* **1990**, *23*, 4347–4350. [CrossRef]
162. Mandelkern, L.; Stack, G.; Mathieu, P. The Melting Temperature of Polymers: Theoretical and Experimental. In *Analytical Calorimetry*; Springer: Berlin, Germany, 1984; Volume 5, pp. 223–237.
163. Wen, J. Heat capacities of polymers. In *Physical Properties of Polymers Handbook*; Springer: Berlin, Germany, 2007; pp. 145–154.
164. Wunderlich, B. The heat capacity of polymers. *Thermochim. Acta* **1997**, *300*, 43–65. [CrossRef]
165. Choy, C. Thermal conductivity of polymers. *Polymer* **1977**, *18*, 984–1004. [CrossRef]
166. dos Santos, W.N.; Mummery, P.; Wallwork, A. Thermal diffusivity of polymers by the laser flash technique. *Polym. Test.* **2005**, *24*, 628–634. [CrossRef]

polymers

MDPI

Article

Predicting Diffusion Coefficients in Nafion Membranes during the Soaking Process Using a Machine Learning Approach

Ivan Malashin [1,*], Daniil Daibagya [1,2], Vadim Tynchenko [1,*], Andrei Gantimurov [1], Vladimir Nelyub [1,3] and Aleksei Borodulin [1]

[1] Artificial Intelligence Technology Scientific and Education Center, Bauman Moscow State Technical University, 105005 Moscow, Russia; agantimurov@emtc.ru (A.G.); vladimir.nelub@emtc.ru (V.N.); alexey.borodulin@emtc.ru (A.B.)

[2] P.N. Lebedev Physical Institute of the Russian Academy of Sciences, 119991 Moscow, Russia

[3] Scientific Department, Far Eastern Federal University, 690922 Vladivostok, Russia

* Correspondence: ivan.p.malashin@gmail.com (I.M.); vadimond@mail.ru (V.T.); Tel.: +7-926-875-7128 (I.M.)

Abstract: Nafion, a versatile polymer used in electrochemistry and membrane technologies, exhibits complex behaviors in saline environments. This study explores Nafion membrane's IR spectra during soaking and subsequent drying processes in salt solutions at various concentrations. Utilizing the principles of Fick's second law, diffusion coefficients for these processes are derived via exponential approximation. By harnessing machine learning (ML) techniques, including the optimization of neural network hyperparameters via a genetic algorithm (GA) and leveraging various regressors, we effectively pinpointed the optimal model for predicting diffusion coefficients. Notably, for the prediction of soaking coefficients, our model is composed of layers with 64, 64, 32, and 16 neurons, employing ReLU, ELU, sigmoid, and ELU activation functions, respectively. Conversely, for drying coefficients, our model features two hidden layers with 16 and 12 neurons, utilizing sigmoid and ELU activation functions, respectively.

Keywords: Nafion; IR spectroscopy; machine learning; diffusion; neural networks; fine-tuning optimization; genetic algorithm

1. Introduction

Nafion [1], a widely used polymer in various applications such as electrochemistry [2], fuel cells [3], and membrane technologies [4], exhibits complex behaviors in saline environments, particularly regarding swelling and drying processes. Gaining insights into these phenomena holds paramount importance for optimizing the performance of Nafion across diverse implementations.

1.1. *Nafion in Salt Solutions*

The investigation of Nafion in salt solutions is extensively covered in many literary sources over the last six decades, encapsulating key insights highlighted in the diagram depicted in Figure 1. One of the first [5] articles in the realm of this field of science investigates the sorption behavior of water and aqueous salt solutions of Nafion and reveals temperature and concentration dependencies of sorption, apparent activation energies for water diffusion in different membrane forms, and a maximum in sorption curves during neutralization, attributed to differences in diffusion coefficients. Additionally, diffusion coefficients of various cations were determined, showing a linear relationship with the charge-to-separation ratio, and water sorption was found to be strongly dependent on both the degree of neutralization and salt type, with a linear correlation between the number of water molecules absorbed and the degree of neutralization for all salts studied.

Figure 1. Retrospective overview of Nafion research in salt solutions: highlights of key topics from top scientific articles from each decade.

Measurements of densities and expansion coefficients of various Nafions in both acid and salt forms were provided in [6], investigating the effects of equivalent weights, moisture contents, and uniaxial orientations. Densities appeared to be independent of the equivalent weight but highly dependent on moisture content, with thin films showing strong uniaxial orientation. Reproducible expansion coefficients were observed after annealing, with distinct breaks in the linear expansion curve correlating with the material's glass transition and mechanical dispersion. The experimental scatter in densities suggests possible microphase separation or partial crystallization of the polymer.

The physical structure of Nafion membranes in [7] was probed using small-angle neutron scattering (SANS) and small-angle X-ray scattering (SAXS). While acid-form samples exhibit two scattering peaks corresponding to crystalline and ion-containing regions, amorphous salt-form samples lack the first peak, allowing for a detailed investigation of the second peak, indicative of water-swollen regions.

Experimental measurements, including zero-current membrane potential, electrolyte sorption, self-diffusion fluxes of co-ions and counter-ions, co-ion fluxes under constant current, and membrane electrical conductance, were conducted in [8] to investigate the factors contributing to the superselectivity of Nafion membranes in NaCl solutions. Swelling was found to influence these factors, with expanded membranes exhibiting higher electrolyte sorption. However, the molarity of the sorbed electrolyte remained unchanged as swelling increased. The loss of superselectivity was attributed to a decrease in the mobility ratio of counter ions to co-ions.

The "cluster network" structure of Nafion membranes was utilized in [9] to develop models for membrane salt solubility and anion transport. Salt uptake was influenced by the size and charge of membrane clusters, while anion transport parameters depended on the membrane tortuosity and electrostatic forces in the interconnected pores. The models accurately predicted cation concentrations and the concentration dependence of anion–membrane interaction parameters.

A molecular-level equilibrium partition coefficient model has been developed in [10] to describe the uptake of single and multicomponent ions by a Nafion 117 cation exchange membrane. The model assumes a cylindrical pore structure and considers changes in ion solvation free energy during partitioning, as well as the orientation of solvent dipoles within the membrane pores due to the strong electric field from fixed charge groups. Membrane structure parameters, derived from experimental data on membrane porosity and X-ray diffraction, are used to predict equilibrium cation concentrations within a 6% accuracy for various aqueous salt mixtures. Analysis of computed electrostatic and hydration forces suggests that ions with a higher surface charge density are excluded from the pore wall region, similar to co-ions.

The overall transport characteristics of various cationic species in Nafion® 117 membranes, including equilibrium salt and solvent uptake to determine membrane chloride concentration, porosity, and water content, were presented in [11]. Conductivity measurements via impedance spectroscopy were conducted for sodium and proton cationic forms, while the electrical mobilities of sodium, nickel, and silver ions were determined using electrophoresis. Additionally, combining these results allowed for the determination of the sodium transport number in the Nafion® 117 membrane equilibrated with NaCl solution.

A sophisticated kinetic model is developed in [12] to study the coupled diffusion of two counterions in an ion-exchange membrane, considering the impact of varying the ionic composition on the membrane water content. Through numerical simulations, diffusion coefficients of alkali metal cations in different Nafion forms are evaluated, revealing a correlation between the membrane-to-aqueous ion diffusion coefficient ratio and the polymer-phase volume fraction. This study suggests that while the polymer phase mainly exhibits a steric effect, differences in the behavior between Nafion forms may be attributed to distinct morphologies.

Analysis of Nafion membranes through a dynamic mechanical analysis and tension film measurements was described in [13], observing a decrease in the initial slope of stress–strain curves with an increasing water content and the addition of certain solvents, and a decrease with increasing temperature. Additionally, the initial slope was found to increase with the replacement of cations, following the order Li^+, Na^+, K^+, Cs^+, and Rb^+. Nafion in salt form typically exhibits an initial slope increase up to around 90 °C, followed by a decrease with rising temperatures.

The dilute solution properties of Nafion in methanol/water mixtures was studied in [14], revealing two aggregation processes: primary aggregation forming smaller, rod-like particles (<103 nm) due to hydrophobic interactions, and secondary aggregation forming larger particles (104 nm) attributed to ionic interactions. Critical concentrations of Cp (1.0 mg/mL) and Cpp (5.0 mg/mL) mark transitions in Nafion aggregation conformations, representing the formation of secondary ionic aggregations and the onset of self-assembly of disordered segments, respectively.

Paper [15] highlighted that the study of transport phenomena across ion-exchange membranes in non-aqueous electrolyte solutions remains limited, despite extensive research in aqueous systems that significantly increased the understanding of transport in aqueous–organic electrolyte solutions, particularly for applications like direct methanol fuel cells. Investigating Nafion membrane behavior in methanol–water electrolytes aims to shed light on its sorption and permeation properties, especially in the context of different electrolyte compositions.

Using the optical method, paper [16] examined how Nafion 112 foil swells anisotropically in methanol–water–inorganic salt solutions. Their findings indicate that even small amounts of inorganic salt in a methanol–water mixture affect both the rate and extent of swelling. We studied the effects of various inorganic salts, including LiCl, NaCl, KCl, CsCl,

$CaCl_2$, $CdCl_2$, K_2CO_3, KNO_3, NH_4Cl, and $AgNO_3$. The swelling kinetics of Nafion in ternary mixtures containing salt exhibited a maximum, suggesting that initially, methanol diffusion is faster than ion transport. This experimental data suggest that the swelling of Nafion decreases with an increase in the ionic radius of the cation.

Employing time-resolved Fourier transform infrared–attenuated total reflectance (FTIR-ATR) spectroscopy to investigate water dynamics in Nafion at low and high humidities was evaluated in [17]. At low humidities, non-Fickian behavior is observed due to a reaction between water and sulfonic acid, while at high humidities, water-induced relaxation in the polymer backbone leads to non-Fickian behavior. A diffusion–reaction model and a diffusion–relaxation model were developed, providing valuable insights into water transport mechanisms and relaxation phenomena in solid-state polymer electrolytes like Nafion.

Water absorption, swelling, and self-diffusivity in 1100 equivalent weight Nafion were analyzed across different temperatures and water activities in [18]. The study revealed a decrease in free volume per water molecule and a transition in the diffusivity rate at a water activity of four. Changes in hydrophilic domain connectivity were observed, informing the determination of interfacial mass transport coefficients and the development of a diffusion model for resolving water activity profiles.

An investigation of the swelling and diffusion behavior of a Nafion® 117 ion-exchange membrane in mixed water–methanol solutions was presented in [19], analyzing the membrane porosity and water/methanol uptake using Raman spectrometry. Permeation experiments, considering adjacent diffusion boundary layers (DBLs) and the methanol diffusion coefficient's dependency on concentration, were conducted under various conditions. Numerical fitting of experimental data determined the DBL thickness and methanol permeability, revealing the significant impact of DBL diffusion resistance on overall permeability, even at high rotation speeds. This study derived an equation relating the apparent non-electrolyte permeability to the true membrane permeability and diffusion layer thickness, following Helfferich's approach.

The optimization of water management in proton exchange membrane fuel cells is crucial for advancing this technology. One study [20] utilized SANS to investigate water sorption processes, revealing flat water concentration profiles across the membrane under different equilibration conditions. The rapid swelling kinetics of a Nafion membrane immersed in liquid water, completing in less than a minute, is also reported for the first time, offering valuable insights into membrane behavior.

Exploration of water uptake and salt transport across different Nafion membranes in various aqueous salt solutions was investigated in [21], showing that water uptake increases with membrane thickness and decreases with cation size. The results also reveal a decrease in the integral permeability coefficient with membrane thickness, while the effect of electrolyte type on this coefficient is minimal for thicker membranes.

The impact of water/alcohol composition on the dispersion of H-Nafion® in water/1-propanol and water/ethanol solutions was described in [22], which is crucial for catalyst layer performance in proton exchange membrane fuel cells (PEMFCs). Utilizing dynamic light scattering (DLS), small-angle X-ray scattering (SAXS), and nuclear magnetic resonance (NMR) spectroscopy, this paper reveals that 1-propanol induces notable changes in rod-like particle characteristics, suggesting implications for enhancing PEMFC performance.

Diffusion within polymer electrolyte membranes is often concurrent with time-dependent processes like swelling and polymer relaxation, limiting their ability to prevent molecular crossover during operation. To explore the coupling of such changes with the permeation process, a stochastic multiscale reaction–diffusion model was developed in [23], simulating methanol uptake and the swelling of hydrated Nafion. Comparison with experimental data suggests that a reaction-limited local response to an increasing methanol concentration best matches the observed behavior, indicating that interactions between methanol and Nafion lead to increased permeation across the membrane.

Coarse-grained molecular dynamics simulations elucidated the self-assembly behavior of Nafion ionomers in 1-propanol (NPA)/water solutions, revealing the formation of cylindrical aggregates with diameters of 2–3 nm [24]. The size and morphology of these aggregates were observed to vary nonlinearly with the ionomer concentration and NPA/water fraction, influenced by electrostatic repulsion among sulfonate groups and modified significantly upon salt addition, forming larger disk-shaped and secondary aggregates.

The hydration behavior of Nafion 117 membranes in alkaline ion forms was examined using high-resolution ^1H NMR [25], revealing different hydration numbers for Li^+, Na^+, and Cs^+ cations [26]. Cation self-diffusion coefficients, measured for the first time using pulsed-field gradient NMR, displayed a trend of $Li^+ \leq Na^+ > Cs^+$, with distinct activation energies indicating differences in diffusion behavior. Ionic conductivities calculated from these coefficients closely matched experimental values obtained via impedance spectroscopy, but with slightly higher estimations.

The study in [27] examines the behavior of Nafion membranes when swelling is limited to spaces smaller than 300 μm, leading to the formation of air cavities due to the expulsion of water from swollen polymer fibers. Investigating the collapse dynamics of these cavities in deionized water [28] and aqueous salt solutions reveals differing characteristic times, suggesting potential implications for pharmaceutical preparation processes requiring standardized dilution protocols.

1.2. Nafion and ML

ML presents a promising approach to optimize PEMFC performance, addressing challenges in cost and efficiency. By extracting patterns from experimental or simulation data, ML can predict outputs and reduce both experimental and computational costs. ML has demonstrated success in tasks such as predicting active electrocatalysts; optimizing membrane electrode assembly (MEA) [29], especially for predicting catalyst utilization [30], activation overpotential [31], maximum power densities, and I-V curves [32]; as an oxygen reduction reaction (ORR) electrocatalyst [33], especially regarding OH adsorption energies [34,35], concentration of surface microstructures and surface facets and clustering for identifying archetypes of Pt NPs [33]; and designing efficient flow channels [36], revolutionizing research in this field. Paper [37] reviews ML applications in optimizing PEMFC performance, offering insights for newcomers and outlining future directions for development. Figure 2 showcases the applications of ML methods in PEMFCs for enhancing performance and efficiency.

Study [38] utilized ML to analyze a database comprising 789 data points from 30 recent publications on polymer electrolyte membrane (PEM) electrolysis. Box whisker plots identified factors such as pure Pt at the cathode surface, Ti at the anode support, and specific compositions at the anode surface that led to higher performance. Principal component analysis (PCA) [39] highlighted risk factors for a low performance, including certain compositions at the cathode and anode surfaces. Classification trees identified the cathode-surface Ni mole fraction and the anode-surface Co mole fraction as crucial variables for electrolyzer performance, while regression trees successfully modeled the polarization behavior with an RMSE value of 0.18.

ML models, including polynomial and logistic regression, were proposed in [40] to predict the optimal design of PEM electrolyzer cells. Trained on 148 samples and validated on 16, the models accurately predicted eleven parameters based on inputs such as the hydrogen production rate and cell design. Hydrogen production rates measured using a custom-made PEM electrolyzer cell closely matched the simulation results, indicating the effectiveness of the models. This approach offers a cost-effective and time-saving solution for developing water electrolyzer cells for future hydrogen production.

Modern industries are increasingly adopting hydrogen as an energy carrier to decarbonize the electricity grid. However, the advancement of PEMFCs, which use hydrogen to produce electricity, is hindered by unpredictable performances and failure events like flooding and dehydration. To address this, paper [41] proposes a machine learning model to predict these failure modes by analyzing cell voltage and current density data. Utilizing advanced regression techniques like support vector machine, Decision Tree Regression, Random Forest Regression, and artificial neural networks, this model accurately forecasts flooding- and dehydration-induced failure events based on features derived from cell polarization data. Validation with real-time test data confirms the model's reliability.

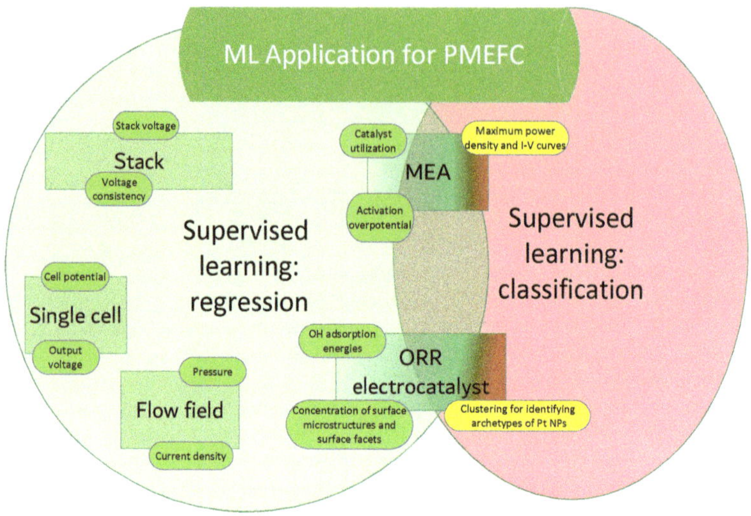

Figure 2. ML methods in PEMFCs.

1.3. Aim of this Study

A significant body of research has been dedicated to investigating Nafion in salt solutions, spanning over several decades and encompassing various aspects such as sorption behavior, structural characterization, and ion transport properties. Despite the extensive use of machine learning (ML) in various aspects of Nafion-based PEMFCs, its application to predict diffusion coefficients for Nafion membranes based on infrared spectroscopy (IR) data remains unexplored. This presents a significant gap in our understanding, as accurate prediction of diffusion behavior is crucial for optimizing the performance and durability of PEMFCs. Therefore, the objective of this research is to fill this gap by leveraging the IR spectra of Nafion membranes to develop ML models capable of predicting diffusion coefficients accurately. By harnessing the information contained within IR spectra, we aim to advance our understanding of Nafion membrane ion-exchange behavior.

2. Materials and Methods

In our study, we employed Nation N117 cast films (Sigma Aldrich, St. Louis, MI, USA) with a thickness of 175 μm and an area of 1 × 1 cm^2. The experimental setup utilized an analytical Fourier spectrometer FSM 2201 (LLC Infraspec, St. Petersburg, Russia), which is also fully described in works [27,42]. This spectrometer featured a total spectral range spanning from 370 to 7800 cm^{-1} (equivalent to 1.3–27 μm), with a spectral resolution of 1.0 cm^{-1} and an absolute error of ±0.05 cm^{-1}.

This investigation was conducted by measuring the infrared (IR) spectra of Nafion membranes in various salt solutions, such as LiCl, KCl, and NaCl, at different concentrations (0.1 M, 0.01 M, 0.001 M). All water-based solutions were made using deionized water with a resistivity of 18 MΩ·cm, obtained from a four-cartridge Nanopure Infinity System (Barnstead, Dubuque, IA, USA). Special attention was paid to the dynamics of the extremum in the range of valence vibrations around 5200 ± 500 cm^{-1} during the soaking and subsequent drying of the Nafion membrane at various time intervals. A time interval of 3–10 min separated each set of measurements. All measurements were conducted at room temperature (23 °C).

Preprocessing of the IR spectra involved baseline correction by identifying local minima within the specified spectral range, which were presumed to represent the underlying baseline. Furthermore, aligning these spectral minima facilitated correction for shifts induced by factors like instrument drift or fluctuations in sample preparation.

Subsequently, the obtained minimum values at different time points were plotted in graphs (Figures 3 and 4) and then fitted with an exponential function; for swelling, this was a decreasing exponential, and for drying, this was a saturation curve. Fitting was conducted using a custom Python 3.10 code, specifically utilizing the module Model from the lmfit library [43]. This module allowed for the implementation of mathematical models to fit the obtained minimum values at different time points. The lmfit library provided robust tools for parameter estimation and curve fitting, enabling precise characterization of the time-dependent behavior observed in the experimental data.

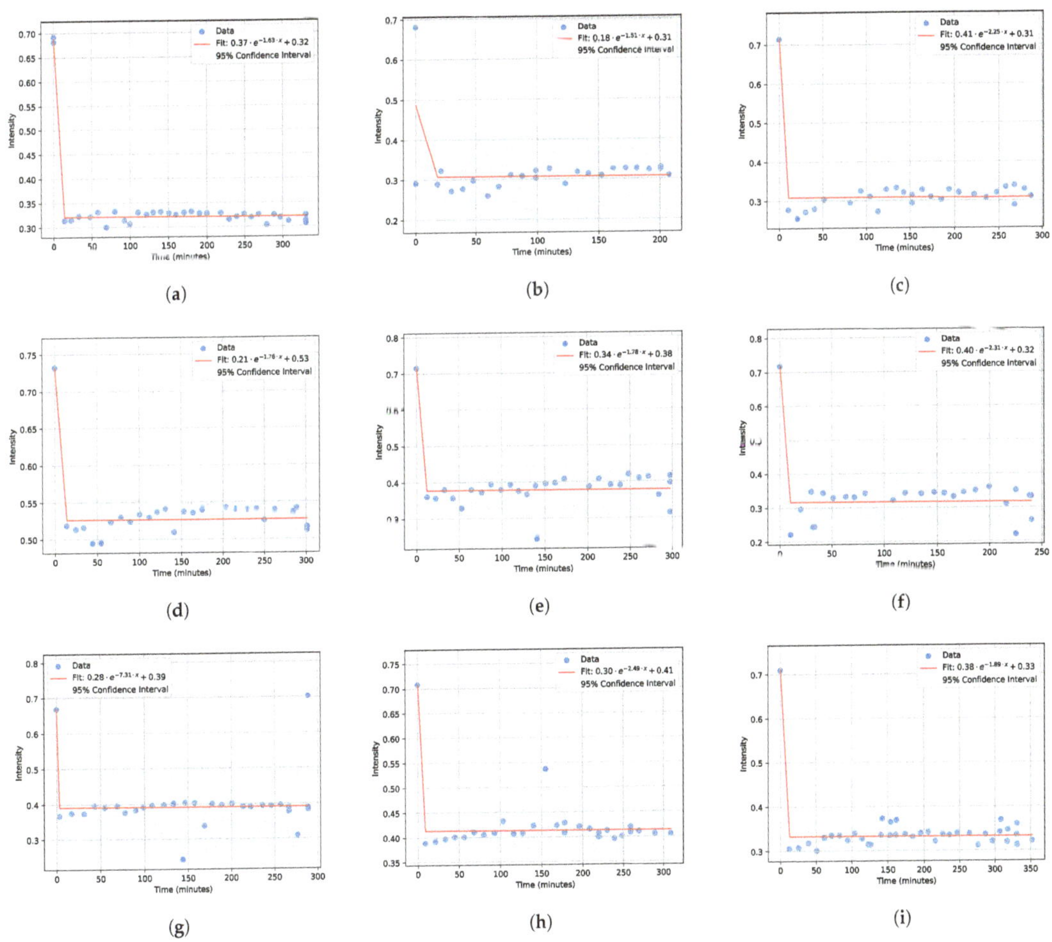

Figure 3. Nafion soaking kinetics for salt solutions of LiCl (**a–c**), KCl (**d–f**), NaCl (**g–i**) at salt concentrations of 0.1 M, 0.01 M, and 0.001 M, respectively.

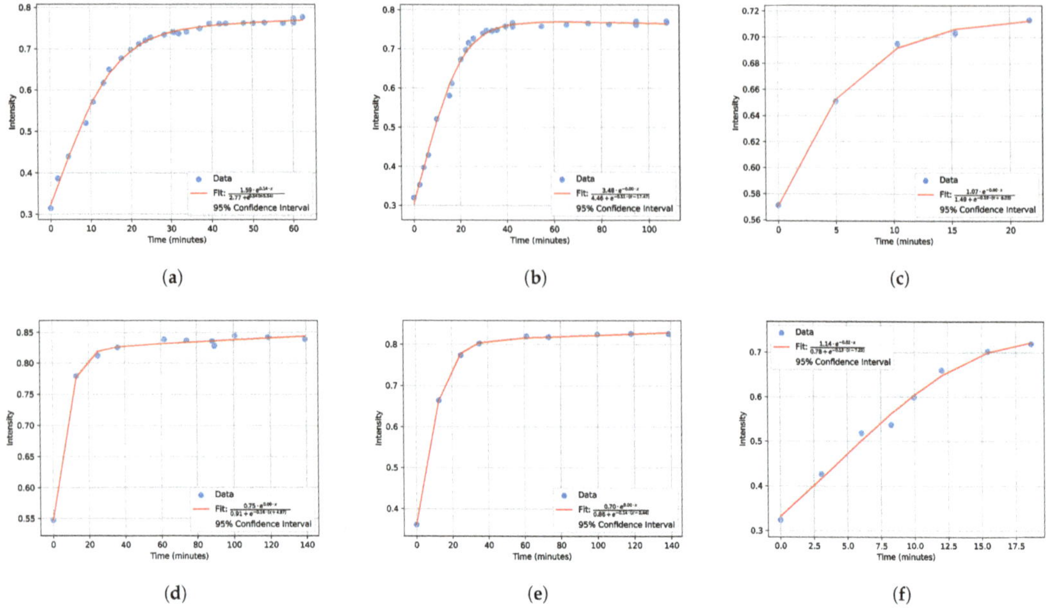

Figure 4. Nafion drying kinetics with saturation approximations for salt solutions of LiCl (**a–c**), KCl (**d–f**) at salt concentrations of 0.1 M, 0.01 M, and 0.001 M, respectively.

Then, ML models were applied to the gathered data. Specifically, the concentration information and salt solution compositions were utilized as input features, while the coefficients of the fitted exponential functions, representing self-diffusion coefficients, served as output (predictive) features.

In modern science, there is a constant need to optimize experimental processes for more efficient data acquisition. It was necessary to conduct numerous measurements to obtain diffusion coefficients (A, B for soaking and C, D, E for drying) at various concentrations of salts and salt solutions. Each measurement required significant time—up to several days—to yield reliable results. The figures presented in our paper (Figures 3 and 4) illustrate only fragments of data collected during the first 6 h of the experiment.

In light of this, we propose an approach based on machine learning [44,45], specifically neural networks (NNs) [46,47], to predict diffusion coefficients based on existing data. This approach could offer an estimate of the diffusion coefficients under various experimental conditions, helping researchers determine the necessity of conducting specific measurements.

The advantages of using NNs in this context include a high flexibility and the ability to adapt to complex nonlinear relationships between input and output data [48]. NNs can extract hidden patterns from data and make accurate predictions even in the presence of noise and incomplete data. The use of NNs in optimizing experimental processes [49] can significantly reduce the time and resources required to obtain valuable data, contributing to faster progress in scientific research.

3. Results

3.1. IR Spectra and Fitting

The obtained results of IR spectra during the swelling and drying processes of Nafion membranes at various time intervals are presented in Figures 5 and 6. Each graph corresponds to Nafion's IR spectra at a specific concentration of a single soaking or drying salt.

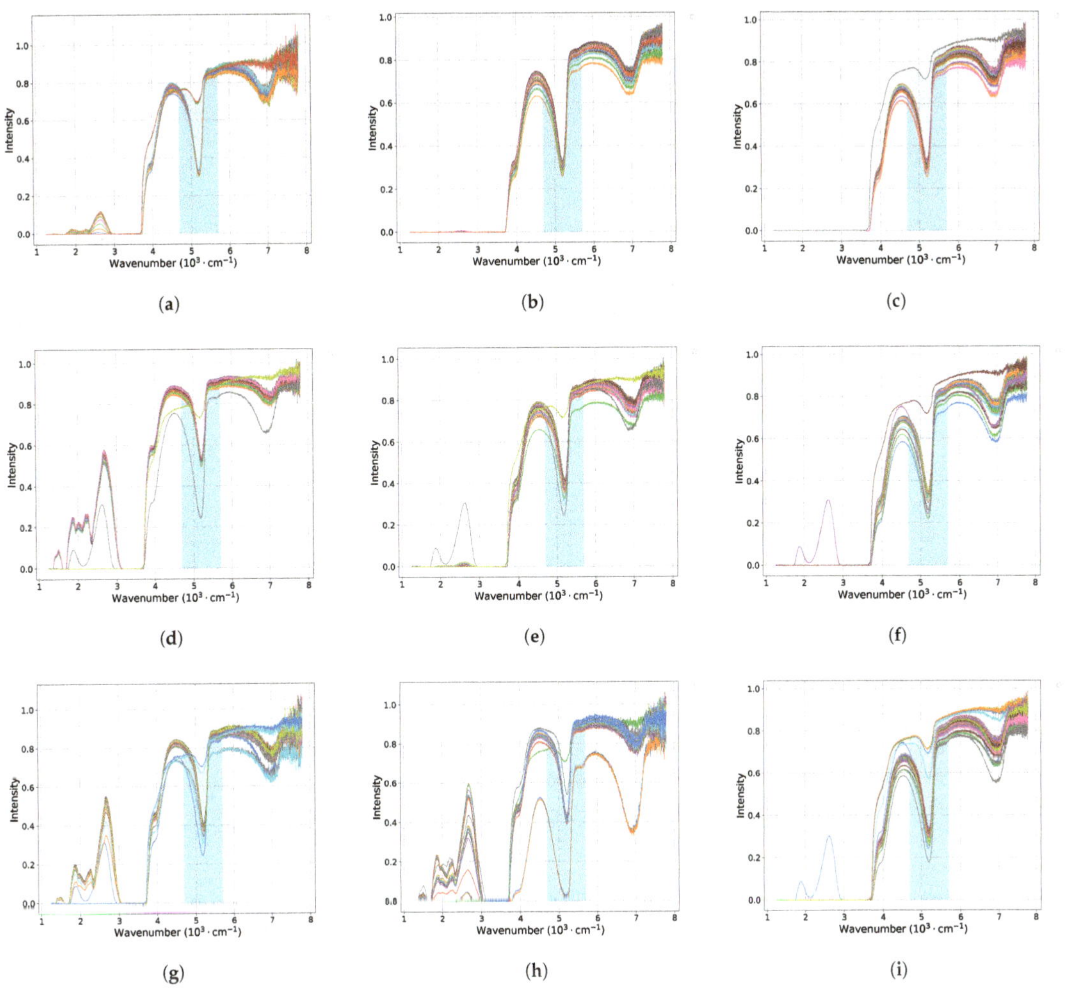

Figure 5. IR spectra of Nafion membranes during soaking in salt solutions of LiCl (**a–c**), KCl (**d–f**), and NaCl (**g–i**) at salt concentrations of 0.1 M, 0.01 M, and 0.001 M, respectively.

The highlighted blue bands in these figures indicate the specific range of interest, within which the identified minima were subsequently fitted over time using exponential functions. The results of these fittings can be observed in Figure 3 for soaking, with approximation function:

$$A \cdot e^{-B \cdot x} + C \tag{1}$$

and for drying, with approximation function:

$$\frac{\widetilde{A} \cdot e^{-\widetilde{B} \cdot x}}{\widetilde{C} + e^{-\widetilde{D} \cdot (x - \widetilde{E})}} \tag{2}$$

in Figure 4. These functions exhibit the most robust behavior suitable for an automated approximation process.

The fitting was performed with a 95% confidence interval [50] which indicates that the range of values obtained from the exponential fitting has a 95% probability of containing the true population parameter.

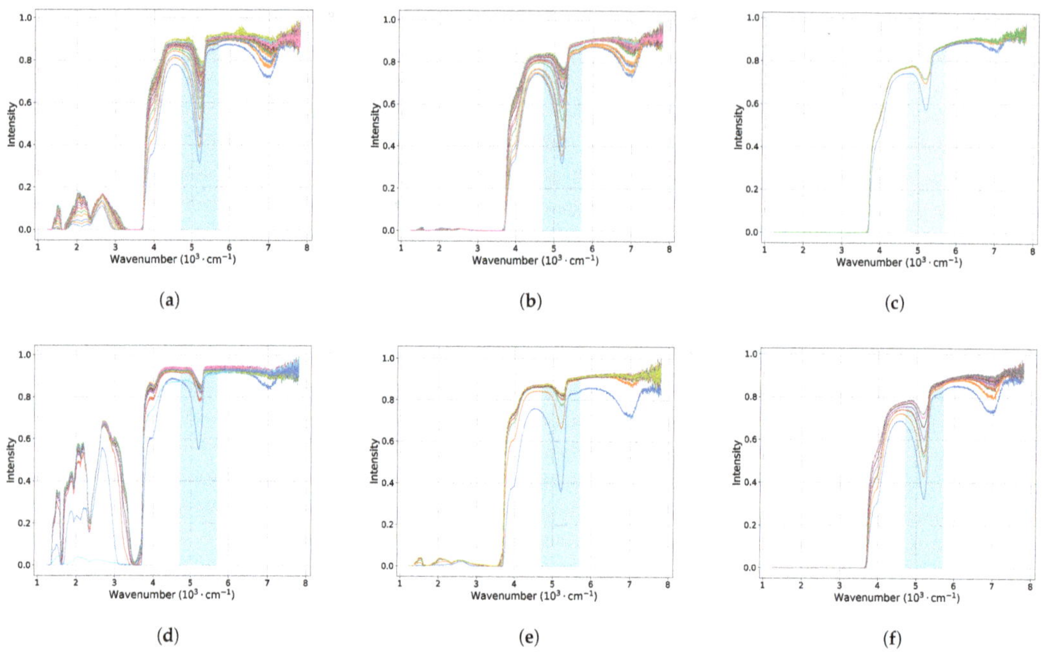

Figure 6. IR spectra of Nafion membranes during drying in salt solutions of LiCl (**a–c**) and KCl (**d–f**) at salt concentrations of 0.1, 0.01, 0.001, respectively.

The coefficients obtained from the fitting graphs are essential for determining the self-diffusion coefficients of ions diffusing into or out of a membrane sample in contact with an equilibrating solution. This approach relies on several assumptions [51]: firstly, the membrane governs the diffusion process; secondly, the self-diffusion coefficient within the membrane remains constant; and thirdly, the diffusion process is one-dimensional.

The validity of the first assumption hinges on ensuring that the concentrations of ions [52] at both boundaries of the membrane match those in the bulk solution, achieved through vigorous stirring of the membrane sample in the equilibrating solution [53]. The concentration profile of radiotracer ions diffusing into or out of the membrane is a function of both time and space, following Fick's second law [54]:

$$\frac{\partial c}{\partial t} = D\frac{\partial^2 c}{\partial x^2}$$

Here, c represents the concentration of ions within the membrane, D denotes the self-diffusion coefficient of the ion within the membrane, and x is the spatial coordinate.

Analytical solution of Fick's second law is given by the expression [55,56]:

$$n(t_k) = n^*\left[1 - \left(\frac{8}{\pi^2}\right)\left\{e^{-D^m_{H_2O}\pi^2 t_k/L^2} + \frac{1}{9}e^{-9D^m_{H_2O}\pi^2 t_k/L^2} + \ldots\right\}\right]$$

where $n(t_k)$ and n^* represent the quantities of ions in the equilibrating solution at fixed times t_k and t_∞, respectively. $D^m_{H_2O}$ denotes the self-diffusion coefficient of water within the membrane, and L represents the thickness of the membrane.

All obtained results of approximations were compiled into Tables 1 and 2 for clarity.

Table 1. Obtained coefficients *A*, *B* and *C* (see Equation (1)) at various concentrations of salts in different solutions from Figure 3.

Salt	Concentration	*A*	*B*	*C*
LiCl	0.1 M	0.37	−1.63	0.32
	0.01 M	0.18	−1.51	0.31
	0.001 M	0.41	−2.25	0.31
KCl	0.1 M	0.21	−1.76	0.53
	0.01 M	0.34	−1.78	0.38
	0.001 M	0.40	−2.31	0.32
NaCl	0.1 M	0.28	0.3	0.38
	0.01 M	−7.31	−2.49	−1.89
	0.001 M	0.39	0.41	0.33

Table 2. Obtained coefficients $\widetilde{A}$, $\widetilde{B}$, $\widetilde{C}$, $\widetilde{D}$, $\widetilde{E}$ (see Equation (2)) at various concentrations of salts in different solutions from Figure 4.

Salt	Concentration	$\widetilde{A}$	$\widetilde{B}$	$\widetilde{C}$	$\widetilde{D}$	$\widetilde{E}$
LiCl	0.1 M	1.59	0.14	2.77	0.14	−5.54
	0.01 M	3.48	0	4.46	−0.11	−17.47
	0.001 M	1.07	0	1.49	−0.19	−5.03
KCl	0.1 M	0.75	0	0.91	−0.16	−4.87
	0.01 M	0.7	0	0.86	−0.14	−0.44
	0.001 M	1.14	−0.02	0.78	−0.13	−7.22

3.2. Neural Network (NN)-Based Model for Predicting Diffusion Coefficients

In this study, we utilized an NN to forecast diffusion coefficients using a dataset encompassing salt concentration, time (during which the approximation was conducted), initial intensity, and intensity measured in the subsequent measurements, which is also referred to as coefficient C. This coefficient essentially represents the average value derived from all measurements taken from the second one onwards.

We have gathered all the necessary experimental parameters, such as concentration and salt type, by conducting measurements including initial measurements, measurements of coefficient C (which represent the average between the second and third measurements), and measurements using time as the interval between the first and third measurement. Our goal is to predict, using a neural network, coefficients *A* and *B* of exponential decay. These coefficients reflect the long-term process of intensity change in IR measurements during the soaking of Nafion membranes in water over time. Before initiating model training, we conducted several steps of data preprocessing. Initially, we employed One-Hot encoding [57] on the 'Salt' column to transform categorical data into binary flags, categorizing salts such as LiCl, KCl, or NaCl accordingly. Then, we scaled numerical features using standardization [58] to ensure similar value ranges.

We employed a genetic algorithm (GA) approach to optimize the parameters of the NN model [59,60] of TensorFlow [61] for predictive analysis. The aim was to develop an accurate model for forecasting the properties of interest based on the input features derived from experimental data. The methodology consisted of several key steps as follows:

1. Definition of a multi-layer NN architecture using the TensorFlow framework. The architecture comprised multiple dense layers with varying numbers of neurons and activation functions, which were treated as parameters to be optimized by the genetic algorithm.
2. GA implementation using the DEAP (Distributed Evolutionary Algorithms in Python) library. We defined the individuals as a combination of integers representing the number of neurons in each dense layer and a categorical variable representing the activation function. The GA aimed to optimize these parameters to minimize the mean squared error (MSE) loss function of the NN model.

3. GA iteratively evolved a population of candidate solutions (NN architectures) over multiple generations. Each candidate solution was evaluated by training the corresponding NN model on the training dataset and computing its MSE loss on the validation dataset. The fitness of each solution was determined by its validation MSE.
4. After a predefined number of generations, the GA selected the best-performing individual (NN architecture) based on its validation MSE. We split the data into training and testing sets in an 80/20 ratio. This architecture was then used to train a final NN model on the entire training dataset.
5. The performance of the final NN model was evaluated on an independent testing dataset to assess its generalization ability and predictive accuracy. The MSE loss on the test dataset was calculated as a measure of model performance.

The results of hyperparameter optimization for the neural network using a genetic algorithm (GA) are depicted in Figure 7, showcasing the performance metrics, including loss and validation loss, for the best-performing architecture obtained through optimization.

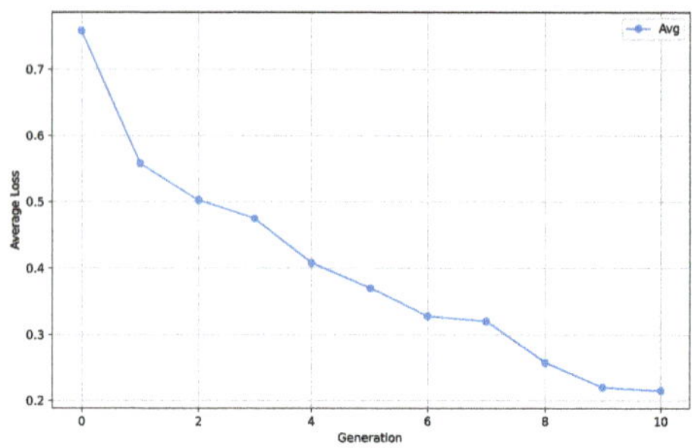

(a)

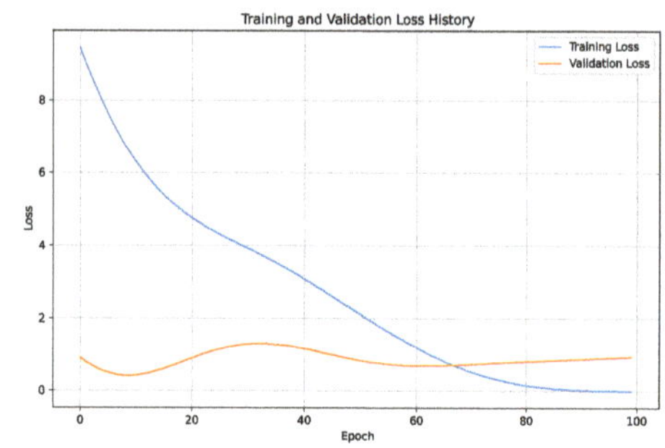

(b)

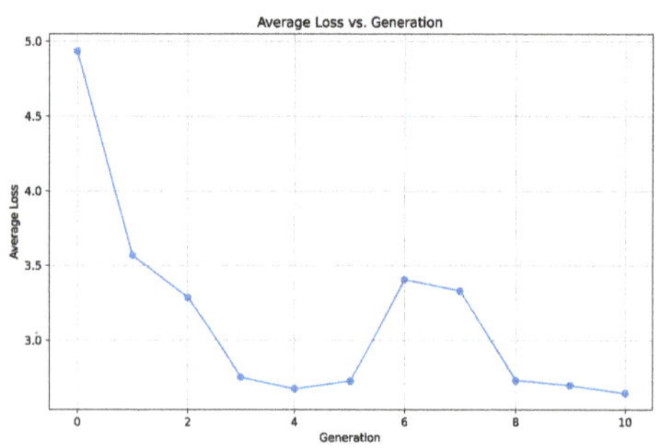

(c)

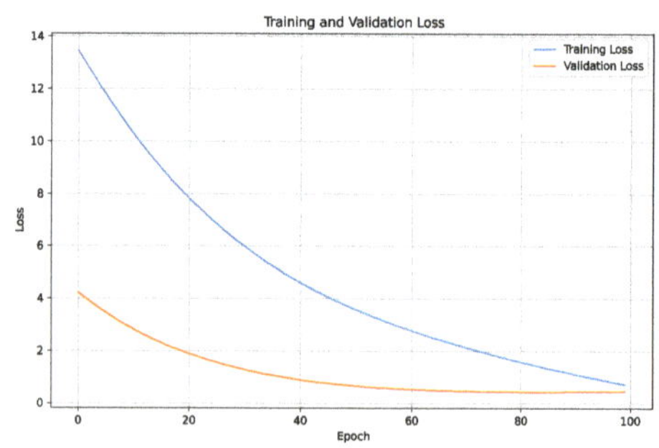

(d)

Figure 7. Average loss across generations during genetic algorithm optimization for predicting (**a**) A and B soaking coefficients, (**c**) $\widetilde{A}$, $\widetilde{B}$, $\widetilde{C}$, $\widetilde{D}$ and $\widetilde{E}$ drying coefficients. Training and validation loss history for the best model for predicting (**b**) A and B soaking coefficients, (**d**) $\widetilde{A}$, $\widetilde{B}$, $\widetilde{C}$, $\widetilde{D}$ and $\widetilde{E}$ drying coefficients.

In addition to predicting the coefficients A and B which characterize the soaking process in our study, we extended our model to predict the coefficients $\widetilde{A}$, $\widetilde{B}$, $\widetilde{C}$, $\widetilde{D}$, and $\widetilde{E}$ obtained from infrared (IR) spectroscopy measurements during the drying phase of Nafion membranes after soaking. This expansion allowed us to capture the dynamic changes in material properties during both the soaking and drying stages. Notably, we also employed hyperparameter optimization techniques using the GA to fine-tune the performance of our predictive model. By optimizing these parameters, we aimed to enhance the model's accuracy and robustness, thereby improving its ability to generalize and provide reliable predictions across diverse experimental conditions.

As a result, the best model for predicting soaking coefficients included four hidden layers (64, 64, 32, 16 neurons with ReLU [62], ELU [63], sigmoid [64] and ELU activation functions [65], respectively), as well as an output layer with two neurons for predicting coefficients A and B.

The best model for predicting drying coefficients included two hidden layers (16 and 12 neurons with sigmoid and ELU activation functions, respectively), as well as an output layer with five neurons for predicting coefficients $\widetilde{A}$, $\widetilde{B}$, $\widetilde{C}$, $\widetilde{D}$, and $\widetilde{E}$.

The uilized architecture for both cases of NNs is shown on Figure 8. Here, the input layer's dimension of X signifies the number of entities in the training dataset. With our dataset limited by experimental constraints, utilizing NNs for coefficient estimation becomes necessary. However, the model's parameters relative to the dataset size raise concerns about overfitting. Balancing complexity and the dataset size is crucial for reliable predictions. Further investigation is warranted.

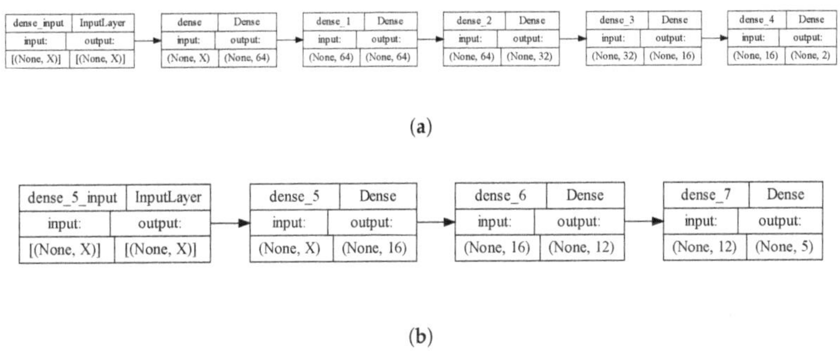

Figure 8. Diagram depicting the architecture of the NN for predicting diffusion coefficients (**a**) for the soaking process and (**b**) for the drying process.

After compiling the model with the Adam optimizer [66] and MSE loss function [67], we trained the model on the training dataset for 100 epochs [68], using a validation set to assess performance and prevent overfitting.

At the end of training, we evaluated the model's performance on the testing dataset and obtained the loss value. We also made predictions for values A and B based on the testing dataset, obtaining a test loss [69] of 0.44. The result is shown in Table 3.

Table 3. Real vs. predicted values for test dataset of diffusion coefficients for soaking Nafion.

Real A	Real B	Predicted A	Predicted B
0.41	−2.25	0.36	−2.32
0.34	−1.78	0.35	−1.42

This approach demonstrates the effective use of NNs for predicting diffusion coefficients based on input parameters. The resulting model can be used to predict the values of A and B with a high accuracy, which is an important step in diffusion research.

The performance of the optimized neural network model was compared against several other regression algorithms, including Linear Regression, Ridge Regression, Lasso Regression, Decision Tree Regressor, Random Forest Regressor, Gradient Boosting Regressor, Support Vector Regressor (SVR), and Multi-layer Perceptron Regressor (MLPRegressor). This comprehensive comparison enabled the assessment of the neural network's predictive capabilities relative to more traditional and state-of-the-art regression techniques, providing valuable insights into its effectiveness across different modeling scenarios. The results are shown in Tables 4 and 5.

Table 4. Mean squared error (MSE) for different regressors for predicting A and B soaking coefficients.

Regressor	A	B
LinearRegression	0.825698	0.011197
Ridge	0.048564	0.127635
Lasso	0.265159	0.589168
DecisionTreeRegressor	0.012850	2.426850
RandomForestRegressor	0.138267	0.145108
GradientBoostingRegressor	0.002541	0.545394
SVR	0.012887	0.134859
MLPRegressor	0.056801	0.514195

Table 5. Mean squared error (MSE) for different regressors for predicting $\widetilde{A}$, $\widetilde{B}$, $\widetilde{C}$, $\widetilde{D}$, and $\widetilde{E}$ drying coefficients.

Regressor	$\widetilde{A}$	$\widetilde{B}$	$\widetilde{C}$	$\widetilde{D}$	$\widetilde{E}$
Linear Regression [70]	0.135808	0.000677	0.272925	0.016502	17.032002
Ridge [71]	0.146546	0.000236	0.445721	0.009340	17.608812
Lasso [72]	0.526937	0.002125	1.763845	0.003050	14.911644
Decision Tree Regressor [73]	0.070900	0.000200	0.009700	0.000650	12.932100
Random Forest Regressor [74]	0.103242	0.002025	0.992413	0.002571	15.126330
Gradient Boosting Regressor [75]	0.304507	0.000200	0.207288	0.001429	11.885889
SVR [76]	0.177954	0.006500	1.016009	0.005524	13.676221
MLPRegressor [77]	0.057762	0.055338	0.031318	0.067580	9.837034

4. Discussion

The derived diffusion coefficients reflect the rate at which different species, such as protons or other ions, diffuse through the membrane [78,79]. Exponential coefficients play a significant role in soaking and drying processes. In the case of soaking, particular attention is given to the coefficient B, while in drying, coefficients $\widetilde{B}$ and $\widetilde{D}$ are considered.

In fuel cells, for instance, protons must efficiently transport across the membrane from the anode to the cathode for electrochemical reactions to occur. The diffusion coefficients help quantify how effectively this transport process happens. Higher diffusion coefficients indicate faster movement of species through the membrane, which can lead to an improved cell performance, a higher power output, and a better overall efficiency.

Moreover, these coefficients can provide insights into the membrane's structural and chemical properties. Changes in diffusion coefficients under different operating conditions or with modifications to the membrane composition can reveal how alterations affect ion transport and membrane performance. This understanding is crucial for designing Nafion membranes tailored to specific applications, optimizing their performance and addressing challenges such as durability, efficiency, and cost-effectiveness in fuel cells and other electrochemical devices.

Here are some possible research directions in the application of neural networks for predicting diffusion coefficients:

- Improving deep learning models: Research on developing more efficient and accurate deep learning models for predicting diffusion coefficients. This may involve using more complex network architectures and optimizing training parameters [80].
- Using recurrent neural networks (RNNs): Research on applying recurrent neural networks to analyze time series data, such as changes in substance concentration over time, to predict diffusion coefficients [81].
- Training on diverse datasets: Research aimed at training models on diverse and larger datasets of diffusion data [82]. This can help improve the models' generalization ability and make them more applicable to various conditions and materials.

- Investigating the influence of material structure: Research on analyzing the influence of the material's structure on its diffusion properties using neural networks [83]. This may involve analyzing the microstructure of the material or its chemical composition.
- Integration of Physical Models: Research aimed at integrating physical diffusion models with neural networks [84,85] to improve prediction accuracy. This can help incorporate the physical laws underlying diffusion into model development.
- Application in multiscale modeling: Research on using neural networks in multiscale diffusion modeling [86,87], allowing for the consideration of different temporal and spatial scales of the process.

These research directions can help advance our understanding of diffusion processes and develop more accurate and applicable models for their prediction.

It is imperative to acknowledge that further improvements in model accuracy necessitate a continuous influx of data [88] by performing more experiments. A larger and more diverse dataset would enable our models to capture a wider range of patterns [89] and nuances inherent in the diffusion process, thereby refining their predictive capabilities and bolstering their applicability across various experimental conditions. Also, it would mitigate the risk of overfitting and enhance the generalizability of our models, ensuring robust performance in real-world scenarios.

5. Conclusions

In conclusion, our study applied various machine learning approaches, including the optimization of neural network hyperparameters using a genetic algorithm, as well as employed different regressors to predict the coefficients governing the IR-based behavior of ion diffusion in Nafion membranes immersed in salt solutions during soaking and drying processes. These methodologies enabled us to enhance our understanding of the complex diffusion processes occurring within Nafion membranes and provide valuable insights into their behavior in different salt environments. Through the integration of advanced analytical techniques with computational methodologies, our research contributes to the ongoing efforts aimed at optimizing the performance of Nafion membranes in various applications, such as fuel cells and electrochemical devices.

Author Contributions: Conceptualization, I.M. and V.T.; Data curation, I.M., D.D. and A.G.; Formal analysis, V.T., A.G., V.N. and A.B.; Funding acquisition, V.T., A.G., V.N. and A.B.; Investigation, A.B.; Methodology, I.M.; Project administration, V.T., A.G., V.N. and A.B.; Resources, I.M.; Software, I.M., D.D. and V.N.; Supervision, V.T., V.N. and A.B.; Validation, I.M. and D.D.; Visualization, I.M. and A.G.; Writing—original draft, I.M., D.D. and V.T.; Writing—review and editing, V.T., V.N. and A.B. All authors have read and agreed to the published version of the manuscript.

Funding: This research received no external funding.

Institutional Review Board Statement: Not applicable.

Informed Consent Statement: Not applicable.

Data Availability Statement: The original contributions presented in the study are included in the article, further inquiries can be directed to the corresponding author.

Conflicts of Interest: The authors declare no conflicts of interest.

References

1. Mauritz, K.A.; Moore, R.B. State of understanding of Nafion. *Chem. Rev.* **2004**, *104*, 4535–4586. [CrossRef] [PubMed]
2. Gerhardt, G.A.; Oke, A.F.; Nagy, G.; Moghaddam, B.; Adams, R.N. Nafion-coated electrodes with high selectivity for CNS electrochemistry. *Brain Res.* **1984**, *290*, 390–395. [CrossRef] [PubMed]
3. Karimi, M.B.; Mohammadi, F.; Hooshyari, K. Recent approaches to improve Nafion performance for fuel cell applications: A review. *Int. J. Hydrogen Energy* **2019**, *44*, 28919–28938. [CrossRef]
4. Banerjee, S.; Curtin, D.E. Nafion® perfluorinated membranes in fuel cells. *J. Fluor. Chem.* **2004**, *125*, 1211–1216. [CrossRef]
5. Takamatsu, T.; Hashiyama, M.; Eisenberg, A. Sorption phenomena in Nafion membranes. *J. Appl. Polym. Sci.* **1979**, *24*, 2199–2220. [CrossRef]

6. Takamatsu, T.; Eisenberg, A. Densities and expansion coefficients of nafion polymers. *J. Appl. Polym. Sci.* **1979**, *24*, 2221–2235. [CrossRef]
7. Roche, E.; Pineri, M.; Duplessix, R.; Levelut, A. Small-angle scattering studies of nafion membranes. *J. Polym. Sci. Polym. Phys. Ed.* **1981**, *19*, 1–11. [CrossRef]
8. Lindheimer, A.; Molenat, J.; Gavach, C. A study of the superselectivity of Nafion perfluorosulfonic membranes. *J. Electroanal. Chem. Interfacial Electrochem.* **1987**, *216*, 71–88. [CrossRef]
9. Capeci, S.W.; Pintauro, P.N.; Bennion, D.N. The Molecular-Level Interpretation of Salt Uptake and Anion Transport in Nafion Membranes. *J. Electrochem. Soc.* **1989**, *136*, 2876. [CrossRef]
10. Bontha, J.; Pintauro, P.N. Water orientation and ion solvation effects during multicomponent salt partitioning in a Nafion cation exchange membrane. *Chem. Eng. Sci.* **1994**, *49*, 3835–3851. [CrossRef]
11. Lehmani, A.; Turq, P.; Périé, M.; Périé, J.; Simonin, J.P. Ion transport in Nafion® 117 membrane. *J. Electroanal. Chem.* **1997**, *428*, 81–89. [CrossRef]
12. Samec, Z.; Trojanek, A.; Langmaier, J.; Samcova, E. Diffusion Coefficients of Alkali Metal Cations in Nafion® from Ion-Exchange Measurements: An Advanced Kinetic Model. *J. Electrochem. Soc.* **1997**, *144*, 4236. [CrossRef]
13. Kawano, Y.; Wang, Y.; Palmer, R.A.; Aubuchon, S.R. Stress-strain curves of Nafion membranes in acid and salt forms. *Polímeros* **2002**, *12*, 96–101. [CrossRef]
14. Lee, S.J.; Yu, T.L.; Lin, H.L.; Liu, W.H.; Lai, C.L. Solution properties of nafion in methanol/water mixture solvent. *Polymer* **2004**, *45*, 2853–2862. [CrossRef]
15. Villaluenga, J.; Barragan, V.; Seoane, B.; Ruiz-Bauza, C. Sorption and permeation of solutions of chloride salts, water and methanol in a Nafion membrane. *Electrochim. Acta* **2006**, *51*, 6297–6303. [CrossRef]
16. Randová, A.; Hovorka, Š.; Izák, P.; Bartovská, L. Swelling of Nafion in methanol–water–inorganic salt ternary mixtures. *J. Electroanal. Chem.* **2008**, *616*, 117–121. [CrossRef]
17. Hallinan, D.T., Jr.; De Angelis, M.G.; Giacinti Baschetti, M.; Sarti, G.C.; Elabd, Y.A. Non-fickian diffusion of water in nafion. *Macromolecules* **2010**, *43*, 4667–4678. [CrossRef]
18. Zhao, Q.; Majsztrik, P.; Benziger, J. Diffusion and interfacial transport of water in Nafion. *J. Phys. Chem. B* **2011**, *115*, 2717–2727. [CrossRef] [PubMed]
19. Chaabane, L.; Dammak, L.; Grande, D.; Larchet, C.; Huguet, P.; Nikonenko, S.; Nikonenko, V. Swelling and permeability of Nafion® 117 in water–methanol solutions: An experimental and modelling investigation. *J. Membr. Sci.* **2011**, *377*, 54–64. [CrossRef]
20. Gebel, G.; Lyonnard, S.; Mendil-Jakani, H.; Morin, A. The kinetics of water sorption in Nafion membranes: A small-angle neutron scattering study. *J. Phys. Condens. Matter* **2011**, *23*, 234107. [CrossRef]
21. Izquierdo-Gil, M.A.; Barragán, V.; Villaluenga, J.; Godino, M. Water uptake and salt transport through Nafion cation-exchange membranes with different thicknesses. *Chem. Eng. Sci.* **2012**, *72*, 1–9. [CrossRef]
22. Yamaguchi, M.; Matsunaga, T.; Amemiya, K.; Ohira, A.; Hasegawa, N.; Shinohara, K.; Ando, M.; Yoshida, T. Dispersion of rod-like particles of nafion in salt-free water/1-propanol and water/ethanol solutions. *J. Phys. Chem. B* **2014**, *118*, 14922–14928. [CrossRef] [PubMed]
23. Soniat, M.; Houle, F.A. Swelling and diffusion during methanol sorption into hydrated nafion. *J. Phys. Chem. B* **2018**, *122*, 8255–8268. [CrossRef]
24. Mabuchi, T.; Huang, S.F.; Tokumasu, T. Dispersion of Nafion ionomer aggregates in 1-propanol/water solutions: Effects of ionomer concentration, alcohol content, and salt addition. *Macromolecules* **2020**, *53*, 3273–3283. [CrossRef]
25. Wang, Z.F.; You, Y.L.; Li, F.F.; Kong, W.R.; Wang, S.Q. Research progress of NMR in natural product quantification. *Molecules* **2021**, *26*, 6308. [CrossRef]
26. Volkov, V.I.; Chernyak, A.V.; Gnezdilov, O.I.; Skirda, V.D. Hydration, self-diffusion and ionic conductivity of Li+, Na+ and Cs+ cations in Nafion membrane studied by NMR. *Solid State Ion.* **2021**, *364*, 115627. [CrossRef]
27. Ninham, B.W.; Bolotskova, P.N.; Gudkov, S.V.; Baranova, E.N.; Kozlov, V.A.; Shkirin, A.V.; Vu, M.T.; Bunkin, N.F. Nafion swelling in salt solutions in a finite sized cell: Curious phenomena dependent on sample preparation protocol. *Polymers* **2022**, *14*, 1511. [CrossRef] [PubMed]
28. Van Meeteren, U.; Van Gelder, H.; Van Ieperen, W. Reconsideration of the use of deionized water as vase water in postharvest experiments on cut flowers. *Postharvest Biol. Technol.* **1999**, *17*, 175–187. [CrossRef]
29. Khajeh-Hosseini-Dalasm, N.; Ahadian, S.; Fushinobu, K.; Okazaki, K.; Kawazoe, Y. Prediction and analysis of the cathode catalyst layer performance of proton exchange membrane fuel cells using artificial neural network and statistical methods. *J. Power Sources* **2011**, *196*, 3750–3756. [CrossRef]
30. Pan, M.; Li, C.; Pan, C.; Lei, H.; Huang, H. A novel predicting method on degree of catalytic reaction in fuel cells. *Int. J. Energy Res.* **2020**, *44*, 6860–6872. [CrossRef]
31. Khajeh-Hosseini-Dalasm, N.; Fesanghary, M.; Fushinobu, K.; Okazaki, K. A study of the agglomerate catalyst layer for the cathode side of a proton exchange membrane fuel cell: Modeling and optimization. *Electrochim. Acta* **2012**, *60*, 55–65. [CrossRef]
32. Ding, R.; Ding, Y.; Zhang, H.; Wang, R.; Xu, Z.; Liu, Y.; Yin, W.; Wang, J.; Li, J.; Liu, J. Applying machine learning to boost the development of high-performance membrane electrode assembly for proton exchange membrane fuel cells. *J. Mater. Chem. A* **2021**, *9*, 6841–6850. [CrossRef]

33. Parker, A.J.; Opletal, G.; Barnard, A.S. Classification of platinum nanoparticle catalysts using machine learning. *J. Appl. Phys.* **2020**, *128*, 014301. [CrossRef]
34. Niu, H.; Wan, X.; Wang, X.; Shao, C.; Robertson, J.; Zhang, Z.; Guo, Y. Single-atom rhodium on defective g-C3N4: A promising bifunctional oxygen electrocatalyst. *ACS Sustain. Chem. Eng.* **2021**, *9*, 3590–3599. [CrossRef]
35. Wang, N.; Bo, X.; Zhou, M. Single-step and room-temperature synthesis of laser-induced Pt/VC nanocomposites as effective bifunctional electrocatalysts for hydrogen evolution and oxygen evolution reactions. *ACS Appl. Mater. Interfaces* **2022**, *14*, 23332–23341. [CrossRef] [PubMed]
36. Seyhan, M.; Akansu, Y.E.; Murat, M.; Korkmaz, Y.; Akansu, S.O. Performance prediction of PEM fuel cell with wavy serpentine flow channel by using artificial neural network. *Int. J. Hydrogen Energy* **2017**, *42*, 25619–25629. [CrossRef]
37. Ding, R.; Zhang, S.; Chen, Y.; Rui, Z.; Hua, K.; Wu, Y.; Li, X.; Duan, X.; Wang, X.; Li, J.; et al. Application of machine learning in optimizing proton exchange membrane fuel cells: A review. *Energy AI* **2022**, *9*, 100170. [CrossRef]
38. Günay, M.E.; Tapan, N.A.; Akkoç, G. Analysis and modeling of high-performance polymer electrolyte membrane electrolyzers by machine learning. *Int. J. Hydrogen Energy* **2022**, *47*, 2134–2151. [CrossRef]
39. Daffertshofer, A.; Lamoth, C.J.; Meijer, O.G.; Beek, P.J. PCA in studying coordination and variability: A tutorial. *Clin. Biomech.* **2004**, *19*, 415–428. [CrossRef]
40. Mohamed, A.; Ibrahem, H.; Yang, R.; Kim, K. Optimization of Proton Exchange Membrane Electrolyzer Cell Design Using Machine Learning. *Energies* **2022**, *15*, 6657. [CrossRef]
41. Zaveri, J.C.; Dhanushkodi, S.R.; Kumar, C.R.; Taler, J.; Majdak, M.; Węglowski, B. Predicting the Performance of PEM Fuel Cells by Determining Dehydration or Flooding in the Cell Using Machine Learning Models. *Energies* **2023**, *16*, 6968. [CrossRef]
42. Bunkin, N.F.; Kozlov, V.A.; Shkirin, A.V.; Ninham, B.W.; Balashov, A.A.; Gudkov, S.V. Dynamics of Nafion membrane swelling in H2O/D2O mixtures as studied using FTIR technique. *J. Chem. Phys.* **2018**, *148*, 124901. [CrossRef] [PubMed]
43. Newville, M.; Stensitzki, T.; Allen, D.B.; Rawlik, M.; Ingargiola, A.; Nelson, A. *LMFIT: Non-Linear Least-Square Minimization and Curve-Fitting for Python;* Astrophysics Source Code Library: Houghton, MI, USA, 2016; p. ascl-1606.
44. Chen, G.; Shen, Z.; Iyer, A.; Ghumman, U.F.; Tang, S.; Bi, J.; Chen, W.; Li, Y. Machine-learning-assisted de novo design of organic molecules and polymers: Opportunities and challenges. *Polymers* **2020**, *12*, 163. [CrossRef] [PubMed]
45. Malashin, I.P.; Tynchenko, V.S.; Nelyub, V.A.; Borodulin, A.S.; Gantimurov, A.P. Estimation and Prediction of the Polymers' Physical Characteristics Using the Machine Learning Models. *Polymers* **2023**, *16*, 115. [CrossRef] [PubMed]
46. Chepurnenko, A. Determining the Rheological Parameters of Polymers Using Artificial Neural Networks. *Polymers* **2022**, *14*, 3977. [CrossRef] [PubMed]
47. Ghaderi, A.; Morovati, V.; Dargazany, R. A physics-informed assembly of feed-forward neural network engines to predict inelasticity in cross-linked polymers. *Polymers* **2020**, *12*, 2628. [CrossRef] [PubMed]
48. Abdolrasol, M.G.; Hussain, S.S.; Ustun, T.S.; Sarker, M.R.; Hannan, M.A.; Mohamed, R.; Ali, J.A.; Mekhilef, S.; Milad, A. Artificial neural networks based optimization techniques: A review. *Electronics* **2021**, *10*, 2689. [CrossRef]
49. Kosarac, A.; Mladjenovic, C.; Zeljkovic, M.; Tabakovic, S.; Knezev, M. Neural-network-based approaches for optimization of machining parameters using small dataset. *Materials* **2022**, *15*, 700. [CrossRef] [PubMed]
50. Smithson, M. *Confidence Intervals;* Number 140; Sage: Thousand Oaks, CA, USA, 2003.
51. Goswami, A.; Acharya, A.; Pandey, A. Study of self-diffusion of monovalent and divalent cations in Nafion-117 ion-exchange membrane. *J. Phys. Chem. B* **2001**, *105*, 9196–9201. [CrossRef]
52. Okada, T.; Møller-Holst, S.; Gorseth, O.; Kjelstrup, S. Transport and equilibrium properties of Nafion® membranes with H+ and Na+ ions. *J. Electroanal. Chem.* **1998**, *442*, 137–145. [CrossRef]
53. Suresh, G.; Scindia, Y.; Pandey, A.; Goswami, A. Isotopic and ion-exchange kinetics in the Nafion-117 membrane. *J. Phys. Chem. B* **2004**, *108*, 4104–4110. [CrossRef]
54. Fick, A. V. On liquid diffusion. *Lond. Edinb. Dublin Philos. Mag. J. Sci.* **1855**, *10*, 30–39. [CrossRef]
55. Suresh, G.; Scindia, Y.; Pandey, A.; Goswami, A. Self-diffusion coefficient of water in Nafion-117 membrane with different monovalent counterions: A radiotracer study. *J. Membr. Sci.* **2005**, *250*, 39–45. [CrossRef]
56. Suresh, G.; Pandey, A.; Goswami, A. Self-diffusion coefficients of water in Nafion-117 membrane with multivalent counterions. *J. Membr. Sci.* **2006**, *284*, 193–197. [CrossRef]
57. Al-Shehari, T.; Alsowail, R.A. An insider data leakage detection using one-hot encoding, synthetic minority oversampling and machine learning techniques. *Entropy* **2021**, *23*, 1258. [CrossRef] [PubMed]
58. Protić, D.; Stanković, M.; Prodanović, R.; Vulić, I.; Stojanović, G.M.; Simić, M.; Ostojić, G.; Stankovski, S. Numerical feature selection and hyperbolic tangent feature scaling in machine learning-based detection of anomalies in the computer network behavior. *Electronics* **2023**, *12*, 4158. [CrossRef]
59. Ali, Y.A.; Awwad, E.M.; Al-Razgan, M.; Maarouf, A. Hyperparameter search for machine learning algorithms for optimizing the computational complexity. *Processes* **2023**, *11*, 349. [CrossRef]
60. Kiliçarslan, S. PSO+ GWO: A hybrid particle swarm optimization and Grey Wolf optimization based Algorithm for fine-tuning hyper-parameters of convolutional neural networks for Cardiovascular Disease Detection. *J. Ambient Intell. Humaniz. Comput.* **2023**, *14*, 87–97. [CrossRef]
61. Novac, O.C.; Chirodea, M.C.; Novac, C.M.; Bizon, N.; Oproescu, M.; Stan, O.P.; Gordan, C.E. Analysis of the application efficiency of TensorFlow and PyTorch in Convolutional Neural Network. *Sensors* **2022**, *22*, 8872. [CrossRef]

62. Hanin, B. Universal function approximation by deep neural nets with bounded width and relu activations. *Mathematics* **2019**, *7*, 992. [CrossRef]
63. Zhang, R.; Zhu, Y.; Ge, Z.; Mu, H.; Qi, D.; Ni, H. Transfer learning for leaf small dataset using improved ResNet50 network with mixed activation functions. *Forests* **2022**, *13*, 2072. [CrossRef]
64. Wang, X.; Ren, H.; Wang, A. Smish: A novel activation function for deep learning methods. *Electronics* **2022**, *11*, 540. [CrossRef]
65. Wang, Y.; Li, Y.; Song, Y.; Rong, X. The influence of the activation function in a convolution neural network model of facial expression recognition. *Appl. Sci.* **2020**, *10*, 1897. [CrossRef]
66. Truong, T.X.; Nhu, V.H.; Phuong, D.T.N.; Nghi, L.T.; Hung, N.N.; Hoa, P.V.; Bui, D.T. A New Approach Based on TensorFlow Deep Neural Networks with ADAM Optimizer and GIS for Spatial Prediction of Forest Fire Danger in Tropical Areas. *Remote Sens.* **2023**, *15*, 3458. [CrossRef]
67. Hayatbini, N.; Kong, B.; Hsu, K.l.; Nguyen, P.; Sorooshian, S.; Stephens, G.; Fowlkes, C.; Nemani, R.; Ganguly, S. Conditional generative adversarial networks (cGANs) for near real-time precipitation estimation from multispectral GOES-16 satellite imageries—PERSIANN-cGAN. *Remote Sens.* **2019**, *11*, 2193. [CrossRef]
68. Tran, T.H.G.; Ressl, C.; Pfeifer, N. Integrated change detection and classification in urban areas based on airborne laser scanning point clouds. *Sensors* **2018**, *18*, 448. [CrossRef] [PubMed]
69. Chen, D.; Hu, F.; Nian, G.; Yang, T. Deep residual learning for nonlinear regression. *Entropy* **2020**, *22*, 193. [CrossRef] [PubMed]
70. James, G.; Witten, D.; Hastie, T.; Tibshirani, R.; Taylor, J. Linear regression. In *An Introduction to Statistical Learning: With Applications in Python*; Springer: London, UK, 2023; pp. 69–134.
71. Barootchi, S.; Tavelli, L.; Majzoub, J.; Stefanini, M.; Wang, H.L.; Avila-Ortiz, G. Alveolar ridge preservation: Complications and cost-effectiveness. *Periodontol. 2000* **2023**, *92*, 235–262. [CrossRef] [PubMed]
72. Zhao, Y.; Huo, X. A survey of numerical algorithms that can solve the Lasso problems. *Wiley Interdiscip. Rev. Comput. Stat.* **2023**, *15*, e1602. [CrossRef]
73. Reddy, P.S.M. Decision tree regressor compared with random forest regressor for house price prediction in mumbai. *J. Surv. Fish. Sci.* **2023**, *10*, 2323–2332.
74. Setiawan, A.; Utami, E.; Ariatmanto, D. Cattle Weight Estimation Using Linear Regression and Random Forest Regressor. *J. RESTI (Rekayasa Sist. Dan Teknol. Inf.)* **2024**, *8*, 72–79. [CrossRef]
75. Rao, N.S.S.V.S.; Thangaraj, S.J.J.; Kumari, V.S. Flight Ticket Prediction Using Gradient Boosting Regressor Compared with Linear Regression. In Proceedings of the 2023 Eighth International Conference on Science Technology Engineering and Mathematics (ICONSTEM), Chennai, India, 6–7 April 2023; IEEE: New York, NY, USA, 2023; pp. 1–6.
76. Luo, C.; Zhu, S.P.; Keshtegar, B.; Niu, X.; Taylan, O. An enhanced uniform simulation approach coupled with SVR for efficient structural reliability analysis. *Reliab. Eng. Syst. Saf.* **2023**, *237*, 109377. [CrossRef]
77. Maqbool, J.; Aggarwal, P.; Kaur, R.; Mittal, A.; Ganaie, I.A. Stock prediction by integrating sentiment scores of financial news and MLP-regressor: A machine learning approach. *Procedia Comput. Sci.* **2023**, *218*, 1067–1078. [CrossRef]
78. Fang, X.; Vitrac, O. Predicting diffusion coefficients of chemicals in and through packaging materials. *Crit. Rev. Food Sci. Nutr.* **2017**, *57*, 275–312. [CrossRef] [PubMed]
79. Tyn, M.T.; Gusek, T.W. Prediction of diffusion coefficients of proteins. *Biotechnol. Bioeng.* **1990**, *35*, 327–338. [CrossRef] [PubMed]
80. Wang, S.; Jin, S.; Bai, D.; Fan, Y.; Shi, H.; Fernandez, C. A critical review of improved deep learning methods for the remaining useful life prediction of lithium-ion batteries. *Energy Rep.* **2021**, *7*, 5562–5574. [CrossRef]
81. Medsker, L.R.; Jain, L. Recurrent neural networks. *Des. Appl.* **2001**, *5*, 2.
82. Donyavi, Z.; Asadi, S. Diverse training dataset generation based on a multi-objective optimization for semi-supervised classification. *Pattern Recognit.* **2020**, *108*, 107543. [CrossRef]
83. Bennemann, C.; Paul, W.; Baschnagel, J.; Binder, K. Investigating the influence of different thermodynamic paths on the structural relaxation in a glass-forming polymer melt. *J. Phys. Condens. Matter* **1999**, *11*, 2179. [CrossRef]
84. Deng, J.; Wu, J.; Zhang, S.; Li, W.; Wang, Y.G. Physical informed neural networks with soft and hard boundary constraints for solving advection-diffusion equations using Fourier expansions. *Comput. Math. Appl.* **2024**, *159*, 60–75.
85. Pajot, A. Incorporating Physical Knowledge into Deep Neural Network. Ph.D. Thesis, Sorbonne Université, Paris, France, 2019.
86. Wang, R.; Xu, X.; Zhang, Y. Multiscale Information Diffusion Prediction with Minimal Substitution Neural Network. *IEEE Trans. Neural Netw. Learn. Syst.* **2023**, 1–12. [CrossRef] [PubMed]
87. Croitoru, F.A.; Hondru, V.; Ionescu, R.T.; Shah, M. Diffusion models in vision: A survey. *IEEE Trans. Pattern Anal. Mach. Intell.* **2023**, *45*, 10850–10869. [CrossRef] [PubMed]
88. Wang, S.; Ren, P.; Takyi-Aninakwa, P.; Jin, S.; Fernandez, C. A critical review of improved deep convolutional neural network for multi-timescale state prediction of lithium-ion batteries. *Energies* **2022**, *15*, 5053. [CrossRef]
89. Krapivin, M.; Autaeu, A.; Marchese, M. *Large Dataset for Keyphrases Extraction*; University of Trento: Trento, Italy, 2009.

polymers

MDPI

Article

Stable 3D Deep Convolutional Autoencoder Method for Ultrasonic Testing of Defects in Polymer Composites

Yi Liu [1], Qing Yu [1], Kaixin Liu [2,*], Ningtao Zhu [3] and Yuan Yao [4,*]

1 Institute of Process Equipment and Control Engineering, Zhejiang University of Technology, Hangzhou 310023, China; yliuzju@zjut.edu.cn (Y.L.); 2112102164@zjut.edu.cn (Q.Y.)
2 The State Key Laboratory for Electronic Testing Technology, North University of China, Taiyuan 030051, China
3 Xi'an Zhanshi Testing & Engineering Co., Ltd., Xi'an 710000, China; zhuningtao270@163.com
4 Department of Chemical Engineering, National Tsing Hua University, Hsinchu 300044, Taiwan
* Correspondence: kxliu@nuc.edu.cn (K.L.); yyao@mx.nthu.edu.tw (Y.Y.); Tel.: +886-3-5713690 (Y.Y.)

Abstract: Ultrasonic testing is widely used for defect detection in polymer composites owing to advantages such as fast processing speed, simple operation, high reliability, and real-time monitoring. However, defect information in ultrasound images is not easily detectable because of the influence of ultrasound echoes and noise. In this study, a stable three-dimensional deep convolutional autoencoder (3D-DCA) was developed to identify defects in polymer composites. Through 3D convolutional operations, it can synchronously learn the spatiotemporal properties of the data volume. Subsequently, the depth receptive field (RF) of the hidden layer in the autoencoder maps the defect information to the original depth location, thereby mitigating the effects of the defect surface and bottom echoes. In addition, a dual-layer encoder was designed to improve the hidden layer visualization results. Consequently, the size, shape, and depth of the defects can be accurately determined. The feasibility of the method was demonstrated through its application to defect detection in carbon-fiber-reinforced polymers.

Keywords: carbon-fiber-reinforced polymer; ultrasonic testing; 3D convolution; deep autoencoder; receptive field

1. Introduction

Carbon-fiber-reinforced polymers (CFRPs) [1] combine the excellent properties of both carbon and fiber materials and are widely used in aerospace, new energy, military applications, and other fields [2]. However, defects may arise during the manufacture and use of these materials. These defects not only affect the comprehensive characteristics of the material and reduce its load-bearing capacity but also have a significant impact on product quality [3]. Therefore, regular nondestructive testing (NDT) of CFRP is crucial to ensure the structural quality of the material. Currently, the mainstream NDT methods include ultrasonic testing (UT) [4], eddy current testing [5], and infrared thermography [6]. UT is widely used because of its nonpolluting and noninvasive nature, ease of operation, accurate positioning, and high reliability [7,8].

However, ultrasound signals are affected by surface echoes, bottom echoes, and noise, which interfere with the judgment of defects, making them difficult to identify [9]. Recently, several methods have been proposed for analyzing ultrasonic inspection data to improve their effectiveness. One-dimensional (1D) signal filtering processing methods, including empirical wavelet transform [10], short-time Fourier transform analysis [11], and Wigner–Ville distribution [12], are mainly used for the analysis of ultrasound signals. These methods can extract the characteristic information of defects but lack visualization. Image defect detection methods include support vector machines [13], manifold learning algorithms [14], and feature-selective clustering [15]. These machine learning methods can improve the identifiability of defects. For ultrasonic data with considerable noise and large echoes, deep learning [16,17] methods are more attractive. Models such as neural networks can automatically extract features and therefore contribute to the development of defect detection.

Recently, the rapid development of deep learning has led to its successful application in several manufacturing and maintenance industries [18]. An autoencoder (AE), a typical unsupervised deep learning model, has shown good performance in extracting potential features from high-dimensional data and eliminating redundant information. Some AE variants have made significant progress in NDT data analysis. Convolutional AE (CAE) [19] can use activation functions to extract nonlinear features, which helps analyze nonlinear, multinoise ultrasonic data [20]. The deep AE (DAE) [21] was successfully employed to analyze infrared thermography defect signals, and the output of the intermediate hidden layer was visualized to accentuate the defects. However, it focuses more on analyzing images and pixel data and often ignores important time-series information. Three-dimensional convolution (3D-conv) [22] can accurately extract deep spatial features from images, significantly outperforming previous feature extraction models based on 1D and two-dimensional (2D) information. However, few studies have applied 3D-conv for defect detection in CFRP.

In this study, a stable 3D deep convolutional autoencoder (3D-DCA) was developed to improve the accuracy and reliability of defect detection. To address the issue of significant ultrasonic noise, a 3D convolutional neural network (CNN) is used to simultaneously extract the spatiotemporal features of the data and reduce noise in the ultrasonic data. The depth-wise receptive field in the hidden layers of the autoencoder maps the defect information to its original depth position, thereby reducing the impact of surface and bottom reflections. Additionally, a two-layer encoder was developed to enhance the visual representation of the hidden layers. The intersection over union (IoU) [23] and contrast-to-noise ratio (CNR) [24] were used to evaluate the effectiveness of the proposed approach.

The remainder of this paper is organized as follows. Section 2 presents the ultrasonic data structure and preprocessing. Section 3 describes the stable 3D-DCA model framework and its implementation. In Section 4, a CFRP specimen is presented and the experimental results are discussed. Finally, the conclusions are presented in Section 5.

2. Data Collection and Preprocessing

2.1. Ultrasound Scanning

Phased-array ultrasonic detectors utilize multichip probes, where each probe functions as a separate transmitter or receiver array. The process operates as shown in Figure 1. The transmitter array directs the emission of the ultrasound signals from each unit to form the desired acoustic beam [25]. Simultaneously, the receiver array manages the timing of each unit to capture reflected signals. The ultrasound waves propagate through the test object and are reflected off the internal defects or interfaces. The strength of the reflected signals correlates with the properties of the materials on either side of the interface [26]. Thus, the ultrasound echoes that return from the workpiece can be used to locate defects within the material.

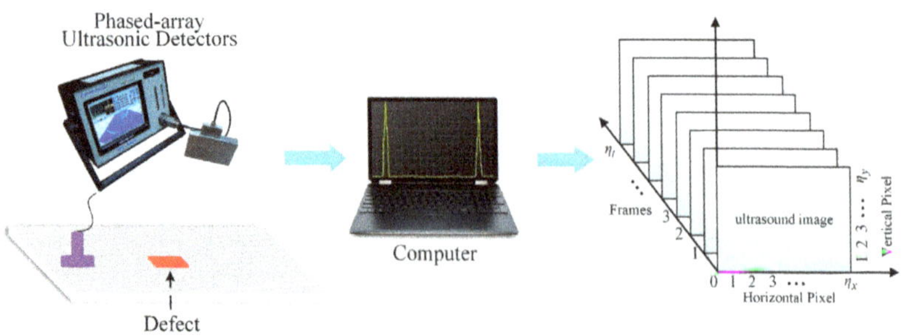

Figure 1. Schematic of a phased-array ultrasonic acquisition system.

The collected ultrasonic data were then converted into a 3D matrix for processing with a stable 3D-DCA. The ultrasonic time series served as the depth dimension, with the horizontal and vertical pixels representing the height and width dimensions of the ultrasound image, respectively. Consequently, the convolution operation extracts feature information from the time series of multiple ultrasound images, thereby reducing the reliance on manual screening operations.

2.2. Data Structures and Preprocessing

The dataset acquired during ultrasonic inspection typically comprises a 3D matrix that displays planar projections of ultrasound echoes at various depths of the specimen. This is achieved by setting a specific depth range to acquire an n_h frame echo signal. Each image frame consists of $n_x \times n_y$ pixels. The 3D ultrasonic matrix $\mathbf{X} \in \mathbf{R}^{T\times M\times N}$ is then transformed into a 2D matrix $\mathbf{X}_1 \in \mathbf{R}^{T\times R}$, where M, N, and T denote the length, width, and number of frames, respectively, and $R = M \times N$.

After converting the 3D ultrasonic matrix into a 2D matrix, it was observed that the amplitude of the echoes at shallow defects was usually larger than that at deep defects, and the amplitudes of the surface and bottom echoes exceeded those of the defect echoes [27]. To mitigate the effects of differences in scale, it is often necessary to preprocess the original ultrasonic data $\mathrm{x}_{i,j}$, where $\mathrm{x}_{i,j}$ is the original value at the jth pixel in column i of $\mathbf{X}_1$. In this study, the robust normalization [28] method was applied, which is calculated as follows:

$$x'_{i,j} = \frac{x_{i,j} - med}{IQR}, \tag{1}$$

where $x'_{i,j}$ represents normalized data. *med* is the median of x_i, and *IQR* is the interquartile range of x_i. The effects of outliers in the data can be reduced using robust normalization techniques.

3. Methodologies

3.1. Description of 3D-DCA

An AE typically consists of encoder and decoder networks. The encoder compresses the data dimensions to extract potential features, and the decoder reconstructs the input data based on these features. As previously mentioned, $\mathbf{X} \in \mathbf{R}^{T\times M\times N}$ represents a set of ultrasonic data, and the feature extraction process using 3D DCA is represented as

$$\begin{aligned} \mathbf{X}_\mathrm{L} &= f(\mathbf{W}_\mathrm{en} \odot \mathbf{X} + \mathbf{b}_\mathrm{en}) \\ \hat{\mathbf{X}} &= f(\mathbf{W}_\mathrm{de} \odot \mathbf{X}_\mathrm{L} + \mathbf{b}_\mathrm{de}) \end{aligned}, \tag{2}$$

where $\mathbf{X}_\mathrm{L} \in \mathbf{R}^{T\times M\times N}$ ($t << T$) denotes the output of the encoder, t is the number of feature maps, $\odot$ represents a 3D-conv operation, $\mathbf{W}_\mathrm{en}$ and $\mathbf{b}_\mathrm{en}$ represent the encoder weight and bias, respectively, $\hat{\mathbf{X}} \in \mathbf{R}^{t\times M\times N}$ denotes the ultrasonic data reconstructed by the decoder, and $\mathbf{W}_\mathrm{de}$ and $\mathbf{b}_\mathrm{de}$ represent the decoder weight and bias.

Compared with 2D convolution, the use of 3D-conv in this model accounts for temporal information. The dimensions of the 2D convolution can be expressed as $k_h \times k_w$, whereas those of 3D-conv are $k_h \times k_w \times k_d$. The formula for 3D-conv is similar to that for 2D convolution, with the primary distinction being that the convolution operation extends across three dimensions—depth, width, and height—that require manipulation in a volumetric space. Thus, 3D-conv can extract both spatial and temporal information, thereby enhancing defect detection accuracy. Similar to the 2D convolutional network structure, 3D-DCA includes an encoder, decoder, and loss function, but with the 2D convolutional kernel replaced by a 3D-conv kernel. Assuming that the feature map of layer l-1 is $\mathrm{A}_{l\text{-}1}$, the representation of the value of the ith feature of layer l at position (x, y, z) is

$$A_{l,i}^{xyz} = f\left(\sum_{c=0}^{C_L-1}\sum_{h=0}^{H_L-1}\sum_{w=0}^{W_L-1}\sum_{d=0}^{D_L-1} W_{l,i}^{chwd} A_{l-1}^{c(x+h)(y+w)(z+d)} + b_{l,i}\right), \tag{3}$$

where $W_{l,i}^{chwd}$ denotes the value of the ith convolution kernel connected to layer l at a position (h, w, and d); c denotes the feature map fed from layer l-1 to layer l; H, W, and D denote the height and width of the 3D-conv kernel and the depth in the time series, respectively; $b_{l,i}$ denotes the bias of the ith feature map in the lth layer; $f(\,)$ denotes the activation function; and the activation function is Leaky Relu (LReLu).

3.2. *Stable 3D-DCA for Ultrasonic Defect Detection*

In this section, the construction steps of the stable 3D-DCA model to improve the visibility of defects are described in detail. A phase-controlled ultrasonic detector was used to acquire ultrasonic data, and a stable 3D-DCA method was applied to extract defective regions from the image. The main steps of the stable 3D-DCA model include 3D-conv for ultrasonic signal denoising, RF-based defect depth prediction, and enhancement of detection performance.

3.2.1. 3D-Conv for Ultrasonic Denoising

The ultrasound images acquired using the phase-controlled ultrasonic detector exhibited significant redundancy. The unfolded data matrix $\mathbf{X_1}$ is preprocessed using the robust normalization method described in Section 2.2 to improve the data presentation. After preprocessing, the 2D data $\mathbf{X_1}$ were converted into a 3D matrix $\mathbf{X} \in \mathbf{R}^{T \times M \times N}$.

The structure of a stable 3D-DCA model is shown in Figure 2. The stable 3D-DCA core architecture comprises two encoders and a decoder.

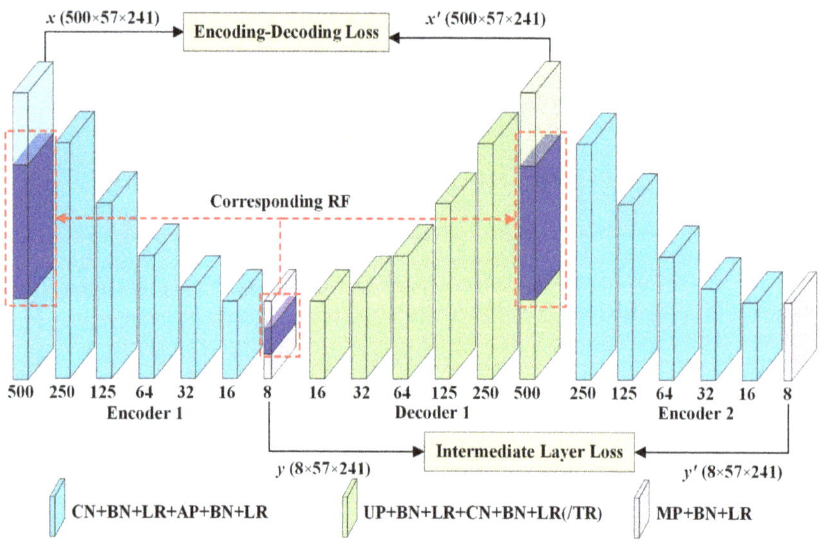

Figure 2. Stable 3D-DCA flowchart. CN represents 3D-conv, BN represents batch normalization, LR represents Leaky ReLU, TH represents Tanh used in the output layer of the decoder, AP represents 3D average pooling, UP represents upsampling, and MP represents 3D maximum pooling.

The encoder has six coded convolutional blocks, each consisting of a convolutional layer and a pooling layer. The 3D filter can move in all three directions of the ultrasonic data and detect features both vertically and horizontally. At each position, a convolved value was calculated to attenuate noise in the original ultrasonic data and clarify the spatial information of the defects. To obtain an enhanced image with low noise, the convolution kernel of the convolutional layer was set to $3 \times 3 \times 3$ and the step size was set to $1 \times 1 \times 1$. Then, 3D maximum pooling was used in the last layer of the encoder, whereas 3D average pooling was used in the other layers. The kernel of the pooling layer was $3 \times 3 \times 3$. Using

3D-conv, the ultrasonic time series can be processed directly in the ultrasound processing tasks without segmenting the 3D data into multiple ultrasound images. Additionally, to accelerate the mapping of the RF and reduce the number of parameters used in the model, the step size of the pooling layer was set to $2 \times 1 \times 1$, focusing only on feature extraction in the depth direction. The Adam optimizer and dropout strategies were used to prevent model overfitting and reduce computation.

The decoder incorporates an upsampling layer to resize the reconstructed map and obtain a noise-reduced map that corresponds one-to-one to the input data. The decoding layer included six convolutional blocks, each featuring an upsampling layer and a convolutional layer. A batch normalization layer and an activation function were added to each layer of neurons, with the tanh activation function [29] used for the output layer of the decoder and LReLU for the other layers.

The ultrasonic data possess a 3D structure. The stable 3D-DCA denoising effect is highlighted by the integration of components such as 3D-conv and 3D pooling.

3.2.2. RF-Based Defect Depth Prediction

In a 3D-conv neural network, the RF refers to the volume of input data that a point on the feature map can perceive. As illustrated in Figure 3, the larger the RF value of the neuron, the wider the range of the original data it can access. The RF of the current layer is calculated as follows:

$$RF_{i+1} = RF_i + (k-1) \times S_i, \tag{4}$$

where RF_{i+1} denotes the RF of the current layer; RF_i denotes the RF of the previous layer; k denotes the size of the convolution kernel; and S_i represents the product of the step sizes of all previous layers (excluding the current layer). The formula used is as follows:

$$S_i = \prod_{i=1}^{i} Stride_i, \tag{5}$$

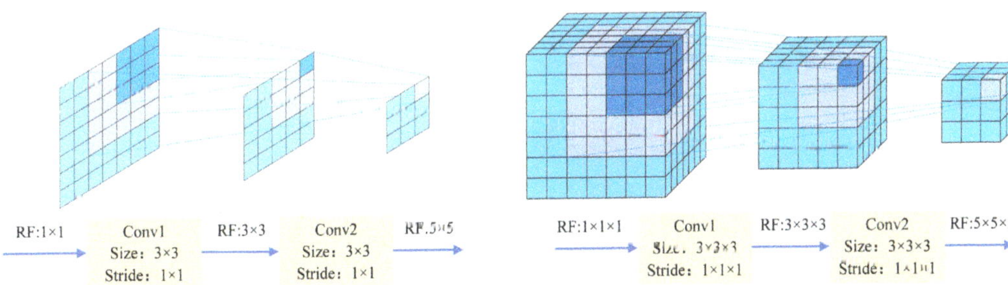

Figure 3. RF mapping of 2D convolution vs. 3D-conv.

Using the data from this study as an example, the network framework and RF calculations are presented in Table 1. The RF of each point on the feature map can be determined using Equation (4): the ultrasonic surface and bottom echoes could be bypassed by adjusting the RF size. Consequently, RF can be utilized to approximate the CFRP defect information for each layer and mitigate the impact of surface and bottom echoes on the results.

Table 1. RF calculations.

Layers	Input	Kernel	Stride	Output	RF
Conv1	$500 \times 57 \times 241$	$3 \times 3 \times 3$	$1 \times 1 \times 1$	$500 \times 57 \times 241$	$3 \times 3 \times 3$
AvgP1	$500 \times 57 \times 241$	$3 \times 3 \times 3$	$2 \times 1 \times 1$	$250 \times 57 \times 241$	$5 \times 5 \times 5$
Conv2	$250 \times 57 \times 241$	$3 \times 3 \times 3$	$1 \times 1 \times 1$	$250 \times 57 \times 241$	$9 \times 7 \times 7$
AvgP2	$250 \times 57 \times 241$	$3 \times 3 \times 3$	$2 \times 1 \times 1$	$125 \times 57 \times 241$	$13 \times 9 \times 9$

Table 1. *Cont.*

Layers	Input	Kernel	Stride	Output	RF
Conv3	125 × 57 × 241	3 × 3 × 3	1 × 1 × 1	125 × 57 × 241	21 × 11 × 11
AvgP3	125 × 57 × 241	3 × 3 × 3	2 × 1 × 1	64 × 57 × 241	29 × 13 × 13
Conv4	64 × 57 × 241	3 × 3 × 3	1 × 1 × 1	64 × 57 × 241	45 × 15 × 15
AvgP4	64 × 57 × 241	3 × 3 × 3	2 × 1 × 1	32 × 57 × 241	53 × 17 × 17
Conv5	32 × 57 × 241	3 × 3 × 3	1 × 1 × 1	32 × 57 × 241	69 × 19 × 19
AvgP5	32 × 57 × 241	3 × 3 × 3	2 × 1 × 1	16 × 57 × 241	85 × 21 × 21
Conv6	16 × 57 × 241	3 × 3 × 3	1 × 1 × 1	16 × 57 × 241	117 × 23 × 23
AvgP6	16 × 57 × 241	3 × 3 × 3	2 × 1 × 1	8 × 57 × 241	149 × 25 × 25
MaxP7	8 × 57 × 241	3 × 3 × 3	1 × 1 × 1	8 × 57 × 241	213 × 27 × 27

3.2.3. Improving the Detection Results

Feature extraction using an encoder compresses numerous ultrasound images into a low-dimensional space, thereby facilitating easier defect detection. However, without stringent parameter constraints, the effectiveness of the extracted features cannot be ensured. In this study, the outputs of the two encoders are used to construct the intermediate layer loss. This adjustment ensured that the defect size and shape closely resembled the ground truth.

$$\text{Loss} = \text{MSELoss}(x, x') + \text{MSELoss}(y, y') \quad (6)$$

Defect recognition is a target detection task that focuses on the location, shape, and number of defects. The raw 3D data encoded by stable 3D-DCA resulted in a new 3D matrix, which was decomposed into several 2D matrices of size $x \times y$. These can be visualized as 2D images for enhanced defect visualization.

A comparison between the stable 3D-DCA and DAE highlighted several advantages. First, a stable 3D-DCA learns the spatial and temporal features of ultrasonic data, whereas the DAE overlooks the depth and potentially original image information. Second, stable 3D-DCA approximates the depth order of the defect information through the mapping between the deep feature, RF, and original image, making defect detection more meaningful. In addition, the loss function designed for a stable 3D-DCA ensures better results.

4. Experimental Results and Discussion

4.1. Specimen and Experiment

In this study, the ultrasonic detection frequency was 5 MHz and a CFRP sample with artificially preset defects was used to demonstrate the effectiveness of the proposed method. The CFRP was produced by vacuum-assisted resin infusion molding. The carbon fiber material was first cut to dimensions of 241.0 mm × 44.5 mm × 20 mm, and the thickness of each layer was 0.26 mm. Subsequently, six Teflon tapes were embedded within the CFRP at designated locations. As illustrated in Figure 4, each Teflon area measured 400 mm^2 with a thickness of 0.5 mm. From left to right, defects h2, h3, and h6 were categorized as shallow, whereas h1, h4, and h5 were considered deep defects. Thus, defects of varying locations and depths were introduced and became invisible after the epoxy resin injection. The details of the defects are listed in Table 2.

After the specimen preparation, data were collected using the pulsed-echo method [30] to capture the ultrasound echoes reflected within the workpiece. If defects are present, the ultrasound echo signals will include not only the bottom and surface echoes but also defect echoes in the middle of their range. Experimentally, 500 ultrasonic images were acquired in which severe background noise was observed, complicating defect detection. Each ultrasound image measured 57 × 241, totaling 13,737 pixels. Figure 5 shows the ultrasound echo diagrams at different locations in the CFRP, where the surface and bottom echoes exhibited relatively high background noise levels. Therefore, raw ultrasonic inspection results are generally unsuitable for direct use, underscoring the need to develop effective ultrasonic defect analysis methods.

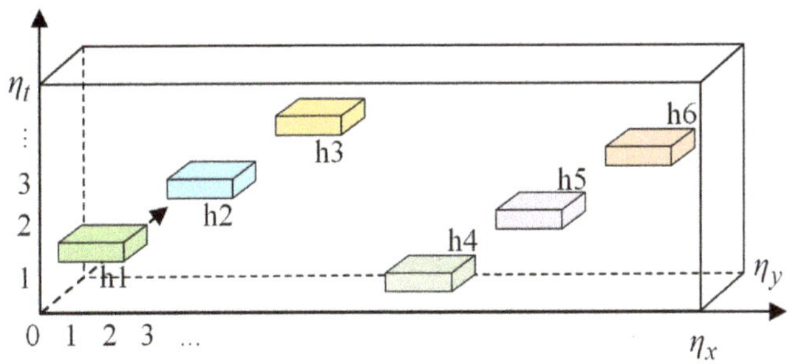

Figure 4. Schematic of the CFRP specimen.

Table 2. Information on defects.

Specimen	Defect	Shape	Area (mm^2)	Layer
CFRP defect specimen	h1	Square	400	50
	h2		400	30
	h3		400	10
	h4		400	60
	h5		400	40
	h6		400	20

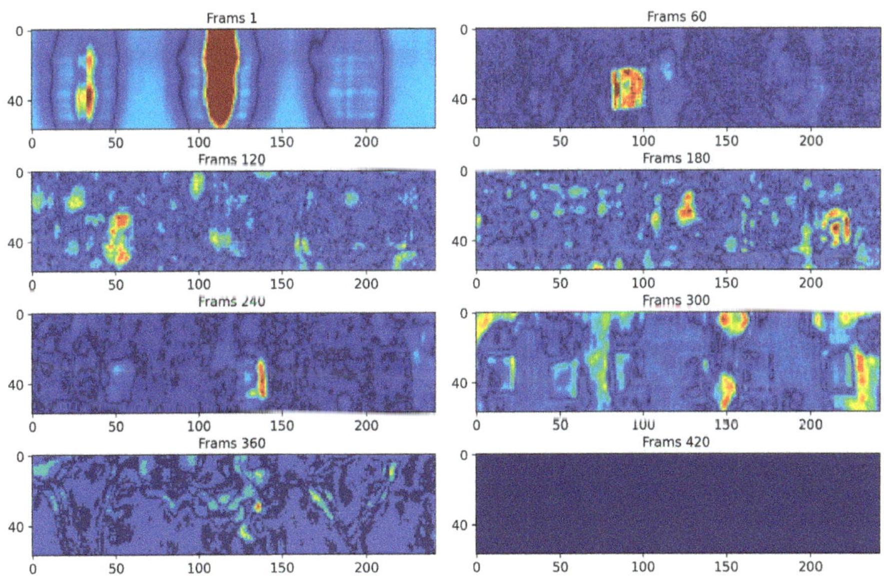

Figure 5. Raw ultrasound images at frames 1, 60, 120, 180, 240, 300, 360, 420.

4.2. Evaluation Metrics

To assess the effectiveness of the stable 3D-DCA for defect location extraction, the IoU, which is commonly used in target detection, was employed. The IoU is the overlapping rate of the predicted frame with the original frame, that is, the ratio of their intersection to their union. This is calculated as follows:

$$\mathrm{IoU} = \frac{A \cap B}{A \cup B}, \quad (7)$$

where A represents the predicted frame and B is the original frame. The formula is used to determine the ratio of the intersecting part of the candidate frame to the labeled frame to measure the correlation between the original frame and the prediction. A larger IoU indicates a closer match between the model predictions and the actual situation.

To objectively evaluate the performance of stable 3D-DCA, CNR, which is commonly used in the image field, was adopted as an assessment index. The formula used is as follows:

$$\mathrm{CNR} = \frac{\left| M_{def} - M_{in} \right|}{\sigma_{in}}, \tag{8}$$

where M_{def} is the average pixel value in the defective region, M_{in} is the average pixel value in the nondefective region, and σ_{in} is the standard deviation of the pixels in the nondefective region. The CNR reflects the contrast between the defective and nondefective regions, with a larger CNR indicating a more discernible defect, particularly when the standard deviation of the nondefective region is small.

To evaluate the performance of stable 3D-DCA in ultrasonic data analysis, the IoU focuses on the accuracy assessment of the defect location, and the CNR focuses on the degree of contrast. A comparison of the experimental results using both IoU and CNR makes the evaluation more reliable.

4.3. Results and Analysis

To validate the effectiveness of the stable 3D-DCA method, principal component analysis (PCA) [31] and DAE [21] were applied to the same dataset.

The PCA transforms raw ultrasonic data into a new coordinate system via a linear transformation to optimize the maximum variance in the data. This dimensionality reduction process aims to concentrate most of the defect information within the principal components visually represented in the loading images. However, this concentration inevitably leads to the loss of defect information. Figure 6 shows the loading images produced by the PCA, with colors indicating the values within the loading vector. Although PCA enhances defect visibility compared to the original ultrasonic images, it falls short in clearly delineating the boundaries of each defect. This lack of definition makes the accurate shape recognition of defects challenging. Moreover, PCA-loading images are marred by considerable background noise, further complicating the task of distinguishing defects from their surrounding environment.

As demonstrated in Figure 7, employing the DAE for feature extraction from ultrasound images and visualizing the encoder's hidden layer features reveal its effectiveness in infrared thermography applications, but not for noisy ultrasonic data. Although the general shape of the defects was discernible, the overwhelming background noise masked the defect information, with some defect shapes partially obscured.

Comparative experiments also assessed the performance of 3D-DCA, which utilizes a single-layer encoder and a conventional autoencoder loss function. The number of iterations, batch size, and learning rate were set to 50, 1, and 0.0005, respectively. The encoder performed feature extraction along the depth dimension by compressing 500 images into eight feature maps, each with an RF size of 213 × 27 × 27. This implied that each pixel on the feature map was derived from the aggregation of 213 × 27 × 27 pixel points from the original data, as processed by the model. Therefore, each feature map represented a distinct sensory region within the raw ultrasonic data. The feature maps generated by 3D-DCA bypass the surface and bottom echo characteristics of ultrasound imaging, making the depth order of the defects readily discernible. For instance, in Figure 8, the first feature map highlights the shallowest defect and the second map reveals the three shallowest defects. Moreover, 3D average pooling was utilized to average out the local area features, which smoothed out the ultrasonic surface and bottom echoes post-boundary filling. Conversely, 3D maximum pooling was employed to retain a greater amount of textural information, thereby enhancing the details captured in the feature maps. Figure 8 shows that while defects are more distinctly extracted and the depth order is apparent, the background noise remains noticeable. This results in poorly defined boundaries between the background and the defects.

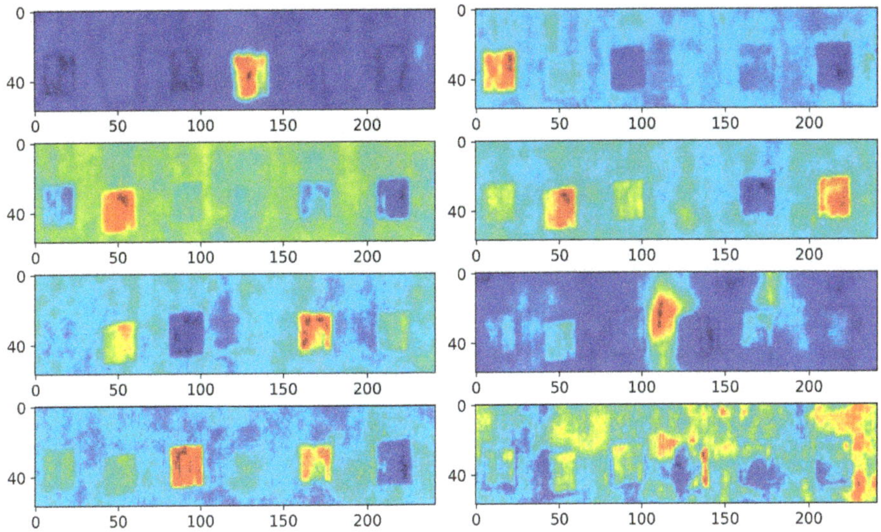

Figure 6. PCA results.

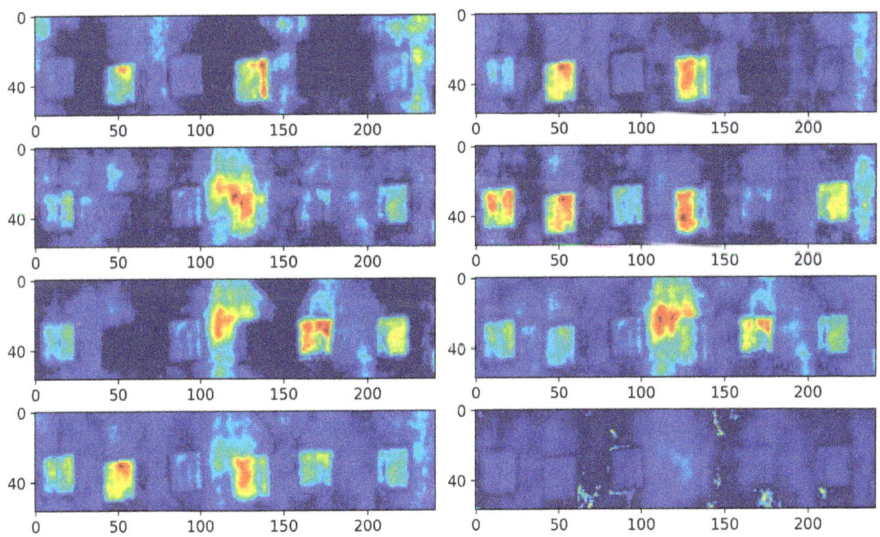

Figure 7. DAE results.

The 3D-DCA technique highlights that feature extraction using a single-layer encoder may not be sufficient to achieve optimal performance. As illustrated in Figure 2, the enhanced approach, termed stable 3D-DCA, employs a dual-layer encoder to introduce intermediate layer loss. This design encourages the model to reconstruct inputs more accurately, compelling the encoder to learn a more discriminative representation. In Figure 9, the background within each feature map is rendered more smoothly and the defect information stands out more sharply. Compared with the original 3D-DCA model, the stable 3D-DCA demonstrated superior depth-order prediction of defects. For instance, the first feature map in Figure 9 highlights the shallowest defect h3. In the subsequent feature map, h3 becomes less noticeable, h6 gains prominence, and h2 appears but remains faint. This pattern suggests that the initial sequence of defects was h3, followed by h6 and h2. Extrapolating further, the sequence continued with defects h5, h1, and h4. This sequencing can also be deduced in reverse from subsequent feature maps, with the depth-order outcomes presented in Table 3.

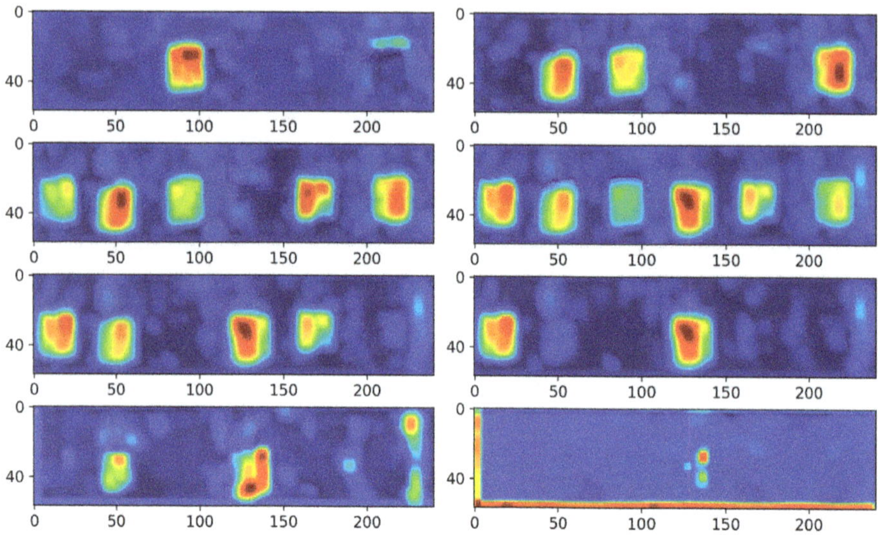

Figure 8. The 3D-DCA results.

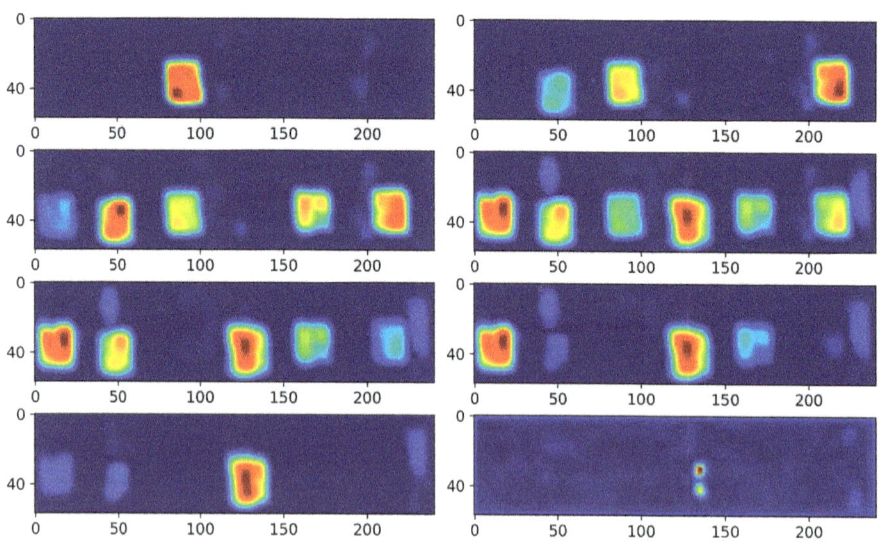

Figure 9. Stable 3D-DCA results.

Table 3. Defect depth order.

	h1	h2	h3	h4	h5	h6
True depth order	5	3	1	6	4	2
Predicted depth order	5	3	1	6	4	2

To compare the defect extraction capabilities of the different methods, Table 4 presents the IoU values for PCA, DAE, 3D-DCA, and stable 3D-DCA across different defect locations. According to Table 4, the IoU values of stable 3D-DCA were notably superior to those of DAE and PCA. This result is consistent with the visualization results, demonstrating the efficacy of the stable 3D-DCA method in enhancing defect detection accuracy.

Table 4. IoU of defects at different locations.

Method	h1	h2	h3	h4	h5	h6	Mean
PCA	0.549	0.663	0.901	0.163	0.802	0.350	0.571
DAE	0.828	0.808	0.675	0.796	0.063	0.315	0.581
3D-DCA	0.790	0.865	0.737	0.712	0.738	0.764	0.768
Stable 3D-DCA	0.839	0.803	0.804	0.684	0.826	0.846	0.800

The CNR index was calculated for each method. Table 5 compares the maximum CNR values achieved by the different methods, indicating that the proposed stable 3D-DCA method outperformed the other methods in detecting each type of defect, showcasing its exceptional performance in defect detection.

Table 5. CNR of defects at different locations.

Method	h1	h2	h3	h4	h5	h6	Mean
PCA	2.188	2.098	2.203	1.812	1.963	2.151	2.069
DAE	5.136	4.472	4.886	3.540	0.001	4.721	3.793
3D-DCA	5.198	4.304	4.818	3.605	3.938	4.636	4.417
Stable 3D-DCA	7.705	6.778	6.972	5.347	5.728	6.749	6.547

5. Conclusions

This paper presents a stable 3D-DCA method, which is a novel approach for analyzing the ultrasonic testing data of composite materials. This method differentiates itself from traditional techniques such as PCA and DAE by accurately capturing the 3D spatiotemporal features embedded in ultrasonic data while simultaneously reducing noise levels. An advantage of a stable 3D-DCA is its improved ability to delineate defect boundaries with enhanced precision, which is a critical factor in ultrasonic data analysis. The shape, location, and depth of the defects were accurately predicted by designing a receptive field to perceive the feature information of the data. Application to a CFRP specimen demonstrates that stable 3D-DCA effectively detects defects, illustrating the feasibility of the method. This method enables earlier defect detection and maintenance, thereby extending the service life of the composite materials. We anticipate that this approach will facilitate the development of various 3D-AE methods for the ultrasonic detection of industrial defects. Further exploration will focus on calculating the depth of the defects using a receptive field with 3D convolutional networks.

Author Contributions: Data curation, Q.Y. and Y.Y.; Funding acquisition, Y.L., K.L. and Y.Y.; Investigation, Y.L., Q.Y., K.L. and N.Z.; Methodology, Y.L., Q.Y. and K.L.; Project administration, Y.L. and N.Z.; Writing—original draft, Y.L. and Q.Y.; Writing—review and editing, Y.L., K.L. and Y.Y. All authors have read and agreed to the published version of the manuscript.

Funding: This research was funded by the National Natural Science Foundation of China (Grant Nos. 62301507 and U23A20328). Yao was supported in part by the National Science and Technology Council, Taiwan, under project number NSTC 112-2221-E-007-105.

Institutional Review Board Statement: Not applicable.

Data Availability Statement: The data presented in this study are available on request from the corresponding author. The data are not publicly available due to privacy.

Conflicts of Interest: Author Ningtao Zhu was employed by the company Xi'an Zhanshi Testing & Engineering Co. Ltd. The remaining authors declare that the research was conducted in the absence of any commercial or financial relationships that could be construed as a potential conflict of interest.

References

1. Shao, W.; Sun, Q.; Xu, X.; Yue, W.; Shi, D. Durability life prediction and horizontal bearing characteristics of CFRP composite piles in marine environments. *Constr. Build. Mater.* **2023**, *367*, 130116. [CrossRef]
2. Rashid, S.M.P.; Bahrami, A. Structural performance of infilled steel–concrete composite thin-walled columns combined with FRP and CFRP: A comprehensive review. *Materials* **2023**, *16*, 1564. [CrossRef] [PubMed]
3. Gañán, P.; Barajas, J.; Zuluaga, R.; Castro, C.; Marín, D.; Tercjak, A.; Builes, D.H. The evolution and future trends of unsaturated polyester biocomposites: A bibliometric analysis. *Polymers* **2023**, *15*, 2970. [CrossRef] [PubMed]
4. Matalgah, K.; Ravindranath, P.; Pulipati, D.; Fleck, T. Automated quantification of interlaminar delaminations in Carbon-Fiber-Reinforced Polymers via high-resolution ultrasonic testing. *Polymers* **2024**, *15*, 4691. [CrossRef] [PubMed]
5. Du, Y.; Li, X.; Xie, S.; Yang, S.; Chen, Z. Reconstruction of cracks in a carbon fiber-reinforced polymer laminate plate from signals of eddy current testing. *J. Compos Mater.* **2020**, *54*, 3527–3536. [CrossRef]
6. Maldague, X.; François, G.; Adel, Z. Advances in pulsed phase thermography. *Infrared Phys. Technol.* **2002**, *43*, 175–181. [CrossRef]
7. Kharrat, M.; Gaillet, L. Non-destructive evaluation of anchorage zones by ultrasonics techniques. *Ultrasonics* **2015**, *61*, 52–61. [CrossRef]
8. Abbas, M.; Shafiee, M. Structural health monitoring (SHM) and determination of surface defects in large metallic structures using ultrasonic guided waves. *Sensors* **2018**, *18*, 3958. [CrossRef] [PubMed]
9. Liu, K.; Yu, Q.; Lou, W.; Sfarra, S.; Liu, Y.; Yang, J.; Yao, Y. Manifold learning and segmentation for ultrasonic inspection of defects in polymer composites. *J. Appl. Phys.* **2022**, *132*, 024901. [CrossRef]
10. Liu, T.; Li, J.; Cai, X.; Yan, S. A time-frequency analysis algorithm for ultrasonic waves generating from a debonding defect by using empirical wavelet transform. *Appl. Acoust.* **2018**, *131*, 16–27. [CrossRef]
11. Le, M.; Kim, J.; Kim, S.; Lee, J. B-scan ultrasonic testing of rivets in multilayer structures based on short-time Fourier transform analysis. *Measurement* **2018**, *128*, 495–503. [CrossRef]
12. Wu, J.; Chen, X.; Ma, Z. A signal decomposition method for ultrasonic guided wave generated from debonding combining smoothed pseudo Wigner-Ville distribution and Vold–Kalman filter order tracking. *Shock Vib.* **2017**, *2017*, 7283450. [CrossRef]
13. Lyu, Y.; Yang, J.; Ge, M.; Xu, L. Wavelet packet energy-based damage detection using guided ultrasonic waves and support vector machine. *Meas. Sci. Technol.* **2023**, *34*, 075107. [CrossRef]
14. Liu, K.; Wang, F.; He, Y.; Liu, Y.; Yang, J.; Yao, Y. Data-augmented manifold learning thermography for defect detection and evaluation of polymer composites. *Polymers* **2023**, *15*, 173. [CrossRef] [PubMed]
15. You, R.; Yao, Y.; Shi, J.; Zheng, K.; Wang, K. Feature-selective clustering for ultrasonic-based automatic defect detection in FRP structures. *Chemometrics Intell. Lab. Syst.* **2016**, *157*, 35–42. [CrossRef]
16. Dai, Y.; Yang, C.; Liu, K.; Liu, A.; Liu, Y. TimeDDPM: Time series augmentation strategy for industrial soft sensing. *IEEE Sens. J.* **2024**, *24*, 2145–2153. [CrossRef]
17. Cheng, X.; Ma, G.; Wu, Z.; Zu, H.; Hu, X. Automatic defect depth estimation for ultrasonic testing in carbon fiber reinforced composites using deep learning. *NDT E Int.* **2023**, *135*, 102804. [CrossRef]
18. Yang, J.; Li, S.; Wang, Z.; Dong, H.; Wang, J.; Tang, S. Using deep learning to detect defects in manufacturing: A comprehensive survey and current challenges. *Materials* **2020**, *13*, 5755. [CrossRef] [PubMed]
19. Wang, B.; Saniie, J. Massive ultrasonic data compression using wavelet packet transformation optimized by convolutional autoencoders. *IEEE Trans. Neural Netw. Learn. Syst.* **2021**, *34*, 1395–1405. [CrossRef] [PubMed]
20. Yan, X.; Wang, H.; Fan, X. Research progress in nonlinear ultrasonic testing for early damage in metal materials. *Materials* **2023**, *16*, 2161. [CrossRef] [PubMed]
21. Liu, K.; Zheng, M.; Liu, Y.; Yang, J.; Yao, Y. Deep autoencoder thermography for defect detection of carbon fiber composites. *IEEE Trans. Ind. Inform.* **2023**, *19*, 6429–6438. [CrossRef]
22. Pintelas, E.; Pintelas, P. A 3D-CAE-CNN model for Deep Representation Learning of 3D images. *Eng. Appl. Artif. Intell.* **2022**, *113*, 104978. [CrossRef]
23. Chen, P.; Zhou, H.; Liu, B.; Liu, P. Shape similarity intersection-over-union loss hybrid model for detection of synthetic aperture radar small ship objects in complex scenes. *IEEE J. Sel. Top. Appl. Earth Observ. Remote Sens.* **2021**, *14*, 9518–9529. [CrossRef]

24. Wang, N.; Yang, C.; Xu, J.; Shi, W.; Huang, W.; Cui, Y.; Jian, X. An improved chirp coded excitation based on compression pulse weighting method in endoscopic ultrasound imaging. *IEEE Trans. Ultrason. Ferroelectr. Freq. Control* **2020**, *68*, 446–452. [CrossRef] [PubMed]
25. She, Y.; Tang, J.; Wang, C.; Wang, Z.; Huang, Z.; Yang, Y. Nano-additive manufacturing and non-destructive testing of nanocomposites. *Nanomaterials* **2023**, *13*, 2741. [CrossRef] [PubMed]
26. Gaydecki, P.A.; Burdekin, F.M.; Damaj, W.; John, D.G. The propagation and attenuation of medium-frequency ultrasonic waves in concrete: A signal analytical approach. *Meas. Sci. Technol.* **1992**, *3*, 126. [CrossRef]
27. Azari, H.; Nazarian, S.; Yuan, D. Assessing sensitivity of impact echo and ultrasonic surface waves methods for nondestructive evaluation of concrete structures. *Constr. Build. Mater.* **2014**, *71*, 384–391. [CrossRef]
28. You, R.; Yao, Y.; Shi, J. Tensor-based ultrasonic data analysis for defect detection in fiber reinforced polymer (FRP) composites. *Chemometrics Intell. Lab. Syst.* **2017**, *163*, 24–30. [CrossRef]
29. Malfliet, W. The tanh method: A tool for solving certain classes of nonlinear evolution and wave equations. *J. Comput. Appl. Math.* **2004**, *164*, 529–541. [CrossRef]
30. Berke, M.; Ballenger, T. Phased array technology for standard ultrasonic testing. *Insight* **2020**, *48*, 218–220. [CrossRef]
31. Maćkiewicz, A.; Ratajczak, W. Principal components analysis (PCA). *Comput. Geosci.* **1993**, *19*, 303–342. [CrossRef]

polymers

MDPI

Article

A Multi-Objective Optimization of Neural Networks for Predicting the Physical Properties of Textile Polymer Composite Materials

Ivan Malashin [1,*], Vadim Tynchenko [1,*], Andrei Gantimurov [1], Vladimir Nelyub [1,2] and Aleksei Borodulin [1]

1 Artificial Intelligence Technology Scientific and Education Center, Bauman Moscow State Technical University, 105005 Moscow, Russia; agantimurov@emtc.ru (A.G.); vladimir.nelub@emtc.ru (V.N.); alexey.borodulin@emtc.ru (A.B.)
2 Scientific Department, Far Eastern Federal University, 690922 Vladivostok, Russia
* Correspondence: ivan.p.malashin@gmail.com (I.M.); vadimond@mail.ru (V.T.); Tel.: +7-926-875-7128 (I.M.)

Abstract: This paper explores the application of multi-objective optimization techniques, including MOPSO, NSGA II, and SPEA2, to optimize the hyperparameters of artificial neural networks (ANNs) and support vector machines (SVMs) for predicting the physical properties of textile polymer composite materials (TPCMs). The optimization process utilizes data on the physical characteristics of the constituent fibers and fabrics used to manufacture these composites. By employing optimization algorithms, we aim to enhance the predictive accuracy of the ANN and SVM models, thereby facilitating the design and development of high-performance textile polymer composites. The effectiveness of the proposed approach is demonstrated through comparative analyses and validation experiments, highlighting its potential for optimizing complex material systems.

Keywords: multi-objective optimization; neural networks; textile polymer composites; physical properties' prediction

1. Introduction

Polymer composite materials (PCMs) are composite materials consisting of two or more components, where the final mixture exhibits properties superior to those of each individual constituent material [1]. PCMs are indispensable in contemporary engineering and industrial applications owing to their exceptional characteristics, including but not limited to lightweight design [2], high strength [3], and resistance to corrosion [4,5]. However, the development of effective polymer composites necessitates the consideration of diverse factors [6], notably encompassing composition and structure.

Attaining the desired attributes of polymers mandates the application of multifaceted approaches, including choosing optimal raw components [7], with the necessary chemical and physical properties being a primary consideration. This may involve selecting different polymers [8], fillers [9], reinforcing materials [10], and additives [11]. Careful tuning of the composition and proportions [12] of components in the material can significantly impact its properties. This could involve altering the concentration of the polymer components [13,14]. Additionally, the chemical modification of the polymers [15,16] can be used to change their properties. This may involve introducing functional groups [17] and changing the molecular weight [18] and crystalline structure [19] of the polymer. Various treatments such as extrusion [20], injection molding [21], compression molding [22], and filament winding [23] can be used to shape and enhance the properties of polymer materials. Furthermore, the addition of nanoparticles [24] or nanotubes [25] to polymer composites can significantly improve their mechanical, thermal, and electrical properties [26].

The specific physical characteristics of polymers become imperative to meet precise functional requisites across various applications [27]. These characteristics may encompass mechanical properties such as tensile strength [28], flexibility [29], and impact resistance [30], alongside thermal attributes like heat resistance [31] and thermal conductivity [32]. Thus, a comprehensive understanding of the polymer composite composition and structure is pivotal, facilitating the customization of the properties to align with distinct application demands.

One of the modern approaches to achieving the desired properties of PCMs involves the application of machine learning (ML) techniques. ML algorithms, such as neural networks [33], support vector machines [34], and random forests [35], can analyze complex datasets comprising the material composition, processing parameters, and desired properties to identify intricate relationships and patterns. By leveraging these relationships, ML models can predict the properties of PCMs [36], optimize material formulations, and expedite the development process. This approach has been actively discussed and extensively researched in the scientific literature.

Fontes et al. [37] showcased the effectiveness of Deep Neural Networks (DNNs) in crafting a data-driven failure model for Fiber-reinforced Polymer (FRP) composite materials. Using experimental failure data from the literature, a fully connected DNN with 20 input units and one output unit was trained. The inputs included the laminate layup sequence, the lamina properties, and the loading conditions, while the output was the failure vector length. Comparative analysis with conventional theories such as Tsai–Wu [38], Cuntze [39], and Pinho theory [40] indicated the superior performance of the DNN in fitting the experimental data. Its ability to handle higher order polynomials makes it a valuable tool for predicting FRP composite laminate failure.

Fahem et al. [41] investigated the impact of porosity on the mechanical properties of Glass Fiber-Reinforced Polymer (GFRP) [42,43] through experimental and numerical analyses. The material characterization included a three-point bending test, while the finite-element modeling explored various air bubble scenarios. The results revealed a significant reduction in load as the bubble size increased. Additionally, an artificial neural network-Enhanced Jaya Algorithm (ANN-E JAYA) [44] predicted the tensile load reduction based on the crack lengths from an Extended finite-element method (XFEM) [45]. A comparison with other algorithms, including the Jaya Algorithm (JAYA) [46] and particle swarm optimization (PSO) [47], demonstrated the superior accuracy of ANN-E JAYA.

Nguen et al. [48] explored the impact of the cure-induced size effect on stress development and tensile transverse failure response in thermoset composite materials. Thick structures, combined with low polymer conductivity, may result in spatially varying temperature fields, affecting the property distribution and residual stress. The study employed a thermo-chemo-mechanical finite-element framework integrated with a crack band model. The cure model considered the kinetics and heat generation, while a neural network-based constitutive model captured the matrix mechanical property evolution.

Gupta et al. [49] proposed an ML model to precisely quantify the mechanical properties of FRP composites for optimal structural design. Using microstructural images as the input, the model visualizes the stress components, specifically S_{11}, with high accuracy. The training data were obtained from the FEM analysis of short carbon fiber-filled specimens using a Representative Area Element (RAE) approach [50]. The study demonstrated the robustness of a pix2pix [51] deep learning Convolutional Neural Network (CNN) model in predicting the stress fields. By focusing on the chronological development of the CNN model, the paper outlined a promising approach to efficiently predict full-field stress maps in fiber-reinforced composite specimens, reducing the time and cost associated with traditional methods.

El [52] used a Deep Recurrent CNN (DCRN) [53] to predict the full non-linear response of composite materials. The framework relies on a Representative Volume Element [54] (RVE) database, encompassing the composite layups, defects, and loading conditions. It incorporates various sources of material non-linearity, including matrix damage, delamination, fiber failure, and shear non-linearity. The proposed DCRN architecture combines convolutional layers for spatial feature detection with Long Short-Term Memory layers for the material loading history dependencies.

Zhang et al. [55] focused on enhancing the thermal protection performance of PCMs for re-entry vehicles. A thermal response model without surface recession was developed to simulate the ablation process. Using ML, the relationship between the piecewise porosity distribution and bondline temperature was explored based on simulated data. Optimal porosity distributions were obtained, leading to a reduction in the bondline temperature by 17.61 K, highlighting the potential of rational porosity optimization to improve material utilization rates.

Song et al. [56] presented an approach using digital material twins to analyze the mechanical performance of woven composites, particularly damage and failure behaviors. Addressing challenges in segmenting low-contrast digital images and reconstructing 3D braided structures, a ResL-U-Net CNN was proposed. The network incorporates the leaky-ReLU [57] activation function for efficiency and employs residual structures to enhance robustness and accuracy. The results demonstrated that the simulations accurately depicted the mechanical performance of GFRP, including the damage locations and material failure patterns.

Li et al. [58] introduced a DL fusion model for predicting the material properties of carbon fiber monofilaments [59] by simultaneously analyzing textual and visual data. Utilizing the greedy-based generation [60] (GBG) algorithm, 1200 stochastic microstructures were generated, and the statistical representations were determined using two-point statistics. The macroscopic properties were calculated via micro-scale finite-element simulation. The developed hybrid CNN-MLP fusion model achieved impressive average testing R^2 values for various mechanical properties of carbon fibers.

Doddashamachar et al. [61] predicted the dielectric properties of polypropylene composites reinforced with banana fiber using an ANN. Composites were prepared according to ASTM standards with varying banana fiber volume fractions [62]. The dielectric characteristics were determined using an impedance analyzer. The ANN, trained with the ReLU activation function, showed accurate prediction of the dielectric properties.

Amor at al. [63] provided an overview of computational intelligence (CI) modeling methods for lightweight composite materials (LWCMs). CI facilitates material data science tasks such as imaging, feature identification, prediction, and design optimization, enhancing LWCM quality through constituent optimization.

Mukhopadhyay [64] explored the use of an ANN in predicting the mechanical properties and behaviors of textile composite materials, including the static and dynamic mechanical properties, time-dependent properties like creep and stress relaxation, fatigue prediction, wear simulation, and crack detection. The discussion highlighted recent developments and applications of ANNs in the field of fiber-reinforced composites, emphasizing the importance of accurately modeling composite properties for engineering applications.

This article aims to bridge the existing gap in knowledge by predicting the physical characteristics of textile polymer composite materials (TPCMs) [65] based on a dataset encompassing the mechanical properties of fabrics and yarns in both the warp and weft directions, which constitute these fabrics. Through this analysis, the study delves into optimization methods aimed at fine-tuning the hyperparameters and selecting appropriate architectures for machine learning (ML) models. The primary focus is on exploring the effectiveness of optimization methods such as multi-objective particle swarm optimization [66] (MOPSO), Non-dominated Sorting Genetic Algorithm II [67] (NSGA-II) [67], and Strength Pareto Evolutionary Algorithm 2 [68] (SPEA2) in optimizing ML models like support vector machines [69] (SVMs) and ANNs to maximize accuracy and minimize inference time.

2. Materials and Methods

2.1. Dataset Description

To evaluate the tensile, compressive, and other properties of TPCMs, we prepared specially designed specimens in the form of strips measuring 10 × 20 cm. These specimens were securely mounted in a universal testing machine, the QUASAR 50 universal testing machine (Galdabini, Cardano al Campo, Italy), which applies mechanical forces to the material.

During a tensile test, the specimen was subjected to gradually increasing axial force until it ruptured, allowing us to assess its tensile strength and elongation properties. Conversely, in a compressive test, the specimen was compressed along its length by applying opposing forces at its ends, enabling the characterization of its compressive strength and modulus.

Figure 1 illustrates histograms depicting the distribution of the physical characteristics of textile PCMs to showcase key physical properties such as the tensile, compressive, and bending strengths, the percentage of elongation, and the modulus of elasticity in tension along the warp and weft directions, respectively, for each mentioned characteristic.

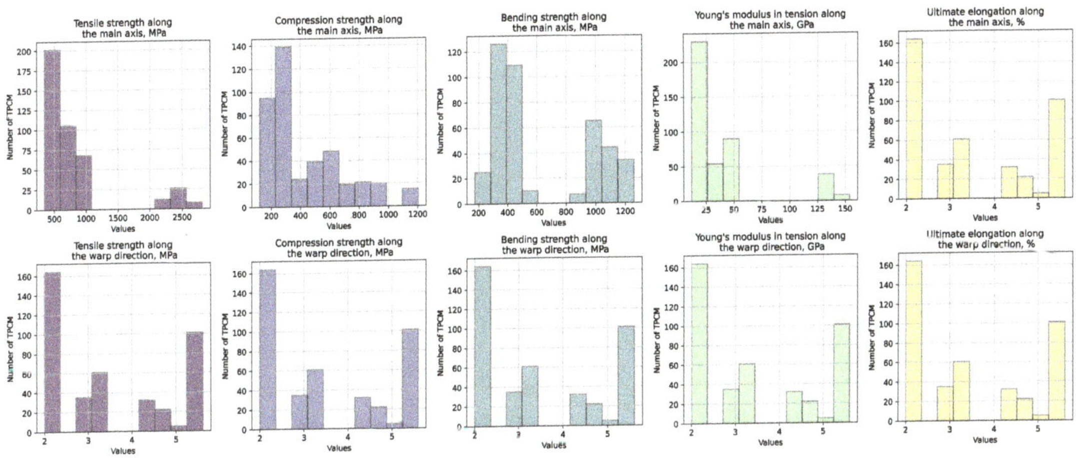

Figure 1. Physical properties of textile PCMs such as the tensile, compressive, and bending strengths and the modulus of elasticity in tension along the warp and weft directions, respectively, for the considered specimens.

Figure 2 showcases histograms representing the spread of physical attributes like the interlaminar shear modulus, coefficient of linear thermal expansion along the warp direction, and density.

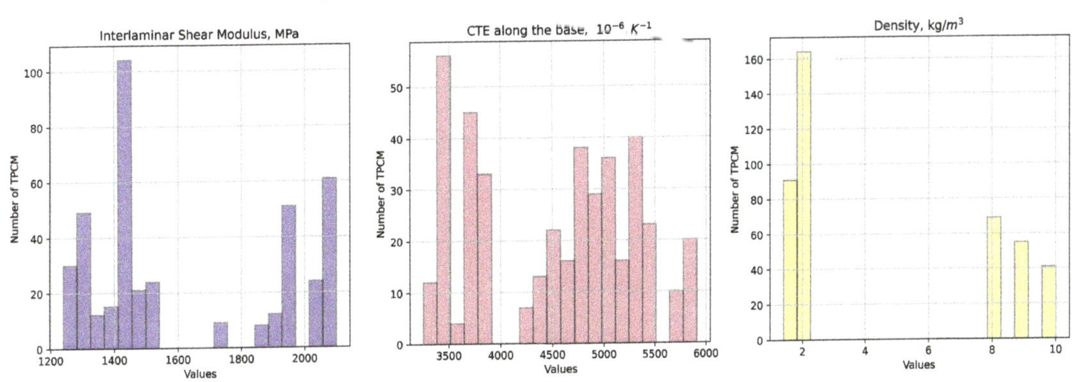

Figure 2. Distribution of the interlaminar shear modulus, coefficient of linear thermal expansion (CTE) along the warp direction, and density for the considered specimens.

The study involved conducting a total of 420 measurements on samples to investigate 11 mentioned physical characteristics across various types of TPCMs and their corresponding constituents. Specifically, the materials under investigation included the following:

1. Basalt plastic, designated as TBK-100 [70,71], is a composite material comprising basalt fibers [72] as the reinforcement phase embedded in a polymer matrix. Basalt fibers are derived from natural volcanic rock [73] and have high tensile strength and resistance to temperature variations. TBK-100 finds application in construction [74]. The weave pattern of these samples is canvas.
2. Fiberglass-reinforced plastic, commonly known as fiberglass [75], is a composite material composed of fine glass fibers embedded in a polymer matrix, typically epoxy or polyester resin [76]. This material exhibits a high strength-to-weight ratio [77], excellent corrosion resistance [78,79], and dimensional stability [80], making it suitable for applications requiring durability and structural integrity [81]. We considered types such as T-10 [82], T-13 [83], T-11 [84], T-SU 8/3(VMP)-78 [85], and T-25 [86]. The fabric construction of these samples predominantly consisted of canvas [87] and satin weaves [88].
3. Carbon fiber-reinforced plastic, also known as carbon fiber composite or carbon composite [89], consists of carbon fibers infused in a polymer matrix, often epoxy resin. This material offers exceptional strength, stiffness, and lightweight properties, making it ideal for aerospace [90], automotive [91], and sporting goods [92] applications where weight reduction and high performance are critical. We considered CC245 [93], CC206 [94], T700SC [95], UMT49 [96], UT-900-3 [97], HTS45 [98], and IMS65 [99]. The weave pattern observed in these samples primarily included twill [100] and unidirectional [101] weaves.
4. Aramid fiber-reinforced plastic, or aramid composite [102], incorporates aramid fibers, such as Kevlar® [103], as the reinforcing component in a polymer matrix. Aramid fibers are known for their exceptional strength, stiffness, and resistance to impact and abrasion [104]. Aramid composites offer high tensile strength, heat resistance, and low weight, making them suitable for ballistic protection [105]. We considered varieties like T-43-76 [106], Satin 5/3 [107], Satin 8/3 [108], T-42-78 [109], and T-42/1-76 [110]. For this type of TPCM, the weave patterns also included canvas and satin.

To analyze the distribution of the key physical characteristics across various types of TPCMs, we utilized histograms. These materials were categorized into basalt plastic, fiberglass, carbon plastic, and aramid plastic. Each histogram in Figure 3 showcases the distribution of specific characteristics, such as the tensile strength [111], compression strength [112], bending strength [113], Young's modulus [114], and ultimate elongation [115], along both the main axis and the warp direction.

Figure 4 illustrates the variations in the distribution of the interlaminar shear modulus [116], CTE [117], and density [118].

Unlike Figures 1 and 2, Figures 3 and 4 feature histograms with variable bar widths. This variation in bar thickness aims to highlight the distribution density of different data points. Thicker bars denote a higher number of TPCMs or the concentration of data within a range, while thinner bars represent a lower number of TPCMs. This method offers a more nuanced understanding of the data distribution compared to uniform bar widths.

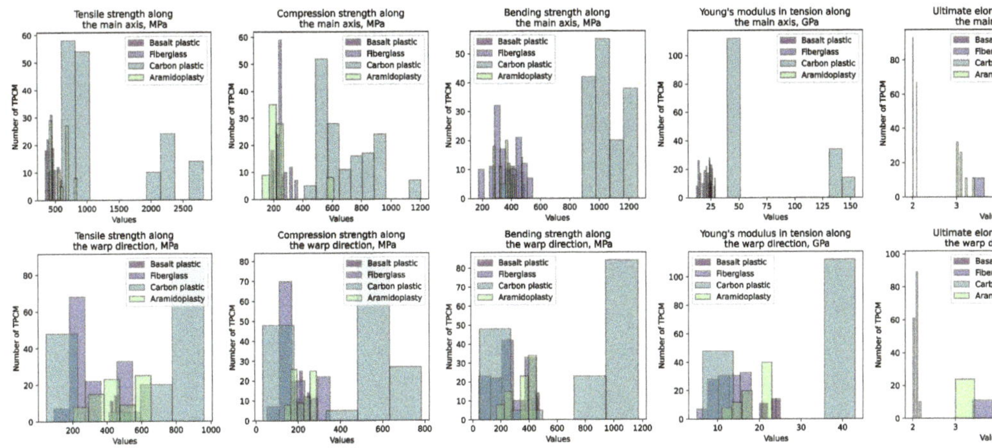

Figure 3. Histograms depicting the distribution of the physical characteristics for TPCMs grouped by type: basalt plastic, fiberglass, carbon plastic, and aramid plastic. Each subplot illustrates the distribution of the following characteristics along the main and warp directions: tensile strength, compression strength, bending strength, Young's modulus, and ultimate elongation.

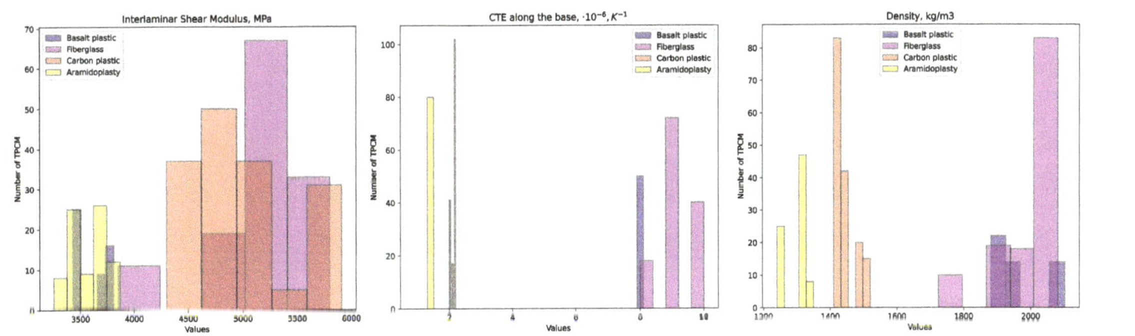

Figure 4. Histograms of physical attributes for TPCMs grouped by type: basalt plastic, fiberglass, carbon plastic, and aramid plastic for the distributions of the interlaminar shear modulus, CTE, and density.

Additionally, an analysis of the correlation matrix was conducted, reflecting the relationship between the physical characteristics of the TPCMs and the properties of the fabrics from which these samples are made (Figure 5). These relationships could be the basis for developing predictive models that estimate the behavior of composite materials based on the fabric properties and can aid in quality control during the manufacturing process. Figure 5 highlights the numerical input features of the dataset in green and the output features in red, as also depicted in Figure 6. From this correlation matrix, one can observe patterns such as the correlation between the parameters related to the yarns, such as the tensile strength along the base and warp, compression strength, bending strength, and Young's modulus. These correlations exhibit a chessboard-like pattern, indicating that the values along the base correlate more strongly with each other than with the values along the warp. This pattern is similarly observed for the output parameters, such as those pertaining to finished TPCM products, with the same names, but for finished products.

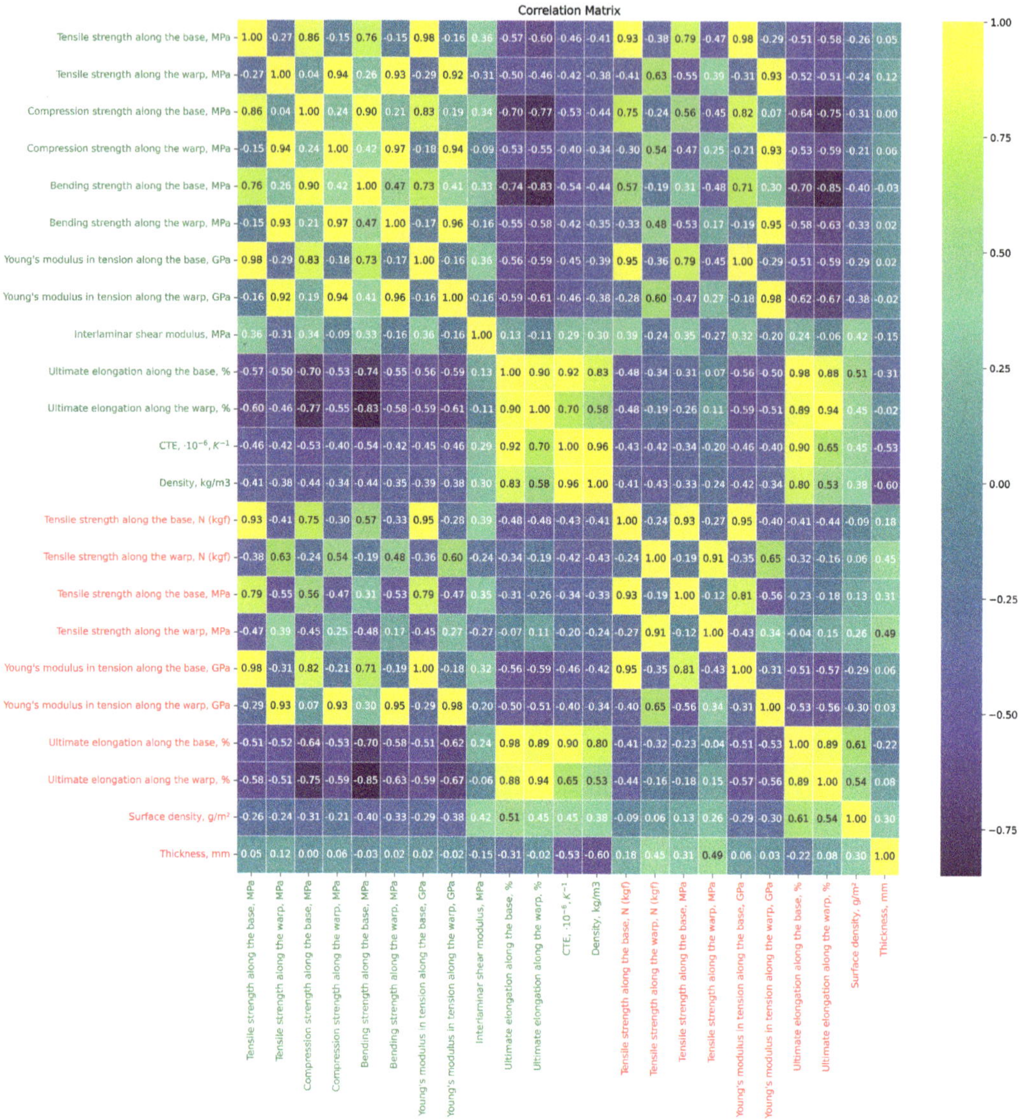

Figure 5. Correlation matrix of physical properties of TPMCs (highlighted in green) and fabric properties (highlighted in red), from which the samples are derived.

To build a model for predicting the physical characteristics of polymer composite materials, we utilized neural networks, which allow modeling complex non-linear relationships between the material components and their properties. Neural networks consist of multiple layers of neurons that process input data and provide predictions.

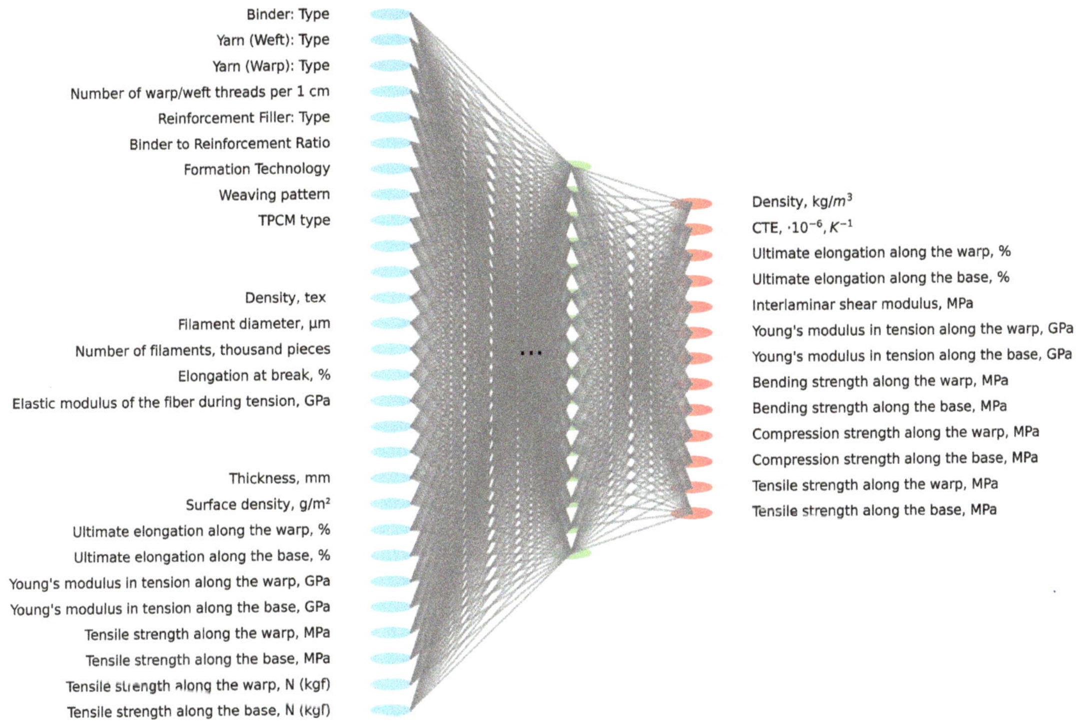

Figure 6. Illustration of a hypothetical neural network architecture designed to predict the physical characteristics of TPCMs: blue dots as inputs, red dots as outputs

To optimize the performance of neural networks, we applied hyperparameter optimization methods such as Grid Search, Random Search, and Bayesian Optimization. These methods enabled us to automatically find optimal values for the model hyperparameters such as the number of hidden layers, the number of neurons in each layer, the learning rate, and others.

2.2. Model Development

To predict the physical characteristics of the TPCMs, an SVM and ANN were selected as the optimization methods. The SVM is capable of handling both linearly and non-linearly separable data and is resistant to overfitting. Due to the small number of hyperparameters, this simplifies the optimization process and parameter tuning. The ANN possesses the ability to adapt to various types of data and tasks, achieving high accuracy when properly configured.

To further address the task at hand, it is necessary to determine the selected optimization methods. NSGA-2 is used for multi-objective optimization problems, aiming to find a set of optimal solutions known as the Pareto set [119,120]. NSGA-2 is designed for problems with multiple optimization criteria, ensuring diversity in the population to prevent premature convergence and achieve uniform coverage of the Pareto set.

SPEA-2 is an evolutionary algorithm also utilized for multi-criteria optimization. It ranks solutions based on their strength and distance from other solutions to effectively select the best solutions for constructing the Pareto set [121,122].

MOPSO is a variant of the particle swarm optimization (PSO) method, also applied to multi-criteria optimization. It iteratively updates the position and velocity of particles in the parameter space [123] based on the best solutions and information exchange between particles.

Python was chosen for implementing the algorithm due to its powerful capabilities in machine learning and optimization. We utilized the scikit-learn library for SVM model handling and PyGMO for the MOPSO algorithm. Data preprocessing was conducted using pandas and numpy, followed by splitting into training and testing sets. The optimization space encompassed the linear [124], polynomial [125], or RBF [124] kernel types and a regularization parameter [126] for the SVM. For multi-objective optimization, we defined minimizing the inference time and maximizing the model accuracy as the objectives. The inference time is computed as the average inference time on the test dataset for each SVM model. Accuracy is computed as the accuracy on the test dataset for each SVM model. The MOPSO algorithm was implemented using PyGMO [127], involving initialization, updating, and evaluating the solutions. The solution updating followed the principles of dominance and best solution selection. The algorithm continued until reaching a specified number of iterations or stopping criteria, after which the best solution was selected based on multi-objective optimization.

Figure 6 shows the schematic representation of a potential neural network architecture for predicting the physical properties of TPCMs.

The input features include the properties of both the fabric and thread. The fabric properties encompass parameters such as the elastic modulus of the fiber during tension (GPa), elongation at break (%), number of filaments (thousand pieces), filament diameter (μm), and density (tex). Additionally, thread properties include the tensile strength along the base and warp (N and MPa), Young's modulus in tension along the base and warp (GPa), ultimate elongation along the base and warp (%), surface density (g/m^2), and thickness (mm). Categorical features such as the TPCM type (like T-43-76, Satin 5/3, Satin 8/3, etc.), weaving pattern, formation technology, binder-to-reinforcement ratio, reinforcement filler type, number of warp/weft threads per 1 cm, yarn type for warp and weft, and binder type were also included as inputs. The network predicts the parameters of the final TPCM product, including the tensile strength, compression strength, bending strength, Young's modulus, interlaminar shear modulus, ultimate elongation, and CTE.

3. Results

Our study investigates the efficacy of employing metaheuristic optimization algorithms, specifically MOPSO, SPEA2, and NSGA-II, for hyperparameter tuning in the ANN and SVM for predicting the physical properties of TPCM samples based on their fabricated components. In Figure 7, we present the dynamic evolution of the loss curves for the five best models for each optimization case alongside the exploration of the optimization parameter space.

This figure encapsulates the iterative optimization process, illustrating how these algorithms navigate the complex landscape of hyperparameters to achieve optimal performance in both the ANN and SVM models. Such analysis provides valuable insights into the comparative effectiveness of these optimization techniques, shedding light on their suitability for enhancing the predictive capabilities of machine learning models.

The ANN model architecture was constructed based on the genetic algorithm's (GA) [128] individual representation for the optimizers NSGA-II and SPEA2. The architecture consisted of densely connected layers with leaky-ReLU [57] activation functions. The number of layers was dynamically determined, but capped at a predefined maximum. Each layer's neuron count was constrained within specified bounds. The model was trained using k-fold cross-validation [129] to mitigate overfitting, and its performance was evaluated based on the root-mean-squared error (RMSE) [130] metric. Key parameters such as the population size [131] specified the number of individuals (ANN architectures) in each generation of the genetic algorithm. In our case, the population size was set to 50. The crossover probability [132] determined the likelihood of crossover occurring between two parent individuals during reproduction. We set a value of 0.7; there was a 70% chance of crossover. The mutation probability [133] represents the probability of mutation, which introduces small random changes to individual genomes. We set a value of 0.3, indicating a 30% chance of mutation. The number of generations [134] indicates how the GA will evolve the population. In our scenario, the algorithm ran for 30 generations. The number of training epochs [135] (iterations over the entire dataset) during the training of each ANN was 100 epochs. The minimum number of neurons allowed in a single layer of the ANN was set to 2 neuron, and the maximum was 32 neurons in increments of 2. We defined the minimum and maximum learning rates as 0.05 and 0.2, respectively, which control the step size during gradient descent optimization. The learning rate [136] typically falls within a predefined range to balance training stability and convergence speed. The maximum number of layers allowed in the ANN architecture was constrained to five.

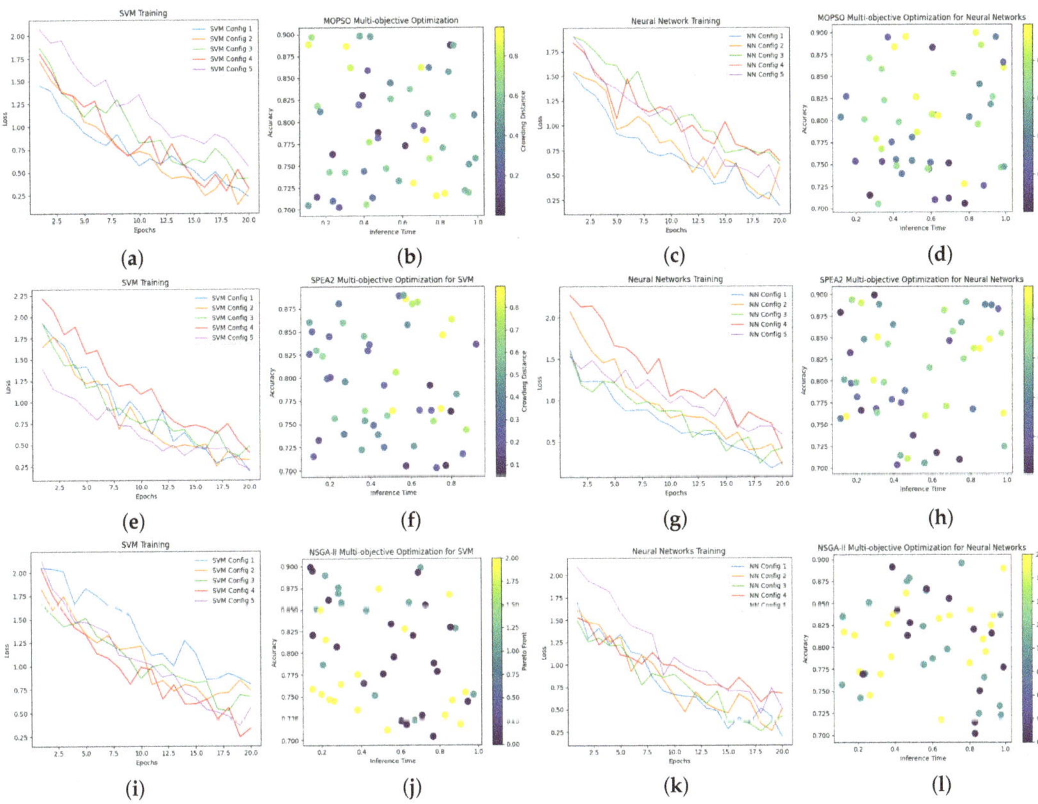

Figure 7. Evolution of loss curves and optimization parameter space for hyperparameter tuning in the ANN and SVM using the MOPSO (**a–d**), SPEA2 (**e–h**), and NGSA-II (**i–l**) optimization methods.

For MOPSO, the fitness function evaluates a solution (particle position) using SVR and the ANN and calculates the RMSE for each target variable. The algorithm runs the MOPSO algorithm for a specified number of iterations as 50 with 20 particles each time, updating the particles' positions and velocities based on their personal best and global best positions.

The predictive performance of the ANN and SVM, optimized using heuristic algorithms, in predicting the physical characteristics of the TPCMs is shown in Figures 8 and 9 for all 13 output features highlighted in red in Figure 6. These Figures 8 and 9 present the median values as whisker plots for some of the polymer grades, while the yellow (for the ANN) and cyan (for the SVM) markers indicate the values proposed by the best ML models, as summarized in Table 1. For convenience, we also provide a table summarizing the optimization parameters for the ANN and SVM using MOPSO, SPEA2, and NSGA-II.

Table 1. Comparison of optimized values for the SVM and ANN using MOPSO, SPEA2, and NSGA-II.

ML Model	Optimized Value	MOPSO	SPEA2	NSGA-II
SVM	accuracy	0.878	0.876	0.899
	inference time, ms	0.88	0.85	0.78
	Parameters (C, γ)	(1.0, 0.1)	(1.2, 0.08)	(0.9, 0.15)
ANN	accuracy	0.902	0.901	0.898
	inference time, ms	0.42	0.36	0.43
	Architecture (layers, neurons, activation)	(4, [4, 28, 20, 12], ReLU)	(4, [2, 16, 8, 16], sigmoid)	(3, [6, 4, 8], tanh)

The findings presented in Figure 9 are a continuation of those shown in Figure 8. Together, they form a series of 13 subfigures, each representing a numerical physical characteristic of the TPCMs distributed across different TPCM grades (types), like TBK-100, T-10 (92), CC245, T-13, T-SU 8/3, T-43-76, Satin 5-3, and Satin 8-3.

The boxplot whisker chart provides a visual representation of the distribution of the values for each physical characteristic across the selected grades of TPCMs for the real dataset. Each boxplot illustrates the median, shown as a line within the colored area of the whisker, and the quartiles, represented by the boundaries of the colored area. Additionally, the chart depicts potential outliers and extreme values as horizontal lines outside the colored area. These elements are presented for each physical characteristic based on the TPCM grade, enabling a detailed comparison of the data distribution across different polymer grades.

Specifically, these grades are those for which experiments were conducted more than 10 times. On each of the whisker plots, the predicted values generated by modeling with the ANN optimized using MOPSO are indicated in yellow digits, and for the SVM optimized using NSGA II represented in cyan digits. These predictions are based on the architecture hyperparameters provided in Table 1. Such analysis offers insights into the efficacy of different machine learning approaches in capturing the complex relationships inherent in the TPCM properties.

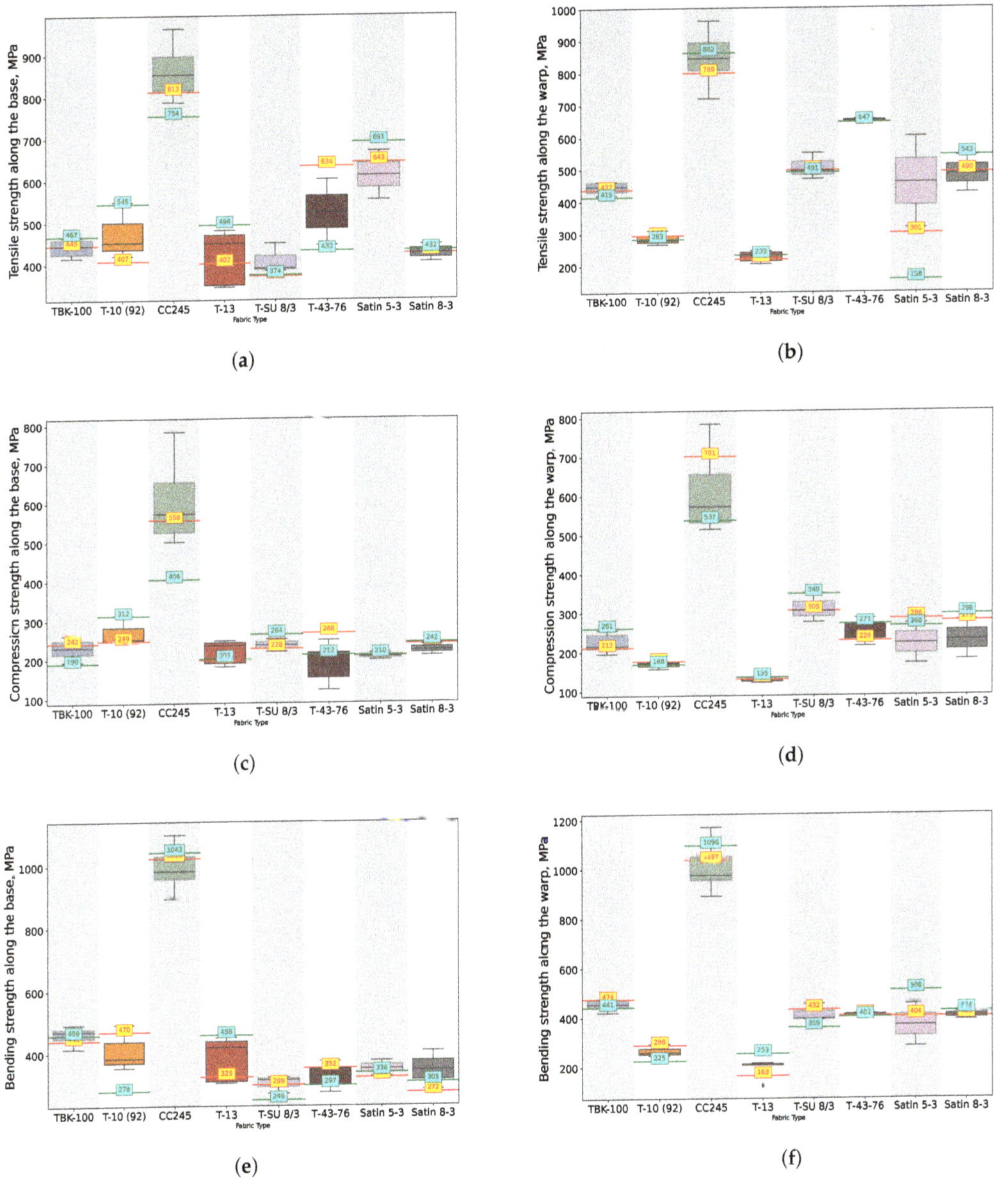

Figure 8. Whisker charts for selected types of TPCMs depending on physical property: (**a**) Tensile strength along the base, (**b**) Tensile strength along the warp, (**c**) Compression strength along the base, (**d**) Compression strength along the warp, (**e**) Bending strength along the base, (**f**) Bending strength along the warp with predictions made by the best architectures of the ANN (digits in yellow) and SVM (digits in cyan), architectures optimized using the MOPSO and NSGA-II algorithms, respectively.

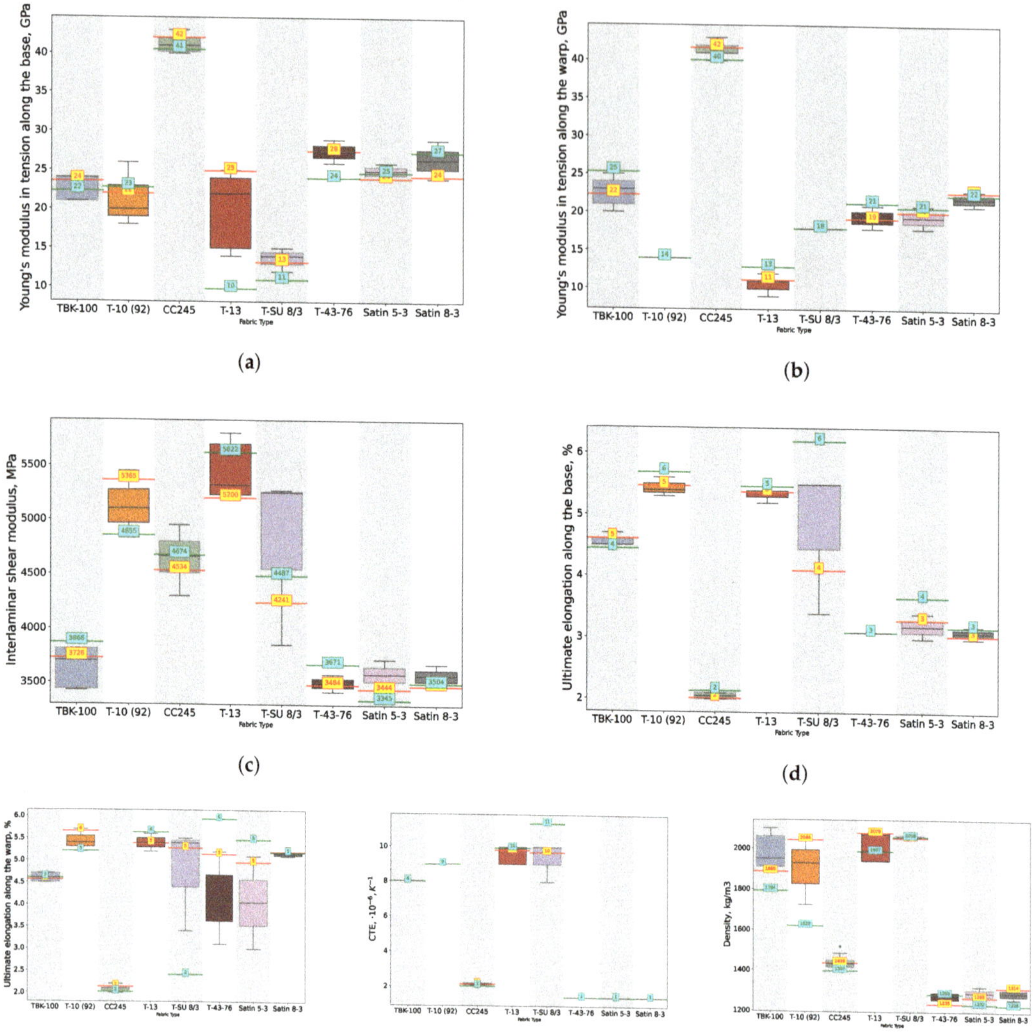

Figure 9. Whisker charts for selected types of TPCMs charts for selected types of TPCMs depending on physical property: (**a**) Tensile strength along the base depending on physical property: (**a**) Young's modulus in tension along the base, (**b**) Young's modulus in tension along the warp, (**c**) Interlaminar shear modulus, (**d**) Ultimate elongation along the base, (**e**) Ultimate elongation along the warp, (**f**) CTE, (**g**) Density with predictions made by the best architectures of the ANN (digits on yellow) and SVM (digits on cyan), architectures optimized using the MOPSO and NSGA-II algorithms, respectively.

4. Discussion

For predicting the physical characteristics of TPCMs, we employed multi-objective optimization algorithms to develop predictive models for the TPCM properties based on the constituents that comprise these materials.

Our choice was motivated by several factors, among which the prominent one was the robustness and extensive application of MOPSO, NSGA II, and SPEA2 in optimization tasks. These algorithms exhibit efficient search capabilities and adeptly manage multiple conflicting objectives [137] concurrently. Furthermore, their evolutionary nature renders them well-suited for navigating high-dimensional parameter spaces. Iteratively exploring these spaces, they refine solutions progressively, converging towards optimal or near-optimal solutions. Also, these algorithms provide a diverse set of solutions, allowing us to explore the trade-offs between different objectives and select the most suitable model configurations based on the specific requirements of the application.

To address classical ML issues such as avoiding local optima [138], heuristic algorithms offer several advantages. They use a population of potential solutions, allowing the exploration of multiple regions of the parameter space simultaneously and reducing the likelihood of becoming stuck in local optima [139]. Additionally, they balance different aspects of model performance, such as accuracy and generalization [140], leading to more reliable solutions. Initializing particles with diverse positions helps cover a broader area of the search space, increasing the chances of finding global optima.

To ensure the robustness of our dataset, we conducted a validation using information obtained from open sources. Several of these sources are listed in Section 2.1. By comparing and verifying the information against established sources, we aimed to confirm the accuracy and reliability of our dataset. Leveraging information from open sources enhances the credibility of our findings and strengthens the validity of our analyses.

Utilizing multi-objective optimization is widely discussed in the scientific literature, as exemplified by Mannodi et al. [141], who developed two Monte Carlo algorithms to pinpoint the Pareto front in the chemical space of dielectric polymers, optimizing both the bandgap and dielectric constant. Using machine learning on a dataset from density functional theory calculations, they created surrogate models for four-block polymers and extended their applicability.

Garcia et al. [142] developed an ANN to predict thermal and electrical conductivity in HDPE–carbon particle composites. ANNs served as objective functions in a multi-objective GA to optimize composite design parameters. The GA generates Pareto-optimal solutions [143] for maximizing thermal conductivity and minimizing electrical conductivity. This approach offers a systematic framework for optimizing polymer composite properties efficiently.

The approach taken in this study showcases several novel aspects, particularly in the optimization of the predictive accuracy and inference time for the physical properties of TPCMs by analyzing the components they are produced from, like the properties of the yarns in the warp and weft directions, as well as the fabrics made from these yarns by utilizing MOPSO, SPEA2, and NSGA-II to achieve a more efficient exploration of the high-dimensional parameter space. This resulted in more accurate predictions and reduced inference times, which is a significant improvement over traditional optimization methods by reducing the need for time-consuming and expensive physical experiments. By accurately predicting the properties of new composite materials based on existing data, we can streamline the development process of new TPCM products.

This method also offers the potential for incorporating multiphysical computational methods, which can further enhance the accuracy and relevance of the predictions. This integration supports the development of comprehensive models that account for various physical phenomena simultaneously.

Despite the promising results, several challenges and limitations must be acknowledged. One notable limitation is the availability and quality of data. While efforts were made to collect comprehensive datasets, variations in the data sources and measurement techniques may introduce inconsistencies and biases.

A trade-off exists between model interpretability and predictive performance. While complex ANNs may achieve superior predictive accuracy, they often lack interpretability, hindering insights into the underlying physical mechanisms. Future research should explore techniques for enhancing model interpretability without compromising performance, such as layerwise relevance propagation [144] and feature attribution methods.

The generalization ability of predictive models is crucial for their practical applicability across diverse TPCM systems and processing conditions. Robust validation strategies, including cross-validation and out-of-sample testing [145], are essential for assessing model generalization and ensuring reliable predictions in real-world scenarios.

Various frameworks exist for multi-physics modeling. For instance, PERMIX [146] is an open-source framework designed for multiscale modeling and simulation of fractures in solids, utilizing the extended finite-element method (XFEM) and integrating with libraries like LAMMPS and ABAQUS. It accommodates both semi-concurrent and concurrent multiscale methods for detailed fracture simulations.

Alternatively, Liu et al. [147] introduced a hybrid ML approach employing an ANN and PSO to predict the thermal conductivity of polymeric nanocomposites (PNCs). By combining the ANN for modeling and PSO for optimization, they achieved superior predictive performance compared to traditional ANNs. Key input parameters included the thermal conductivity of fibers and matrix, Kapitza resistance, volume fraction, and the aspect ratio, with the output being the composite's macroscopic thermal conductivity.

Moreover, N. Vu-Bac [148] integrated molecular dynamics (MD) simulations to examine the impact of the single-walled carbon nanotube (SWCNT) radius, temperature, and pulling velocity on the interfacial shear stress (ISS) of PNCs by assessing the influence of uncertain input parameters on ISS prediction by computing partial derivatives via averaged local sensitivity analysis (SA) and employing surrogate models (polynomial regression, moving least squares, and hybrid models) for computational efficiency.

Ilyani Abu et al. [149] utilized unit cells and evolutionary algorithms to forecast the geometric characteristics and elastic properties of woven fabric composites, through optimizing elastic properties within these unit cells to accurately predict mechanical behavior. TexGen generated the weave patterns; ABAQUS was used to conduct the simulations; finite-element (FE) analysis estimated the effective elastic properties of the yarn. The parameter studies delved into the effects of various geometric parameters, facilitating the selection of an optimal parameter set for composite performance.

The Parametric Deep Energy Method (P-DEM) for elasticity problems incorporating strain gradient effects was suggested in [150]. Utilizing physics-informed neural networks [151] (PINNs), the authors optimized a cost function associated with potential energy, eliminating the need for classical discretization. By defining a parametric/reference space akin to isoparametric finite elements, and leveraging NURBS [152] basis functions, P-DEM achieves efficient computation of the total potential energy.

Accurate predictive models for TPCM properties offer significant implications for materials' design, process optimization, and product performance prediction. By leveraging these models, engineers and designers can expedite material development cycles, optimize manufacturing processes, and tailor materials' properties to meet specific application requirements.

Future research directions may include the incorporation of advanced feature engineering techniques, such as image-based analysis and spectral imaging, to extract rich structural and compositional information from TPCM samples. Additionally, the integration of physics-based models with machine learning approaches could enhance predictive accuracy and facilitate mechanistic understanding of material behavior.

5. Conclusions

In conclusion, our study highlights the significant potential of ML and optimization techniques in advancing the predictive modeling of TPCM properties. By leveraging SVMs and ANNs as powerful modeling tools, we successfully optimized their hyperparameters using state-of-the-art optimization algorithms, including MOPSO, NSGA II, and SPEA 2. Through this approach, we have demonstrated the effectiveness of integrating advanced ML techniques with optimization methodologies to enhance the accuracy, robustness, and applicability of predictive models for TPCM properties. By addressing key challenges and exploring innovative methodologies, researchers can further propel advancements in material science and engineering, ultimately facilitating the development of high-performance textile polymer composites for diverse applications.

Author Contributions: Conceptualization, I.M., V.T. and A.G.; data curation, A.G. and V.N.; funding acquisition, A.G., V.N. and A.B.; investigation, V.T. and V.N.; methodology, I.M.; project administration, A.G., V.N. and A.B.; resources, I.M.; software, I.M., V.T., V.N. and A.B.; supervision, V.T., A.G., V.N. and A.B.; validation, A.B.; visualization, I.M.; writing—original draft, I.M.; writing—review and editing, V.T. and A.B. All authors have read and agreed to the published version of the manuscript.

Funding: This research received no external funding.

Institutional Review Board Statement: Not applicable.

Data Availability Statement: All the data used in this study are available in the TPCM repository. The repository contains a comprehensive dataset comprising the properties of the TPCMs, meticulously compiled from experimental data. Researchers interested in accessing the data can find them in the provided repository.

Conflicts of Interest: The authors declare no conflicts of interest.

References

1. Hsissou, R.; Seghiri, R.; Benzekri, Z.; Hilali, M.; Rafik, M.; Elharfi, A. Polymer composite materials: A comprehensive review. *Compos. Struct.* **2021**, *262*, 113640. [CrossRef]
2. Fan, J.; Njuguna, J. An introduction to lightweight composite materials and their use in transport structures. In *Lightweight Composite Structures in Transport*; Elsevier: Amsterdam, The Netherlands, 2016; pp. 3–34.
3. Qian, D.; Bao, L.; Takatera, M.; Kemmochi, K.; Yamanaka, A. Fiber-reinforced polymer composite materials with high specific strength and excellent solid particle erosion resistance. *Wear* **2010**, *268*, 637–642. [CrossRef]
4. Bazli, L.; Yusuf, M.; Farahani, A.; Kiamarzi, M.; Seyedhosseini, Z.; Nezhadmansari, M.; Aliasghari, M.; Iranpoor, M. Application of composite conducting polymers for improving the corrosion behavior of various substrates: A Review. *J. Compos. Compd.* **2020**, *2*, 228–240. [CrossRef]
5. Dobrzański, L.; Drak, M.; Trzaska, J. Corrosion resistance of the polymer matrix hard magnetic composite materials Nd–Fe–B. *J. Mater. Process. Technol.* **2005**, *164*, 795–804. [CrossRef]
6. Kangishwar, S.; Radhika, N.; Sheik, A.A.; Chavali, A.; Hariharan, S. A comprehensive review on polymer matrix composites: Material selection, fabrication, and application. *Polym. Bull.* **2023**, *80*, 47–87. [CrossRef]
7. Ali, B.A.; Sapuan, S.; Zainudin, E.; Othman, M. Implementation of the expert decision system for environmental assessment in composite materials selection for automotive components. *J. Clean. Prod.* **2015**, *107*, 557–567.
8. Sapuan, S.; Kho, J.; Zainudin, E.; Leman, Z.; Ali, B.; Hambali, A. Materials selection for natural fiber reinforced polymer composites using analytical hierarchy process. *Indian J. Eng. Mater. Sci.* **2011**, *18*, 255–267.
9. DeArmitt, C.; Rothon, R. Particulate fillers, selection and use in polymer composites. In *Encyclopedia of Polymers and Composites*; Springer: Berlin/Heidelberg, Germany, 2017; pp. 1–19.
10. Rajak, D.K.; Pagar, D.D.; Kumar, R.; Pruncu, C.I. Recent progress of reinforcement materials: A comprehensive overview of composite materials. *J. Mater. Res. Technol.* **2019**, *8*, 6354–6374. [CrossRef]
11. Kamath, S.S.; Chandrappa, R.K. Additives used in natural fibre reinforced polymer composites—A review. *Mater. Today Proc.* **2022**, *50*, 1417–1424. [CrossRef]
12. Lhamo, P.; Mahanty, B. Tuning material properties of microbially synthesized poly (3-hydroxybutyrate-co-hydroxyvalerate) and their composites for various applications. *J. Polym. Environ.* **2023**, *31*, 4641–4661. [CrossRef]
13. Xue, T.; He, T.; Peng, L.; Syzgantseva, O.A.; Li, R.; Liu, C.; Sun, D.T.; Xu, G.; Qiu, R.; Wang, Y.; et al. A customized MOF-polymer composite for rapid gold extraction from water matrices. *Sci. Adv.* **2023**, *9*, eadg4923. [CrossRef]
14. Krishnan, M.R.; Omar, H.; Almohsin, A.; Alsharaeh, E.H. An overview on nanosilica–polymer composites as high-performance functional materials in oil fields. *Polym. Bull.* **2024**, *81*, 3883–3933. [CrossRef]
15. Yadav, R.; Singh, M.; Shekhawat, D.; Lee, S.Y.; Park, S.J. The role of fillers to enhance the mechanical, thermal, and wear characteristics of polymer composite materials: A review. *Compos. Part A Appl. Sci. Manuf.* **2023**, *175*, 107775. [CrossRef]
16. Hasan, K.F.; Al Hasan, K.N.; Ahmed, T.; György, S.T.; Pervez, M.N.; Bejó, L.; Sándor, B.; Alpár, T. Sustainable bamboo fiber reinforced polymeric composites for structural applications: A mini review of recent advances and future prospects. *Case Stud. Chem. Environ. Eng.* **2023**, *8*, 100362. [CrossRef]
17. Fu, X.; Lin, J.; Liang, Z.; Yao, R.; Wu, W.; Fang, Z.; Zou, W.; Wu, Z.; Ning, H.; Peng, J. Graphene oxide as a promising nanofiller for polymer composite. *Surfaces Interfaces* **2023**, *37*, 102747. [CrossRef]
18. Xu, J.; Sun, J.; Zhao, J.; Zhang, W.; Zhou, J.; Xu, L.; Guo, H.; Liu, Y.; Zhang, D. Eco-friendly wood plastic composites with biomass-activated carbon-based form-stable phase change material for building energy conversion. *Ind. Crop. Prod.* **2023**, *197*, 116573. [CrossRef]

19. Huseynov, O.; Hasanov, S.; Fidan, I. Influence of the matrix material on the thermal properties of the short carbon fiber reinforced polymer composites manufactured by material extrusion. *J. Manuf. Process.* **2023**, *92*, 521–533. [CrossRef]
20. Al-Tamimi, A.A.; Tlija, M.; Abidi, M.H.; Anis, A.; Abd Elgawad, A.E.E. Material Extrusion of Multi-Polymer Structures Utilizing Design and Shrinkage Behaviors: A Design of Experiment Study. *Polymers* **2023**, *15*, 2683. [CrossRef] [PubMed]
21. Zou, W.; Zheng, X.; Hu, X.; Huang, J.; Wang, G.; Guo, Z. Recent advances in injection molding of carbon fiber reinforced thermoplastic polymer composites: A review. *ES Gen.* **2023**, *1*, 938. [CrossRef]
22. Barthod-Malat, B.; Hauguel, M.; Behlouli, K.; Grisel, M.; Savary, G. Influence of the Compression Molding Temperature on VOCs and Odors Produced from Natural Fiber Composite Materials. *Coatings* **2023**, *13*, 371. [CrossRef]
23. Błachut, A.; Wollmann, T.; Panek, M.; Vater, M.; Kaleta, J.; Detyna, J.; Hoschützky, S.; Gude, M. Influence of fiber tension during filament winding on the mechanical properties of composite pressure vessels. *Compos. Struct.* **2023**, *304*, 116337. [CrossRef]
24. Mayakannan, S.; Raj, J.B.; Raja, V.; Nagaraj, M. Effectiveness of silicon nanoparticles on the mechanical, wear, and physical characteristics of PALF/sisal fiber–based polymer hybrid nanocomposites. *Biomass Convers. Biorefinery* **2023**, *13*, 13291–13305. [CrossRef]
25. Ali, A.; Koloor, S.S.R.; Alshehri, A.H.; Arockiarajan, A. Carbon nanotube characteristics and enhancement effects on the mechanical features of polymer-based materials and structures—A review. *J. Mater. Res. Technol.* **2023**, *24*, 6495–6521. [CrossRef]
26. dos Anjos, E.G.; Moura, N.K.; Antonelli, E.; Baldan, M.R.; Gomes, N.A.; Braga, N.F.; Santos, A.P.; Rezende, M.C.; Pessan, L.A.; Passador, F.R. Role of adding carbon nanotubes in the electric and electromagnetic shielding behaviors of three different types of graphene in hybrid nanocomposites. *J. Thermoplast. Compos. Mater.* **2023**, *36*, 3209–3235. [CrossRef]
27. Low, K.O.; Wong, K.J. Influence of ball burnishing on surface quality and tribological characteristics of polymers under dry sliding conditions. *Tribol. Int.* **2011**, *44*, 144–153. [CrossRef]
28. Feih, S.; Mouritz, A.; Mathys, Z.; Gibson, A. Tensile strength modeling of glass fiber—polymer composites in fire. *J. Compos. Mater.* **2007**, *41*, 2387–2410. [CrossRef]
29. Chang, S.M.; Hur, S.; Park, J.; Lee, D.G.; Shin, J.; Kim, H.S.; Song, S.E.; Baik, J.M.; Kim, M.; Song, H.C.; et al. Optimization of piezoelectric polymer composites and 3D printing parameters for flexible tactile sensors. *Addit. Manuf.* **2023**, *67*, 103470. [CrossRef]
30. Wen, S.M.; Chen, S.M.; Gao, W.; Zheng, Z.; Bao, J.Z.; Cui, C.; Liu, S.; Gao, H.L.; Yu, S.H. Biomimetic Gradient Bouligand Structure Enhances Impact Resistance of Ceramic-Polymer Composites. *Adv. Mater.* **2023**, *35*, 2211175. [CrossRef] [PubMed]
31. He, L.; Zhang, W.; Liu, X.; Tong, L. Substantial improvement of thermal conductivity and mechanical properties of polymer composites by incorporation of boron nitride nanosheets and modulation of thermal curing reaction. *Polym. Compos.* **2024**, *45*, 2215–2231. [CrossRef]
32. Wang, Z.; Wu, Z.; Weng, L.; Ge, S.; Jiang, D.; Huang, M.; Mulvihill, D.M.; Chen, Q.; Guo, Z.; Jazzar, A.; et al. A roadmap review of thermally conductive polymer composites: Critical factors, progress, and prospects. *Adv. Funct. Mater.* **2023**, *33*, 2301549. [CrossRef]
33. Taye, M.M. Theoretical understanding of convolutional neural network: Concepts, architectures, applications, future directions. *Computation* **2023**, *11*, 52. [CrossRef]
34. Roy, A.; Chakraborty, S. Support vector machine in structural reliability analysis: A review. *Reliab. Eng. Syst. Saf.* **2023**, *233*, 109126. [CrossRef]
35. Hu, J.; Szymczak, S. A review on longitudinal data analysis with random forest. *Briefings Bioinform.* **2023**, *24*, bbad002. [CrossRef]
36. Gao, Y.; Shigidi, I.M.; Ali, M.A.; Homod, R.Z.; Safaei, M.R. Thermophysical properties prediction of carbon-based nano-enhanced phase change material's using various machine learning methods. *J. Taiwan Inst. Chem. Eng.* **2023**, *148*, 104662. [CrossRef]
37. Fontes, A.; Shadmehri, F. Data-driven failure prediction of Fiber-Reinforced Polymer composite materials. *Eng. Appl. Artif. Intell.* **2023**, *120*, 105834. [CrossRef]
38. Groenwold, A.A.; Haftka, R.T. Optimization with non-homogeneous failure criteria like Tsai–Wu for composite laminates. *Struct. Multidiscip. Optim.* **2006**, *32*, 183–190. [CrossRef]
39. Rezasefat, M.; Gonzalez-Jimenez, A.; Giglio, M.; Manes, A. An evaluation of Cuntze and Puck inter fibre failure criteria in simulation of thin CFRP plates subjected to low velocity impact. *Compos. Struct.* **2021**, *278*, 114654. [CrossRef]
40. Pimenta, S.; Pinho, S.T. Recycling carbon fibre reinforced polymers for structural applications: Technology review and market outlook. *Waste Manag.* **2011**, *31*, 378–392. [CrossRef]
41. Fahem, N.; Belaidi, I.; Brahim, A.O.; Noori, M.; Khatir, S.; Wahab, M.A. Prediction of resisting force and tensile load reduction in GFRP composite materials using Artificial Neural Network-Enhanced Jaya Algorithm. *Compos. Struct.* **2023**, *304*, 116326. [CrossRef]
42. Sathishkumar, T.; Satheeshkumar, S.; Naveen, J. Glass fiber-reinforced polymer composites—A review. *J. Reinf. Plast. Compos.* **2014**, *33*, 1258–1275. [CrossRef]
43. Rajak, D.K.; Pagar, D.D.; Menezes, P.L.; Linul, E. Fiber-Reinforced Polymer Composites: Manufacturing, Properties, and Applications. *Polymers* **2019**, *11*, 1667. [CrossRef] [PubMed]
44. Zhang, Y.; Chi, A.; Mirjalili, S. Enhanced Jaya algorithm: A simple but efficient optimization method for constrained engineering design problems. *Knowl.-Based Syst.* **2021**, *233*, 107555. [CrossRef]
45. Cervera, M.; Barbat, G.; Chiumenti, M.; Wu, J.Y. A comparative review of XFEM, mixed FEM and phase-field models for quasi-brittle cracking. *Arch. Comput. Methods Eng.* **2022**, *29*, 1009–1083. [CrossRef]

46. Houssein, E.H.; Gad, A.G.; Wazery, Y.M. Jaya algorithm and applications: A comprehensive review. In *Metaheuristics and Optimization in Computer and Electrical Engineering*; Springer: Cham, Switzerland, 2021; pp. 3–24.
47. Nayak, J.; Swapnarekha, H.; Naik, B.; Dhiman, G.; Vimal, S. 25 years of particle swarm optimization: Flourishing voyage of two decades. *Arch. Comput. Methods Eng.* **2023**, *30*, 1663–1725. [CrossRef]
48. Nguyen, M.H.; D'Mello, R.J.; Waas, A.M. Use of a neural network constitutive model for the size-dependent effects of curing on the deformation response and failure of fiber-reinforced polymer matrix composites. *Arch. Appl. Mech.* **2023**, *93*, 369–387. [CrossRef]
49. Gupta, S.; Mukhopadhyay, T.; Kushvaha, V. Microstructural image based convolutional neural networks for efficient prediction of full-field stress maps in short fiber polymer composites. *Def. Technol.* **2023**, *24*, 58–82. [CrossRef]
50. Sharma, A.; Munde, Y.; Kushvaha, V. Representative volume element based micromechanical modelling of rod shaped glass filled epoxy composites. *SN Appl. Sci.* **2021**, *3*, 1–10. [CrossRef]
51. Kim, S.; Lee, J.; Jeong, K.; Lee, J.; Hong, T.; An, J. Automated door placement in architectural plans through combined deep-learning networks of ResNet-50 and Pix2Pix-GAN. *Expert Syst. Appl.* **2024**, *244*, 122932. [CrossRef]
52. El Said, B. Predicting the non-linear response of composite materials using deep recurrent convolutional neural networks. *Int. J. Solids Struct.* **2023**, *276*, 112334. [CrossRef]
53. Koller, O.; Zargaran, S.; Ney, H. Re-sign: Re-aligned end-to-end sequence modelling with deep recurrent CNN-HMMs. In Proceedings of the IEEE Conference on Computer Vision and Pattern Recognition, Honolulu, HI, USA, 21–26 July 2017; pp. 4297–4305.
54. Reinaldo A.A.; Javier L.M.; Pablo J.S. Multiscale formulation for saturated porous media preserving the representative volume element size objectivity. *Int. J. Numer. Methods Eng.* **2024**, *125*, e7381. [CrossRef]
55. Zhang, H.; Li, B. Improvement of thermal protection performance of polymer composites based on optimized piecewise porosity distribution using a neural network method. *Therm. Sci. Eng. Prog.* **2023**, *43*, 102010. [CrossRef]
56. Song, Y.; Qu, Z.; Liao, H.; Ai, S. Material twins generation of woven polymer composites based on ResL-U-Net convolutional neural networks. *Compos. Struct.* **2023**, *307*, 116672. [CrossRef]
57. Xu, J.; Li, Z.; Du, B.; Zhang, M.; Liu, J. Reluplex made more practical: Leaky ReLU. In Proceedings of the 2020 IEEE Symposium on Computers and Communications (ISCC), Rennes, France, 7–10 July 2020; pp. 1–7.
58. Li, M.; Li, S.; Tian, Y.; Fu, Y.; Pei, Y.; Zhu, W.; Ke, Y. A deep learning convolutional neural network and multi-layer perceptron hybrid fusion model for predicting the mechanical properties of carbon fiber. *Mater. Des.* **2023**, *227*, 111760. [CrossRef]
59. Chen, L.; Hao, L.; Liu, S.; Ding, G.; Sun, X.; Zhang, W.; Li, F.; Jiao, W.; Yang, F.; Xu, Z.; et al. Modulus distribution in polyacrylonitrile-based carbon fiber monofilaments. *Carbon* **2020**, *157*, 47–54. [CrossRef]
60. Li, M.; Zhang, H.; Ma, J.; Li, S.; Zhu, W.; Ke, Y. Greedy-based approach for generating anisotropic random fiber distributions of unidirectional composites and transverse mechanical properties prediction. *Comput. Mater. Sci.* **2023**, *218*, 111966. [CrossRef]
61. Doddashamachar, M.; Sen, S.; Nama Vasudeva Setty, R. A novel Artificial Neural Network-based model for predicting dielectric properties of banana fiber filled with polypropylene composites. *J. Thermoplast. Compos. Mater.* **2023**, *36*, 4106–4123. [CrossRef]
62. Arpitha, G.; Jain, N.; Verma, A. Banana biofiber and glass fiber reinforced hybrid composite for lightweight structural applications: Mechanical, thermal, and microstructural characterization. *Biomass Convers. Biorefinery* **2023**, 1–10.. [CrossRef]
63. Amor, N.; Noman, M.T.; Petru, M.; Sebastian, N.; Balram, D. A review on computational intelligence methods for modelling of light weight composite materials. *Appl. Soft Comput.* **2023**, *147*, 110812. [CrossRef]
64. Mukhopadhyay, S. *Artificial Neural Network Applications in Textile Composites*; Soft Computing in Textile Engineering; Elsevier: Amsterdam, The Netherlands, 2011.
65. Amor, N.; Noman, M.T.; Petru, M. Classification of textile polymer composites: Recent trends and challenges. *Polymers* **2021**, *13*, 2592. [CrossRef]
66. Borhanazad, H.; Mekhilef, S.; Ganapathy, V.G.; Modiri-Delshad, M.; Mirtaheri, A. Optimization of micro-grid system using MOPSO. *Renew. Energy* **2014**, *71*, 295–306. [CrossRef]
67. Verma, S.; Pant, M.; Snasel, V. A comprehensive review on NSGA-II for multi-objective combinatorial optimization problems. *IEEE Access* **2021**, *9*, 57757–57791. [CrossRef]
68. Liu, X.; Zhang, D. An improved SPEA2 algorithm with local search for multi-objective investment decision-making. *Appl. Sci.* **2019**, *9*, 1675. [CrossRef]
69. Cervantes, J.; Garcia-Lamont, F.; Rodríguez-Mazahua, L.; Lopez, A. A comprehensive survey on support vector machine classification: Applications, challenges and trends. *Neurocomputing* **2020**, *408*, 189–215. [CrossRef]
70. Sapozhnikov, S.; Kudryavtsev, O. Modeling of thermoplastic composites used in protective structures. *Mech. Compos. Mater.* **2015**, *51*, 419–426. [CrossRef]
71. Kreutzer-Schmid, C.; Schmid, H.P. The prosomal protein of 27 kDa and a nuclear 38 kDa protein are immunologically related. *FEBS Lett.* **1990**, *267*, 142–146. [CrossRef] [PubMed]
72. Hashim, U.R.; Jumahat, A.; Jawaid, M.; Dungani, R.; Alamery, S. Effects of accelerated weathering on degradation behavior of basalt fiber reinforced polymer nanocomposites. *Polymers* **2020**, *12*, 2621. [CrossRef] [PubMed]
73. Jamshaid, H.; Mishra, R. A green material from rock: Basalt fiber—A review. *J. Text. Inst.* **2016**, *107*, 923–937. [CrossRef]
74. Kozinets, G.L.; Chernov, P.V.; Kolotvina, N.S.; Zotov, D.K.; Kärki, T.; Lahtela, V. Optimization of wooden constructions with basalt-based materials. In Proceedings of the International Seminar, Saint Petersburg, Germany, 25 May 2021; pp. 7–20.

75. Wallenberger, F.T.; Bingham, P.A. *Fiberglass and Glass Technology. Energy-Friendly Compositions And Applications*; Springer: New York, NY, USA, 2010.
76. Patel, A.; Kravchenko, O.; Manas-Zloczower, I. Effect of curing rate on the microstructure and macroscopic properties of epoxy fiberglass composites. *Polymers* **2018**, *10*, 125. [CrossRef] [PubMed]
77. Poopakdee, N.; Thammawichai, W. Improvement on cost-performance ratio of fiberglass/carbon fiber hybrid composite. *J. Met. Mater. Miner.* **2021**, *31*.. [CrossRef]
78. Zhang, W.; Zhang, X.; Qin, Z.; Wu, Y.; Zhang, W.; Yang, R. High-transparency polysilsesquioxane/glycidyl-azide-polymer resin and its fiberglass-reinforced composites with excellent fire resistance, mechanical properties, and water resistance. *Compos. Part B Eng.* **2021**, *219*, 108913. [CrossRef]
79. Trentin, A.; Pakseresht, A.; Duran, A.; Castro, Y.; Galusek, D. Electrochemical characterization of polymeric coatings for corrosion protection: A review of advances and perspectives. *Polymers* **2022**, *14*, 2306. [CrossRef] [PubMed]
80. Thomason, J.; Ali, J. The dimensional stability of glass–fibre reinforced polyamide 66 during hydrolysis conditioning. *Compos. Part A Appl. Sci. Manuf.* **2009**, *40*, 625–634. [CrossRef]
81. Stickel, J.M.; Nagarajan, M. Glass fiber-reinforced composites: From formulation to application. *Int. J. Appl. Glass Sci.* **2012**, *3*, 122–136. [CrossRef]
82. Kalinin, Y.E.; Kudrin, A.; Ovdak, O.; Popov, I. Internal friction in an epoxy polymer and a fiberglass reinforced plastic based on it. *Polym. Sci. Ser. A* **2022**, *64*, 1–9. [CrossRef]
83. Tusnin, A.; Nikolay, L.; Aleksandr, K. The Load-Bearing Capacity and Deformability of Connections of Wooden Elements with Composite Materials Based on Fiberglass. *Buildings* **2023**, *13*, 3063. [CrossRef]
84. Bashkov, O.; Bryansky, A. Cluster analysis of the acoustic emission signals registered during bending deformation of FGRP. *AIP Conf. Proc.* **2023**, *2899*, 020013.
85. Gorev, Y.A.; Rivkind, V. Polyester composites for shipbuilding. *Russ. J. Gen. Chem.* **2010**, *80*, 2098–2114. [CrossRef]
86. Chermoshentseva, A.; Pokrovskiy, A.; Bokhoeva, L. The behavior of delaminations in composite materials-experimental results. *IOP Conf. Ser. Mater. Sci. Eng.* **2016**, *116*, 012005. [CrossRef]
87. Shmoilov, E.; Fedotov, M.; Sharutin, I.; Ilyukhin, R.; Stepanov, S.; Panina, N.; Gurenchuk, L.; Kapyrin, P.; Kabantsev, O.; Kornev, O. Polymer Composites for External Reinforcement of Building Structures. *Int. J. Comput. Civ. Struct. Eng.* **2024**, *20*, 21–34.
88. Liang, B.; Zhao, Z.; Cheng, H.; Boisse, P.; Zhang, K.; Luo, B. A combined method for analyzing the effective thermal conductivity evolution of satin weave thermoset prepregs during preforming process. *Int. J. Therm. Sci.* **2022**, *177*, 107574. [CrossRef]
89. Zhang, Y.; Zhang, Q.; Chen, G. Carbon and carbon composites for thermoelectric applications. *Carbon Energy* **2020**, *2*, 408–436. [CrossRef]
90. Srinivasan, V.; Kunjiappan, S.; Palanisamy, P. A brief review of carbon nanotube reinforced metal matrix composites for aerospace and defense applications. *Int. Nano Lett.* **2021**, *11*, 321–345. [CrossRef]
91. Ahmad, H.; Markina, A.; Porotnikov, M.; Ahmad, F. A review of carbon fiber materials in automotive industry. *IOP Conf. Ser. Mater. Sci. Eng.* **2020**, *971*, 032011. [CrossRef]
92. Sharma, S.; Sudhakara, P.; Misra, S.; Singh, J. A comprehensive review of current developments on the waste-reinforced polymer-matrix composites for automotive, sports goods and construction applications: Materials, processes and properties. *Mater. Today Proc.* **2020**, *33*, 1671–1679. [CrossRef]
93. Garifullin, A.; Krasina, I.; Skidchenko, E.; Shaekhov, M.; Tikhonova, N. Modification of carbon fabrics by radio-frequency capacitive discharge at low pressure to regulate mechanical properties of carbon fiber reinforced plastics based on it. *J. Phys. Conf. Ser.* **2017**,*789*, 012014. [CrossRef]
94. Ghiringhelli, G.; Terraneo, M.; Vigoni, E. Improvement of structures vibroacoustics by widespread embodiment of viscoelastic materials. *Aerosp. Sci. Technol.* **2013**, *28*, 227–241. [CrossRef]
95. Wang, Z.; Dong, S.; Ding, Y.; Zhang, X.; Zhou, H.; Yang, J.; Lu, B. Mechanical properties and microstructures of Cf/SiC–ZrC composites using T700SC carbon fibers as reinforcements. *Ceram. Int.* **2011**, *37*, 695–700. [CrossRef]
96. Slovikov, S.; Babushkin, A.; Gusina, M. Nonlinearity of compression behavior of 3D-epoxy reinforced with carbon fibers composites. *Frat. Integrità Strutt.* **2023**, *17*, 311–321. [CrossRef]
97. Gordeev, Y.I.; Binchurov, A.; Yasinskii, V.; Pikalov, Y.Y.; Vakulin, M.; Kazakov, I. Influence of Cutting Conditions and End Mill Geometry on the Surface Quality in High-Speed Machining of Carbon Composites. *Russ. Eng. Res.* **2023**, *43*, 592–597. [CrossRef]
98. Käppler, I.; Hund, R.D.; Cherif, C. Surface modification of carbon fibres using plasma technique. *Autex Res. J.* **2014**, *14*, 34–38. [CrossRef]
99. Kravchuk, L.; Buiskikh, K.; Derevyanko, I.; Potapov, O. Load-bearing capacity of elements of composite shell structures in rocket and space engineering made of composite materials. *Strength Mater.* **2022**, *54*, 613–621. [CrossRef]
100. Chairman, C.A.; Jayasathyakawin, S.; Babu, S.K.; Ravichandran, M. Mechanical properties of basalt fabric plain and twill weave reinforced epoxy composites. *Mater. Today Proc.* **2021**, *46*, 9480–9483. [CrossRef]
101. Ahmed, U.; Tariq, A.; Nawab, Y.; Shaker, K.; Khaliq, Z.; Umair, M. Comparison of mechanical behavior of biaxial, unidirectional and standard woven fabric reinforced composites. *Fibers Polym.* **2020**, *21*, 1308–1315. [CrossRef]
102. Fan, Y.; Li, Z.; Wei, J. Application of aramid nanofibers in nanocomposites: A brief review. *Polymers* **2021**, *13*, 3071. [CrossRef] [PubMed]

103. Zhao, Y.; Li, X.; Shen, J.; Gao, C.; Van der Bruggen, B. The potential of Kevlar aramid nanofiber composite membranes. *J. Mater. Chem. A* **2020**, *8*, 7548–7568. [CrossRef]
104. Bijwe, J.; Awtade, S.; Ghosh, A. Influence of orientation and volume fraction of Aramid fabric on abrasive wear performance of polyethersulfone composites. *Wear* **2006**, *260*, 401–411. [CrossRef]
105. Farias-Aguilar, J.; Ramírez-Moreno, M.; Gonzalez-García, D.; Téllez-Jurado, L.; Balmori-Ramirez, H. Evaluation of the ballistic protection level of (glass-fiber reinforced polyamide 6)-aramid fabric sandwich composite panels. *J. Mater. Res. Technol.* **2021**, *12*, 1606–1614. [CrossRef]
106. Dos Santos, D.; Carbas, R.; Marques, E.; Da Silva, L. Reinforcement of CFRP joints with fibre metal laminates and additional adhesive layers. *Compos. Part B Eng.* **2019**, *165*, 386–396. [CrossRef]
107. Xiaobo, Y.; Binjie, X.; Baciu, G.; Jinlian, H. Fourier-analysis based satin fabric density and weaving pattern extraction. *Res. J. Text. Appar.* **2007**, *11*, 71.
108. Saiman, M.; Wahab, M.; Wahit, M. The effect of fabric weave on the tensile strength of woven kenaf reinforced unsaturated polyester composite. In *Proceedings of the International Colloquium in Textile Engineering, Fashion, Apparel and Design 2014 (ICTEFAD 2014)*; Springer: Singapore, 2014; pp. 25–29.
109. Lobkovskiy, S.; Shaydurova, G.; Zubarev, S. Study of waste utilization technology generated during the production of rocket engine casings from polymer composite materials. *Ecol. Ind. Russ.* **2016**, *20*, 10–15.
110. Barannikov, A.A.; Veshkin, E.A.; Postnov, V.I.; Strelnikov, S.V. On the issue of producing PCM floor panels for aircraft (review article). *Proc. Samara Sci. Cent. Russ. Acad. Sci.* **2017**, *19*, 198–213.
111. Karamzadeh, N.S.; Aliha, M.; Karimi, H.R. Investigation of the effect of components on tensile strength and mode-I fracture toughness of polymer concrete. *Arab. J. Geosci.* **2022**, *15*, 1213. [CrossRef]
112. Mohammed, A.; Mahmood, W.; Ghafor, K. TGA, rheological properties with maximum shear stress and compressive strength of cement-based grout modified with polycarboxylate polymers. *Constr. Build. Mater.* **2020**, *235*, 117534. [CrossRef]
113. Iqbal, M.; Aminanda, Y.; Firsa, T.; Nazaruddin, N.; Nasution, I.S.; Erawan, D.F.; Saputra, D.A.; Nasution, A.R. The effect of fiber content and fiber orientation on bending strength of abaca fiber reinforce polymer composite fabricated by press method. *AIP Conf. Proc.* **2023**, *2643*, 050055.
114. Kim, S.; Lee, Y.; Lee, M.; An, S.; Cho, S.J. Quantitative visualization of the nanomechanical Young's modulus of soft materials by atomic force microscopy. *Nanomaterials* **2021**, *11*, 1593. [CrossRef] [PubMed]
115. Akagi, Y.; Katashima, T.; Sakurai, H.; Chung, U.i.; Sakai, T. Ultimate elongation of polymer gels with controlled network structure. *RSC Adv.* **2013**, *3*, 13251–13258. [CrossRef]
116. Yavas, D.; Zhang, Z.; Liu, Q.; Wu, D. Interlaminar shear behavior of continuous and short carbon fiber reinforced polymer composites fabricated by additive manufacturing. *Compos. Part B Eng.* **2021**, *204*, 108460. [CrossRef]
117. Kamarian, S.; Bodaghi, M.; Isfahani, R.B.; Shakeri, M.; Yas, M. Influence of carbon nanotubes on thermal expansion coefficient and thermal buckling of polymer composite plates: Experimental and numerical investigations. *Mech. Based Des. Struct. Mach.* **2021**, *49*, 217–232. [CrossRef]
118. Wei, J.; Zhu, L. Intrinsic polymer dielectrics for high energy density and low loss electric energy storage. *Prog. Polym. Sci.* **2020**, *106*, 101254. [CrossRef]
119. Zouambi, M.; Dhaenens, C.; Jacques, J. An Alternative Pareto-based Approach to Multi-objective Neural Architecture Search. In Proceedings of the 2023 IEEE Congress on Evolutionary Computation (CEC), Chicago, IL, USA, 1–5 July 2023; pp. 1–8.
120. Ngo, S.T.; Jaafar, J.; Aziz, I.A.; Tong, G.T.; Nguyen, G.H.; Bui, A.N. Different Approaches of Evolutionary Algorithms to Multiple Objective RCPSP. In Proceedings of the 7th International Conference on Big Data and Computing, Yogyakarta, Indonesia, 7–11 November 2022; pp. 58–66.
121. Wan, L.; Deng, K.; Li, X.; Zhao, L.; Long, J. Multi-objective optimization strategy for industrial catalytic cracking units: Kinetic model and enhanced SPEA-2 algorithm with economic, CO_2, and SO_2 emission considerations. *Chem. Eng. Sci.* **2023**, *282*, 119331. [CrossRef]
122. Mohanty, R.; Das, S.K.; Mohanty, M. Shear Wave Velocity-Based Liquefaction Susceptibility of Soil Using Extreme Learning Machine (ELM) with Strength Pareto Evolutionary Algorithm (SPEA 2). In *Earthquake Geotechnics: Select Proceedings of 7th ICRAGEE 2021*; Springer: Singapore, 2022; pp. 33–44.
123. Zhang, X.; Wang, J.; Zhou, Y.; Wang, H.; Xie, N. A multi-objective optimization method for enclosed-space lighting design based on MOPSO. *Build. Environ.* **2024**, *250*, 111185. [CrossRef]
124. Liu, Y.; Yang, Z.; Liu, X.; Dan, H.; Xiong, W.; Ling, T.; Su, M. Weighting factor design based on SVR–MOPSO for finite set MPC operated power electronic converters. *J. Power Electron.* **2022**, *22*, 1085–1099. [CrossRef]
125. Guo, X.; Zhu, C.; Hao, J.; Zhang, S. Multi-step wind speed prediction based on an improved multi-objective seagull optimization algorithm and a multi-kernel extreme learning machine. *Appl. Intell.* **2023**, *53*, 16445–16472. [CrossRef]
126. Vapnik, V.; Izmailov, R. Reinforced SVM method and memorization mechanisms. *Pattern Recognit.* **2021**, *119*, 108018. [CrossRef]
127. Tettelaar, K. Towards Multi-Objective Bayesian Global Optimization for Space Missions. Master's Thesis, Leiden Institute of Advanced Computer Science, Leiden, The Netherlands, 2020.
128. Hassanat, A.; Almohammadi, K.; Alkafaween, E.; Abunawas, E.; Hammouri, A.; Prasath, V.S. Choosing mutation and crossover ratios for genetic algorithms—A review with a new dynamic approach. *Information* **2019**, *10*, 390. [CrossRef]

129. Nti, I.K.; Nyarko-Boateng, O.; Aning, J. Performance of machine learning algorithms with different K values in K-fold cross-validation. *Int. J. Inf. Technol. Comput. Sci.* **2021**, *13*, 61–71.
130. Hodson, T.O. Root mean square error (RMSE) or mean absolute error (MAE): When to use them or not. *Geosci. Model Dev. Discuss.* **2022**, *15*, 5481–5487. [CrossRef]
131. Abdul-Muneer, P. Application of microsatellite markers in conservation genetics and fisheries management: Recent advances in population structure analysis and conservation strategies. *Genet. Res. Int.* **2014**, *2014*, 691759. [CrossRef] [PubMed]
132. Zhang, X. Differential Evolution without the Scale Factor and the Crossover Probability. *J. Math.* **2023**, *2023*, 8973912. [CrossRef]
133. Li, Q.; Ma, Z. A hybrid dynamic probability mutation particle swarm optimization for engineering structure design. *Mob. Inf. Syst.* **2021**, *2021*, 1–32. [CrossRef]
134. Angelova, M.; Pencheva, T. Tuning genetic algorithm parameters to improve convergence time. *Int. J. Chem. Eng.* **2011**, *2011*. [CrossRef]
135. Manjula Devi, R.; Kuppuswami, S.; Suganthe, R. Fast linear adaptive skipping training algorithm for training artificial neural network. *Math. Probl. Eng.* **2013**, *2013*. [CrossRef]
136. Johny, A.; Madhusoodanan, K. Dynamic learning rate in deep CNN model for metastasis detection and classification of histopathology images. *Comput. Math. Methods Med.* **2021**, *2021*, 5557168. [CrossRef] [PubMed]
137. Lalwani, S.; Singhal, S.; Kumar, R.; Gupta, N. A comprehensive survey: Applications of multi-objective particle swarm optimization (MOPSO) algorithm. *Trans. Comb.* **2013**, *2*, 39–101.
138. Klincewicz, J.G. Avoiding local optima in the p-hub location problem using tabu search and GRASP. *Ann. Oper. Res.* **1992**, *40*, 283–302. [CrossRef]
139. Lin, Q.; Li, J.; Du, Z.; Chen, J.; Ming, Z. A novel multi-objective particle swarm optimization with multiple search strategies. *Eur. J. Oper. Res.* **2015**, *247*, 732–744. [CrossRef]
140. Qasem, S.; Shamsuddin, S. Generalization improvement of radial basis function network based on multi-objective particle swarm optimization. *J. Artif. Intell* **2010**, *3*, 1–16. [CrossRef]
141. Mannodi-Kanakkithodi, A.; Pilania, G.; Ramprasad, R.; Lookman, T.; Gubernatis, J.E. Multi-objective optimization techniques to design the Pareto front of organic dielectric polymers. *Comput. Mater. Sci.* **2016**, *125*, 92–99. [CrossRef]
142. García-Carrillo, M.; Espinoza-Martínez, A.B.; Ramos-de Valle, L.F.; Sánchez-Valdés, S. Simultaneous optimization of thermal and electrical conductivity of high density polyethylene-carbon particle composites by artificial neural networks and multi-objective genetic algorithm. *Comput. Mater. Sci.* **2022**, *201*, 110956. [CrossRef]
143. Kong, Y.; Mei, Y.; Wang, X.; Ben, Y. Solution selection from a pareto optimal set of multi-objective reservoir operation via clustering operation processes and objective values. *Water* **2021**, *13*, 1046. [CrossRef]
144. Montavon, G.; Binder, A.; Lapuschkin, S.; Samek, W.; Müller, K.R. Layer-wise relevance propagation: An overview. In *Explainable AI: Interpreting, Explaining and Visualizing Deep Learning*; Springer: Cham, Switzerland, 2019; pp. 193–209.
145. Pickering, B.; Choudhary, R. Quantifying resilience in energy systems with out-of-sample testing. *Appl. Energy* **2021**, *285*, 116465. [CrossRef]
146. Talebi, H.; Silani, M.; Bordas, S.P.; Kerfriden, P.; Rabczuk, T. A computational library for multiscale modeling of material failure. *Comput. Mech.* **2014**, *53*, 1047–1071. [CrossRef]
147. Liu, B.; Vu-Bac, N.; Rabczuk, T. A stochastic multiscale method for the prediction of the thermal conductivity of Polymer nanocomposites through hybrid machine learning algorithms. *Compos. Struct.* **2021**, *273*, 114269. [CrossRef]
148. Vu-Bac, N.; Lahmer, T.; Zhang, Y.; Zhuang, X.; Rabczuk, T. Stochastic predictions of interfacial characteristic of polymeric nanocomposites (PNCs). *Compos. Part B Eng.* **2014**, *59*, 80–95. [CrossRef]
149. Bakar, I.A.A.; Kramer, O.; Bordas, S.; Rabczuk, T. Optimization of elastic properties and weaving patterns of woven composites. *Compos. Struct.* **2013**, *100*, 575–591. [CrossRef]
150. Nguyen-Thanh, V.M.; Anitescu, C.; Alajlan, N.; Rabczuk, T.; Zhuang, X. Parametric deep energy approach for elasticity accounting for strain gradient effects. *Comput. Methods Appl. Mech. Eng.* **2021**, *386*, 114096. [CrossRef]
151. Cuomo, S.; Di Cola, V.S.; Giampaolo, F.; Rozza, G.; Raissi, M.; Piccialli, F. Scientific machine learning through physics–informed neural networks: Where we are and what's next. *J. Sci. Comput.* **2022**, *92*, 88. [CrossRef]
152. Dimas, E.; Briassoulis, D. 3D geometric modelling based on NURBS: A review. *Adv. Eng. Softw.* **1999**, *30*, 741–751. [CrossRef]

polymers

MDPI

Article

AI-Aided Crystallization Elution Fractionation (CEF) Assessment of Polyolefin Resins

Lorenzo Brighel [1,2], Gabriella Maria Lucia Scuotto [1], Giuseppe Antinucci [1,2], Roberta Cipullo [1,2] and Vincenzo Busico [1,2,*]

1 Department of Chemical Sciences, Federico II University of Naples, via Cinthia, 80126 Napoli, Italy; l.brighel@studenti.unina.it (L.B.); gab.scuotto@studenti.unina.it (G.M.L.S.); giuseppe.antinucci@unina.it (G.A.); rcipullo@unina.it (R.C.)
2 Dutch Polymer Institute, 5600 AX Eindhoven, The Netherlands
* Correspondence: busico@unina.it

Abstract: Artificial Intelligence (AI) tools and methods are dramatically innovating the application protocols of most polymer characterization techniques. In this paper, we demonstrate that, with the aid of custom-made and properly trained machine learning algorithms, analytical Crystallization Elution Fractionation (aCEF) can be changed from an ancillary to a standalone approach usable to identify and categorize commercially relevant polyolefin materials without any prior information. The proposed protocols are fully operational for monomaterials, whereas for multimaterials, integration with AI-aided ^{13}C NMR is a realistic intermediate step.

Keywords: crystallization elution fractionation; artificial intelligence; machine learning; polyolefin characterization; mechanical recycling

1. Introduction

The spread of Artificial Intelligence (AI) is revolutionizing—inter alia—the analytical approach to polymeric materials. Ultrafast classification and property evaluation methods are essential for applications in high-throughput experimentation and to sort post-consumer mixtures for recycling purposes [1–4]. In the latter domain, rapid automated identifications and separations of the various components in multimaterial waste streams, primarily based on Near-Infrared (NIR) spectroscopic fingerprinting [5–8], are now fully operational, and for several polymer classes (e.g., polyesters and polyamides), the fraction of recycled products is close to that of paper, glass, and some common metals [9,10]. A most notable exception is represented by polyolefins (POs), because the lack of idiosyncratic functional groups in the polymer chains complicates the NIR categorization of the various polyethylene (PE) and polypropylene (PP) grades [1,8]. Unfortunately, these amount to about 50% by weight of the polymer market [11–13], and are very difficult to re-use in mixtures because thermodynamic incompatibility results in phase separation issues [14].

^{13}C NMR is the only spectroscopy potentially able to finely unravel the microstructural details of PO blends [15–18]. Compared with NIR, ^{13}C NMR is a much more demanding technique in terms of Capex and Opex. On the other hand, the introduction of high-temperature cryoprobes has greatly reduced the time for the acquisition of quantitative PO spectra (from several hours down to a few minutes) [16–18]; therefore, a spectrometer equipped with one such probe can process up to several hundred samples per day. In ref. [15], we introduced AI methods capable of automating ^{13}C NMR PO data analysis downstream of spectral acquisition; applications can open the door to higher-value-added solutions for mechanical recycling, because a previous microstructural assessment enables the rational design of compatibilizers' packages [19–22].

The integration of ^{13}C NMR with average molar mass information would represent a further important step forward. To the best of our knowledge, hyphenating NMR spectroscopy with Gel Permeation Chromatography (GPC) has been achieved successfully (albeit not commercially) for ^{1}H NMR, but not yet for ^{13}C NMR [23,24]. Moreover, GPC curves of PO mixtures are poorly resolved and therefore difficult to analyze. A potentially better option is to integrate ^{13}C NMR with analytical Crystallization Elution Fractionation (aCEF) data [25,26]. The aCEF technique was introduced to determine the distribution of crystallizable sequences in a PO sample from its solvent elution profile in a temperature ramp, but modern aCEF equipment can also provide the intrinsic viscosity of the material at each point of the elution curve [25,27]; moreover, the resolution is superior to GPC (Figure 1).

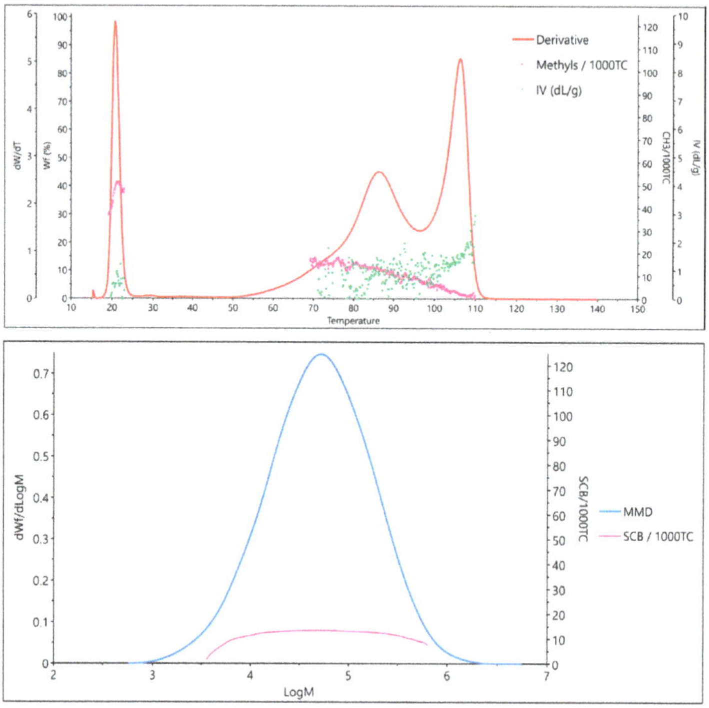

Figure 1. Characterization of a typical LDPE sample by aCEF (**top**) and GPC (**bottom**).

In the present paper, we demonstrate that machine learning methods can be implemented to categorize a PO material *based on aCEF information alone*. This is an important first achievement for the automated application of this technique in the realm of fast polyolefin analytics. The approach readily applies to monomaterials, whereas extension to multimaterials is feasible, but complex, and requires further work, because the contributions from the individual components of a PO mixture to the overall aCEF trace are not necessarily additive due to cocrystallization effects. Along with covariance issues associated with limitations in resolution, this calls for a more sophisticated training of the machine learning algorithm; we will elaborate on this question in the final section.

2. Materials and Methods

2.1. Dataset Description and Pre-Processing

In this study, a diverse set of PO monomaterials was analyzed by aCEF to develop a machine learning pipeline for recognition and classification purposes from the elution profiles. For PO sample classification, we adopted the scheme proposed in Ref. [15] (Supporting Information (SI), Table S1). The monomaterial constraint was essential to accurately establish the baseline analytical behavior and elution characteristics of each PO class without confounding effects due to overlapping or interacting signals from multiple classes.

The data were obtained from the output files generated by a Polymer Char aCEF [28] setup, which, for each given sample, provides the elution profile hyphenated with the point-by-point signals from a FTIR detector (in a limited spectral region transparent to the eluent, and yet usable to determine the ratio between methyl and total carbon atoms), and a dual capillary viscometry detector providing the value of intrinsic viscosity.

Four dedicated Python scripts were implemented to automate the analysis pipeline, each addressing a specific stage of the data workflow. The first script processes raw output files from the Polymer Char aCEF measurement system and reorganizes them into structured Excel spreadsheets tailored for downstream analysis. For each PO (sub-)class, the script extracts key analytical parameters directly from the aCEF traces, including:

- The elution signal profile ("Derivative Norm");
- The ratio between methyl and total carbon atoms;
- The intrinsic viscosity.

The elution signal profile provides a detailed distribution of crystallinity within the sample. The ratio between methyl and total carbon atoms is an essential microstructural marker for quantifying short-chain branching, and can be used to distinguish, e.g., HDPE from LLDPE and iPP class. Moreover, intrinsic viscosity serves as an indirect measure of molecular weight distribution. The combination of these parameters can give information about the composition, polymer microstructure, and crystallinity of the polymer materials that contribute to its identification and categorization.

In addition, the script integrates external metadata files that contain sample identifiers and comonomer composition in case of copolymers; this is critical for determining the specific makeup of each sample (SI, Table S2). To ensure data integrity, a filtering step was implemented to exclude any samples lacking the aforementioned data.

2.2. Data Processing Pipeline

The core of the analysis is an automated, Python-based (ver. 3.11.8) data processing pipeline specifically developed for the comprehensive analysis of elution peaks in aCEF traces of PO samples. This custom script handles the complete analytical workflow, from importing pre-processed elution data to generating structured descriptors of the elution signal that are optimized for statistical interpretation and machine learning tasks. The pipeline is tailored to the unique characteristics of PO aCEF data and integrates advanced signal processing, robust peak detection, and physically meaningful parametric fitting.

The process begins by importing the elution traces previously generated during the initial pre-processing stage. Each trace undergoes a smoothing operation based on spline interpolation [29], which effectively reduces high-frequency noise and baseline fluctuations while preserving the shape, asymmetry, and subtle features of the elution peaks. This step is critical, as accurate identification and modeling of elution features depend on the integrity of these signals. After smoothing, the traces are normalized with respect to their total integral, ensuring comparability across samples by providing a unified intensity scale.

Following normalization, a custom elution peak detection function is applied. This algorithm locates local maxima based on the first derivative of the smoothed signal and enforces dynamic thresholding criteria involving peak prominence and local peak-to-baseline contrast. Only those peaks that satisfy strict significance conditions, indicative of true elution contributions, are retained. This selection step is essential, especially for samples containing overlapping or noisy signals, where artifacts or 'micropeaks' might otherwise contaminate the descriptor space and reduce the reliability of downstream analyses.

For each selected elution peak, a non-linear least-squares fitting procedure is performed using a linear combination of Exponentially Modified Gaussian (EMG) functions. The EMG model is particularly well-suited for describing aCEF peaks, as it captures the asymmetric tailing effects resulting from mass transport and other diffusion related phenomena during sample elution. Each EMG is defined by four parameters (Equation (1)), namely amplitude (A), center (c), standard deviation (σ), and exponential decay constant (τ), allowing for a flexible, yet physically interpretable, reconstruction of the elution peak shape.

$$EMG(x) = Aexp\left(\frac{1}{2}\left(\frac{\sigma}{\tau}\right)^2 - \frac{x-c}{\tau}\right)erfc\left(\frac{1}{\sqrt{2}}\left(\frac{\sigma}{\tau}\right)\right) \quad (1)$$

The fitting process uses bounded minimization routines under constraints that enforce physical realism and avoid overfitting.

Fit accuracy is validated by comparing the area under the combined EMG curve to the corresponding segment of the experimental trace. A fit is accepted if it captures at least 90% of the local area under the curve in the elution peak region. If this criterion is not met, an iterative fitting improvement process is initiated: at each iteration, peak detection is repeated with an increased minimum standard deviation requirement, effectively filtering out noise and allowing for additional EMG components to be added only when justified. This refinement continues until the accuracy threshold is achieved and no significant residuals remain (SI, Figure S1).

Once the fitting process is completed, a set of quantitative descriptors is extracted for each EMG component in the elution trace. These include shape parameters, peak area, and structural features, such as methyl per thousand carbon atoms and average intrinsic viscosity measured at the peak location. The resulting descriptors are saved on an Excel file in a structured format, mapped to the original sample ID and trace filename. To improve sample classification, statistical averaging of descriptors is performed across all detected peaks, with each peak weighted by its intensity. This yields a representative elution fingerprint for each sample type. The feature matrix serves as the foundation for further analyses, including dimensionality reduction, clustering, and supervised classification.

To support transparency and quality assurance, the pipeline also generates diagnostic visualizations for each sample. These plots overlay the raw elution trace, the smoothed baseline, and the fitted elution EMG components, providing a visual summary of the decomposition and fit quality. Such outputs are saved automatically for every sample, facilitating manual review and the identification of potential outliers or fitting anomalies. The entire pipeline is designed for scalability, enabling batch processing of large sample sets, and is easily extensible to incorporate new polymer classes, updated analytical parameters, or evolving experimental protocols (SI, Figure S2 and Table S3).

2.3. Data Augmentation and Visualization

To overcome the significant class imbalance in the original dataset, primarily due to the uneven distribution of samples across different PO (sub-)classes, a targeted oversampling strategy was implemented. This step was critical to prevent statistical and machine learning models from being biased toward the most represented (sub-)classes, thereby improving the robustness, generalizability, and predictive accuracy of the classification tasks. The core of the strategy relied on data augmentation techniques designed specifically for the elution peak domain, ensuring the generation of realistic and chemically meaningful synthetic examples that could enrich the feature space without introducing noise or distortion.

Data augmentation was performed using the SMOTE (Synthetic Minority Oversampling Technique) algorithm [30] via a custom Python-based script that operates directly on the parametric descriptors of the EMG functions (Equation (1)) previously obtained from the elution peak fitting routine. For each PO (sub-)class, synthetic elution traces were created by perturbing the EMG parameters, namely amplitude, center, width, and decay constant, within empirically derived bounds. These ranges were carefully selected based on observed intra-class variability, ensuring that the generated traces remained faithful to the physical and chemical characteristics typical of each PO (sub-)class. The aim was to simulate natural variability found within the same (sub-)class while preserving class-specific fingerprint features related to elution behavior.

Each synthetic trace was reconstructed as a linear combination of modified EMG components and normalized in the same manner as the original traces, preserving comparability across the dataset. Once generated, statistical averaging was again performed on the descriptors of the synthetic peaks to derive class-specific fingerprints, allowing these new data points to be seamlessly integrated into the analytical pipeline. The number of synthetic instances per class was selected to approximately equalize the sample count across all PO (sub-)classes in Table S1, producing a more balanced dataset suited for training supervised learning models.

In parallel, a set of synthetic "negative" traces was also generated to enrich the model with representative counterexamples. Rather than simulating only baseline or noise, these samples were created by randomly sampling across the entire multidimensional feature space defined by the extracted EMG parameters. As such, they span a wide range of plausible values, including regions that may overlap with polymer-containing signals. This approach ensures that the model learns to identify polymer-specific patterns robustly, rather than simply detecting deviations from a baseline, thereby reducing the risk of overfitting and improving generalization to ambiguous or borderline cases.

To further guide feature selection for classification, the augmented dataset was explored through a series of diagnostic scatterplots, which mapped key EMG-averaged parameters against the elution peak position (SI, Figure S3). Parameters such as total peak area, total amplitude, average variance, methyl per thousand carbons, and intrinsic viscosity at the elution peak were plotted to identify separable regions in the feature space. These visual tools provided insight into how distinct polymer (sub-)classes distribute across the elution profile and helped to prioritize the most discriminative descriptors for machine learning. As a complement, Figure S4 reports the same parameter scatterplots for the synthetic negative samples, randomly generated across the feature space.

Overall, the augmentation framework significantly improved the coverage and balance of the dataset. By simulating intra-class variability and introducing chemically plausible synthetic data, while maintaining focus on the elution peak features, the approach enabled a more comprehensive representation of PO (sub-)classes and enhanced the performance and resilience of downstream predictive models.

2.4. Machine Learning Models: Generalities

The final step of the workflow involves the application of supervised machine learning models for the recognition and classification of PO monomaterials based on their aCEF profiles and associated analytical parameters. This step was designed to address three distinct prediction tasks, each of which builds upon the previous one: binary classification of PO presence, multiclass classification of PO identity, and regression-based estimation of comonomer composition in case of copolymers. Each model was trained on the augmented dataset created during the earlier stages of data processing, which included both real and synthetic samples. This dataset was constructed from a uniform vector of quantitative descriptors derived from signal fitting, averaging, and metadata extraction.

All machine learning models share a common input representation consisting of analytical descriptors extracted from the aCEF signal and the corresponding EMG fitting parameters. These features include:

- Total normalized area of all EMG peaks;
- Average EMG peaks position;
- Total peak amplitude summed across all EMGs;
- Average variance of the EMGs;
- Average methyl per 1000 carbons value at the peak positions;
- Average intrinsic viscosity at the peak positions.

Each sample, whether experimental or synthetic, was represented as a single vector containing these six features, ensuring full compatibility with standard classifiers and regressors available in scikit-learn. Additionally, each sample was pre-labeled with the corresponding target outputs required for the specific prediction task. For classification tasks, the target labels were either binary or multilabel, depending on the task type. In the case of copolymers, for regression tasks, the target was comonomer content (in mol%), computed from sample metadata or assigned during synthetic generation. The final dataset was split into training (80%) and testing (20%) subsets using a fixed random seed to ensure reproducibility. Although stratified sampling was not explicitly applied, the dataset had been pre-balanced through data augmentation, resulting in approximately uniform class distributions in both subsets. Class proportions were verified post-split to confirm the absence of significant imbalances.

The dataset used for training and testing the machine learning models consisted of a total of 142 samples per polyolefin (PO) class, combining both real and synthetically augmented data. Additionally, a set of 142 synthetic "no-polymer" samples was created to simulate baseline conditions, bringing the overall dataset size to 1278 samples. The choice of 142 samples per class was guided by practical and statistical considerations for Random Forest classifiers [31]. According to empirical guidelines in machine learning literature, although not explicitly stated as a rule, a sample size of approximately 20–30 times the number of features per class is generally considered a reasonable lower bound for stable model performance. Given the limited number of input features in the present work, a total of over 1200 polymer-positive samples provides a sufficiently large and diverse dataset to avoid overfitting, ensure meaningful model training, and support generalization.

2.5. *Binary, Classification, and Regression Models*

The first predictive model addresses the binary task of determining whether a given sample is a true polymeric material. This classifier was trained on a dataset composed of both real polymer samples and synthetic negative controls, which were generated during the previous step using uniformly sampled baseline parameters with Gaussian noise. The model was implemented using a Random Forest classifier [31,32], with hyperparameters optimized via grid search cross-validation. The output of the model is a probability score ranging from 0 to 1, indicating the likelihood of polymer presence. To facilitate practical use, thresholds were defined to categorize samples into two binary categories:

- "No polymer detected" (score < 0.7);
- "Polymer detected" (score ≥ 0.7).

These cutoffs were determined empirically by analyzing the precision–recall curve, optimizing the balance between sensitivity and specificity to minimize false positives and false negatives (SI, Figure S5).

For samples classified as polymer-positive (score $\geq$ 0.7 in the binary classification), a second predictive model was implemented to identify the specific polymer type(s) present. Although this model was initially applied to the classification of PO monomaterials, it was designed with future applications involving PO mixtures in mind. Therefore, in light of the possibility of multiple PO components in a given multimaterial sample (e.g., iPP + EPR, or LDPE + LLDPE), this task was framed as a multilabel classification problem. The model was built using a MultiOutputClassifier [33] wrapper around a RandomForestClassifier, which enabled independent binary classifiers to be trained for each polymer class present in the dataset. Each classifier outputs the probability of presence for its respective polymer, allowing the model to predict multiple coexisting polymer classes without assuming mutual exclusivity. To interpret these probabilities, the following thresholds were applied:

- Polymer not present (score < 0.3);
- Polymer present with low confidence (score between 0.3 and 0.5);
- Polymer present with moderate confidence (score between 0.5 and 0.8);
- Polymer present with high confidence (score > 0.8)

The polymer classification results are stored in a structured Excel file, where each row corresponds to a real or synthetic sample. For each sample, the file includes the outcome of the binary classification, the actual polymer class, and the model's interpretation, which lists all detected polymer classes with probabilities above a 0.3 threshold. It also reports the most likely polymer identified by the model, its associated confidence level, and the probability estimates for the presence of each polymer in the material.

The adoption of a two-stage approach (first detecting the presence of a polymer, then classifying its type) was aimed at improving both performance and interpretability. The presence detection step treats the "no polymer" condition as a distinct binary task, reducing false positives and preventing forced classifications when no meaningful signal is detected. Separating the tasks allows for each model to specialize, facilitates threshold optimization, and offers greater modularity and transparency for future refinements.

The third and final model was designed to estimate the composition of samples classified as copolymers, namely LLDPE, raco-PP, and EPR. This task was approached using regression modeling with a RandomForestRegressor [31]. As with the previous models, the same set of six analytical descriptors served as the input features. The regressor was trained on both real and synthetic data, with the target output being the known comonomer content in mol%, in the ranges of commercial relevance indicated in Table S1. The predictions were saved alongside the classifier outputs, and for each sample, both the real and predicted copolymer composition were added to the structured Excel file (SI, Table S4). This allows for users to cross-reference the qualitative presence/absence information with the quantitative composition estimates.

Specifically, the regression model was applied exclusively to samples that had been correctly classified in terms of copolymer (sub-)class, ensuring that composition estimates were only derived from confidently identified materials.

Random Forest (RF) models were selected for both binary/classification and regression tasks due to their robustness, ease of use, and strong performance on structured datasets such as the one used in this study. RF classifiers are particularly well-suited for scenarios involving a limited number of features and relatively small sample sizes, as in our case. They effectively capture nonlinear patterns, are robust to noise, and require minimal hyperparameter tuning [31].

Importantly, Random Forests are straightforward to interpret and analyze, for example, through feature importance scores, and are easy to integrate into modular and scalable pipelines. This was especially beneficial in our two-stage architecture, where separate models were used for presence detection and polymer classification. For regression, Random Forests can capture complex, nonlinear relationships between spectral features and polymer composition without relying on rigid assumptions about the underlying data distribution.

3. Results and Discussion

3.1. Binary and Multiclass Classification—Polymer Presence and Type

The performance of the developed machine learning pipeline, which integrates a binary classifier for polymer presence detection and a multi-label classifier for polymer-type (PO (sub-)class) identification, is summarized in the confusion matrix reported in Figure 2. For the binary classification task (polymer presence vs. absence), the model was evaluated using an identification threshold of 0.7 on the output probabilities. Under this condition, the classifier demonstrated perfect performance on the test set, achieving 100% precision and 100% recall. All 32 "No Polymer" samples were correctly classified, with no instances of false positives or false negatives. This level of accuracy confirms the robustness of the model in discriminating samples containing polymers from empty or irrelevant signals, which is a crucial step in automating the screening of post-consumer or mixed-material samples.

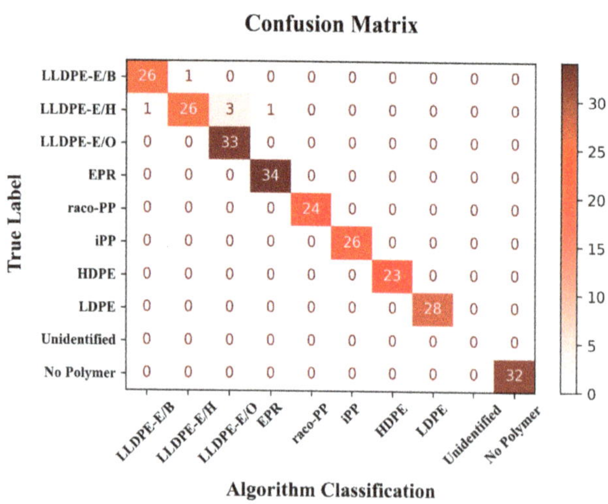

Figure 2. Confusion matrix displaying the performance of the binary presence/absence and polymer-type classification models.

The polymer-type classification model, in turn, demonstrated excellent performance across the main polymer classes included in the dataset. All samples belonging to the HDPE, LDPE, iPP, raco-PP, and EPR classes were correctly classified (with only one misassignment (false positive) in the EPR case. Some minor limitations were observed in the classification of LLDPE samples, most notably within the LLDPE-E/H sub-class, due to an inherent variability; as a matter of fact, the fingerprint parameter scatterplots in Figure S5 displayed substantial dispersion across nearly all descriptors for reasons that still need to be clarified.

Notably, the system yielded no "Unidentified" outputs: in all cases where the binary model identified the presence of at least one polymer component, the classification model was able to provide a corresponding polymer identity. This outcome confirms the internal consistency between the two stages of the model and suggests that the classification model is well-calibrated with respect to the presence predictions.

To further assess model performance, classification results were grouped according to the confidence level associated with each prediction, based on the maximum class probability returned by the classifier. Table 1 summarizes the number of samples and the corresponding classification accuracy across four confidence intervals.

Table 1. Summary of classification results based on confidence levels, with the number of samples and corresponding accuracy reported for each category.

Category	Samples	Accuracy (%)
Unidentified	0	0
No Polymer	32	100.0
Low Confidence	6	66.7
Moderate Confidence	22	81.8
High Confidence	198	100.0

As shown in Table 1, the majority of classified samples (198 out of 258, approximately 77%) fell within the high-confidence category, where the accuracy reached 100%. Only a limited number of cases (approximately 11%) fell within the low-confidence or moderate-confidence categories, where misclassifications were more likely to occur.

These results demonstrate that confidence filtering may serve as a practical and reliable post-processing step, allowing analysts to tailor the balance between sensitivity and precision depending on the application. Specifically, filtering based on prediction confidence offers a pathway to mitigate uncertainties associated with certain ethylene-rich polymers, without discarding correctly classified high-confidence predictions. This approach is especially valuable for screening large datasets or automated pipelines, where interpretability and trust in the output are critical.

3.2. *Regression—Composition Estimation*

The performance of the regression model is illustrated in Figure 3, which displays the predicted versus actual comonomer content for all correctly classified LLDPE-E/B, LLDPE E/H, LLDPE-E/O, raco-PP and EPR samples. To facilitate interpretation, the plot includes relative error bands at ±5% and ±10%, which serve as visual benchmarks for assessing prediction accuracy.

The distribution of points within these bands reveals the strong agreement between the predicted and actual values across most PO samples, supporting the overall reliability of the model's quantitative estimates. The plot is particularly useful in highlighting both the tight clustering of data points around the parity line and the rare deviations from expected values. One notable exception is an outlier within the LLDPE-E/H class, which falls significantly outside the confidence intervals and contributes disproportionately to the lower overall performance of the model for such PO sub-classes.

A more detailed breakdown of the regression performance is provided in Table 2, which reports the coefficient of determination (R^2) for each polymer class, stratified by the confidence level of the classification.

As the table shows, the regression model performed exceptionally well across most PO (sub-)classes, with R^2 values approaching or equal to 1.00 for LLDPE-E/B, LLDPE-E/O, Raco-PP, and EPR. The model for LLDPE-E/O is particularly noteworthy for its robustness: it maintained excellent accuracy across all confidence levels, including moderate and low-confidence classifications. The lower overall R^2 observed for the LLDPE-E/H class

(0.52) is primarily due to one single outlier sample with a substantial prediction error, also observed in the scatterplot. Importantly, the model achieves a perfect fit (R^2 = 1.00) for high-confidence LLDPE-E/H samples, indicating that the drop in total performance is not systematic, but rather the result of isolated variability.

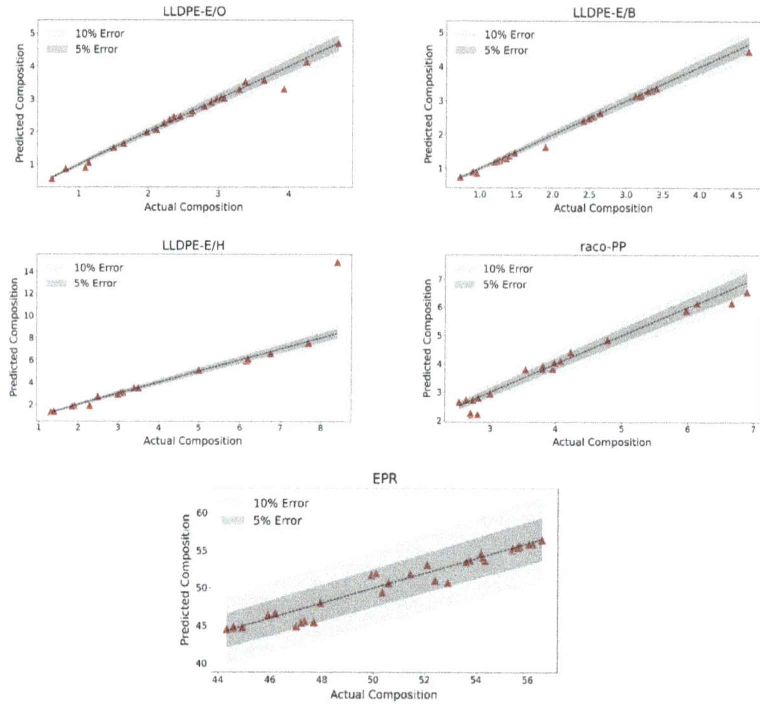

Figure 3. Scatterplot showing the predicted versus actual comonomer content for all correctly classified copolymer samples. The plot includes 5% and 10% relative error bands, indicating the accuracy of the regression model.

Table 2. Coefficient of determination (R^2) for each polymer class at different levels of classification confidence for the validation set.

Polymer	R^2 (Total)	R^2 (High)	R^2 (Moderate)	R^2 (Low)
EPR	0.94	0.94	-	-
LLDPE-E/O	0.98	0.99	1.00	0.97
LLDPE-E/B	1.00	1.00	-	-
LLDPE-E/H	0.52	1.00	−0.15	-
raco-PP	0.97	0.97	-	-

To evaluate the consistency and generalizability of the regression model, the coefficient of determination (R^2) was also calculated on the training set. The R^2 values obtained for each polymer class in the training set, particularly for predictions associated with high confidence levels, were comparable to those observed in the test set. This consistency confirms the model's stability and suggests minimal overfitting. Detailed R^2 values by polymer class and confidence level in the training set are reported in Table 3.

These results demonstrate that, when classification is reliable, the regression model can accurately estimate the intrinsic composition of polyolefins. This capability is crucial for downstream tasks such as compositional fingerprinting, quality control, and process optimization in both research and industrial recycling workflows.

Table 3. Coefficient of determination (R^2) for each polymer class at different levels of classification confidence for the training set.

Polymer	R^2 (Total)	R^2 (High)	R^2 (Moderate)	R^2 (Low)
EPR	0.99	0.99	-	-
LLDPE-E/O	0.99	0.99	0.98	-
LLDPE-E/B	0.62	0.99	−0.26	-
LLDPE-E/H	0.85	0.92	0.14	-
raco-PP	0.97	0.98	0.82	-

In order to better understand the internal decision process of the Random Forest regression model, a feature importance analysis was carried out based on the importance scores provided by the trained estimator.

As shown in Figure 4, the most influential feature in predicting comonomer composition was the average methyl per 1000 carbons value at the peak positions, which accounted for nearly 50% of the model's decision weight. This was followed by the total peak amplitude and average elution peak position, highlighting the importance of chemical structure and elution profile in determining copolymer microstructure. Features related to intrinsic viscosity and elution peak variance were also relevant, though with lower contributions.

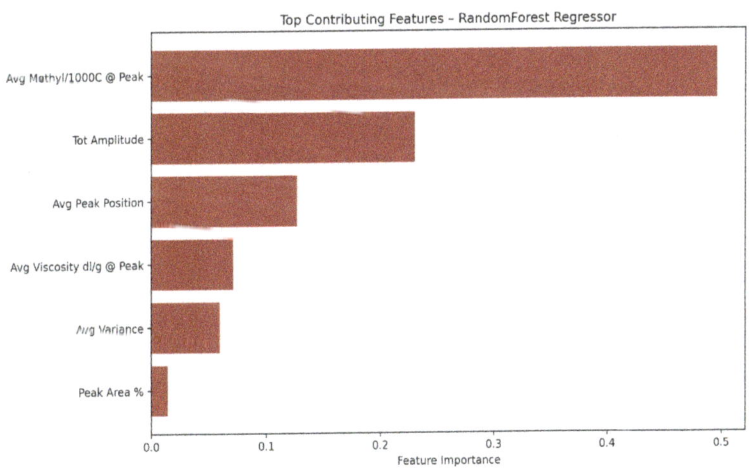

Figure 4. Relative importance of the six input features used in the Random Forest regression model for predicting comonomer content in PO copolymers.

4. Conclusions

Analytical Crystallization Elution Fractionation (aCEF) is an important technique for the characterization of PO materials. Originally introduced as a faster and more efficient version of Temperature Rising Elution Fractionation (TREF) [25,26], it has progressively gained further applications thanks to the smart utilization of FTIR and dual-capillary viscometry detectors. As a matter of fact, in addition to the distribution of crystallizable sequences in any PO samples, it can also provide semiquantitative information on average molar masses via intrinsic viscosity measurements and, with proper calibrations, on comonomer contents for copolymer samples of known nature.

In this work, we demonstrated that, with the aid of AI methods, it is possible to move further important steps forward. In fact, custom-made machine learning algorithms can be trained on aCEF traces to identify and categorize practically all commercially relevant PO monomaterials *without any prior information*, with average molar masses and comonomer contents for copolymers added as a bonus.

Extension to PO mixtures, including multimaterial waste streams, is a most desirable—albeit most complex too—ultimate objective. In our opinion, integrating automated AI-aided aCEF and ^{13}C NMR characterizations is a realistic intermediate step that can already enable higher-value-added solutions in the mechanical recycling of post-consumer PO blends.

Supplementary Materials: The following supporting information can be downloaded at: https://www.mdpi.com/article/10.3390/polym17121597/s1, Table S1. Proposed microstructural categorization of commercial PO monomaterials; Table S2. Structure of the Excel sheet corresponding to a single LDPE sample (as a representative example). The data structure includes the sample ID, polymer class, composition data (for copolymers) retrieved from external metadata, the CEF elution signal, the point-by-point FTIR-calculated value of methyl carbons per 1000 carbons, and the point-by-point value of intrinsic viscosity; Figure S1. Illustration of the iterative EMG fitting process applied to the elution region of a iPP sample (for exemplification purposes); Figure S2. Example of EMG-based elution peak fitting for multiple raco-PP samples (for exemplification purposes). Each plot displays the original CEF trace (black), the detected elution peaks (cross markers), the individual EMG components (colored dashed lines), and the resulting composite fit (solid red line); Table S3. Overview of the Excel file summarizing the extracted EMG fit descriptors for raco-PP samples (for exemplification purposes). Each row represents a sample and includes its aggregated statistical descriptors derived from the fitted crystalline components; Figure S3. Scatterplots of key EMG-derived descriptors across the augmented dataset (excluding synthetic negative samples). Each point represents a PO sample, with colors indicating PO (sub-)classes. Parameters such as total elution peak area, cumulative amplitude, average variance, methyl per thousand carbons, and intrinsic viscosity are plotted against the peak center position; Figure S4. Scatterplots of key parameters for synthetic "No Polymer" samples, randomly distributed across the multidimensional parameter space; Figure S5. Precision–Recall curve for the classification model. Precision represents the percentage of true positives among all instances classified as positive, while recall indicates the percentage of true positives which were classified as positives. A higher precision means lower numbers of false positives, while higher recall means a lower number of false negatives; Table S4. Structure of the Excel file containing the results of the machine learning model for polymer classification and composition estimation. The presence probabilities for each polymer, not reported in the figure, are also included in the file for a detailed interpretation of the results.

Author Contributions: Conceptualization, R.C. and V.B.; methodology, L.B. and G.M.L.S.; software, L.B. and G.M.L.S.; validation, L.B., G.A. and R.C.; formal analysis, L.B., G.A., R.C. and V.B.; investigation, L.B. and G.M.L.S.; data curation, G.A. and R.C.; writing—original draft preparation, L.B. and V.B.; writing—review and editing, L.B., G.M.L.S., G.A., R.C. and V.B.; funding acquisition, V.B. All authors have read and agreed to the published version of the manuscript.

Funding: This research was financially supported by DPI under project #863. R.C. and V.B. acknowledge financial support under the National Recovery and Resilience Plan (NRRP), Mission 4, Component 2, Investment 1.1, Call for tender No. 1409 published on 14 September 2022 by the Italian Ministry of University and Research (MUR), funded by the European Union—NextGenerationEU–Project—CUP: E53D23016000001 Grant Assignment Decree No. 1187 adopted on 27 July 2023 by the Italian Ministry of Ministry of University and Research (MUR).

Institutional Review Board Statement: Not applicable.

Data Availability Statement: The original contributions presented in this study are included in the article and/or Supporting Information. Further inquiries can be directed to the corresponding authors.

Acknowledgments: This work forms part of the research program of DPI, project #863.

Conflicts of Interest: The authors declare no conflicts of interest.

References

1. Lubongo, C.; Alexandridis, P. Assessment of Performance and Challenges in Use of Commercial Automated Sorting Technology for Plastic Waste. *Recycling* **2022**, *7*, 11. [CrossRef]
2. Wilts, H.; Garcia, B.R.; Garlito, R.G.; Gómez, L.S.; Prieto, E.G. Artificial Intelligence in the Sorting of Municipal Waste as an Enabler of the Circular Economy. *Resources* **2021**, *10*, 28. [CrossRef]
3. Bernat, K. Post-Consumer Plastic Waste Management: From Collection and Sortation to Mechanical Recycling. *Energies* **2023**, *16*, 3504. [CrossRef]
4. Lubongo, C.; Bin Daej, M.A.A.; Alexandridis, P. Recent Developments in Technology for Sorting Plastic for Recycling: The Emergence of Artificial Intelligence and the Rise of the Robots. *Recycling* **2024**, *9*, 59. [CrossRef]
5. Huth-Fehre, T.; Feldhoff, R.; Kantimm, T.; Quick, L.; Winter, F.; Cammann, K.; van den Broek, W.; Wienke, D.; Melssen, W.; Buydens, L. NIR—Remote Sensing and Artificial Neural Networks for Rapid Identification of Post Consumer Plastics. *J. Mol. Struct.* **1995**, *348*, 143–146. [CrossRef]
6. van den Broek, W.H.A.; Wienke, D.; Melssen, W.J.; Buydens, L.M.C. Plastic Material Identification with Spectroscopic near Infrared Imaging and Artificial Neural Networks. *Anal. Chim. Acta* **1998**, *361*, 161–176. [CrossRef]
7. Signoret, C.; Caro-Bretelle, A.-S.; Lopez-Cuesta, J.-M.; Ienny, P.; Perrin, D. Alterations of Plastics Spectra in MIR and the Potential Impacts on Identification towards Recycling. *Resour. Conserv. Recycl.* **2020**, *161*, 104980. [CrossRef]
8. Neo, E.R.K.; Yeo, Z.; Low, J.S.C.; Goodship, V.; Debattista, K. A Review on Chemometric Techniques with Infrared, Raman and Laser-Induced Breakdown Spectroscopy for Sorting Plastic Waste in the Recycling Industry. *Resour. Conserv. Recycl.* **2022**, *180*, 106217. [CrossRef]
9. Haque, F.M.; Ishibashi, J.S.A.; Lidston, C.A.L.; Shao, H.; Bates, F.S.; Chang, A.B.; Coates, G.W.; Cramer, C.J.; Dauenhauer, P.J.; Dichtel, W.R.; et al. Defining the Macromolecules of Tomorrow through Synergistic Sustainable Polymer Research. *Chem. Rev.* **2022**, *122*, 6322–6373. [CrossRef]
10. Jubinville, D.; Esmizadeh, E.; Saikrishnan, S.; Tzoganakis, C.; Mekonnen, T. A Comprehensive Review of Global Production and Recycling Methods of Polyolefin (PO) Based Products and Their Post-Recycling Applications. *Sustain. Mater. Technol.* **2020**, *25*, e00188. [CrossRef]
11. IEA. *The Future of Petrochemicals: Towards More Sustainable Plastics and Fertilisers*; OECD: Paris, France, 2018.
12. Statista. Polyethylene Market Volume Worldwide 2015–2029. Available online: https://www.statista.com/statistics/1245162/polyethylene-market-volumeworldwide/ (accessed on 28 June 2024).
13. Statista. Polypropylene Market Volume 2015–2029. Available online: https://www.statista.com/statistics/1245169/polypropylene-market-volumeworldwide/ (accessed on 28 June 2024).
14. Vogt, B.D.; Stokes, K.K.; Kumar, S.K. Why Is Recycling of Postconsumer Plastics so Challenging? *ACS Appl. Polym. Mater.* **2021**, *3*, 4325–4346. [CrossRef]
15. Giampaolo, F.; Cipullo, R.; Cuomo, S.; Piccialli, F.; Busico, V. Automated Ultra-Fast ^{13}C NMR Analysis of Polyolefin Materials. *Anal. Chem.* **2025**, *97*, 2503–2510. [CrossRef] [PubMed]
16. Zhou, Z.; Kümmerle, R.; Stevens, J.C.; Redwine, D.; He, Y.; Qiu, X.; Cong, R.; Klosin, J.; Montañez, N.; Roof, G. ^{13}C NMR of Polyolefins with a New High Temperature 10 mm Cryoprobe. *J. Magn. Reson.* **2009**, *200*, 328–333. [CrossRef]
17. Zhou, Z.; Kuemmerle, R.; Rau, N.; Eldred, D.; Moreno, A.; Czarniecki, B.; Qiu, X.; Cong, R.; Gies, A.P.; Fan, L.; et al. Polyolefin Analyses with a 10 mm Multinuclear NMR Cryoprobe. *Anal. Chem.* **2020**, *92*, 15596–15603. [CrossRef] [PubMed]
18. Antinucci, G.; Vittoria, A.; Cipullo, R.; Busico, V. Regioirregular Monomeric Units in Ziegler–Natta Polypropylene: A Sensitive Probe of the Catalytic Sites. *Macromolecules* **2020**, *53*, 3789–3795. [CrossRef]
19. Self, J.L.; Zervoudakis, A.J.; Peng, X.; Lenart, W.R.; Macosko, C.W.; Ellison, C.J. Linear, Graft, and Beyond: Multiblock Copolymers as Next-Generation Compatibilizers. *JACS Au* **2022**, *2*, 310–321. [CrossRef]
20. Bovey, F. *Chain Structure and Conformation of Macromolecules*; Elsevier: Amsterdam, The Netherlands, 2012.
21. Busico, V.; Cipullo, R. Microstructure of Polypropylene. *Prog. Polym. Sci.* **2001**, *26*, 443–533. [CrossRef]
22. Busico, V.; Cipullo, R.; Segre, A.L. Advances in the ^{13}C NMR Characterization of Ethene/Propene Copolymers, 1. *Macromol. Chem. Phys.* **2002**, *203*, 1403–1412. [CrossRef]
23. Hiller, W.; Pasch, H.; Macko, T.; Hofmann, M.; Ganz, J.; Spraul, M.; Braumann, U.; Streck, R.; Mason, J.; Van Damme, F. On-Line Coupling of High Temperature GPC and ^{1}H NMR for the Analysis of Polymers. *J. Magn. Reson.* **2006**, *183*, 290–302. [CrossRef]
24. Cudaj, M.; Guthausen, G.; Hofe, T.; Wilhelm, M. Online Coupling of Size-Exclusion Chromatography and Low-Field ^{1}H NMR Spectroscopy. *Macromol. Chem. Phys.* **2012**, *213*, 1933–1943. [CrossRef]

25. Monrabal, B.; Romero, L.; Mayo, N.; Sancho-Tello, J. Advances in Crystallization Elution Fractionation. *Macromol. Symp.* **2009**, *282*, 14–24. [CrossRef]
26. Monrabal, B.; Romero, L. Separation of Polypropylene Polymers by Crystallization and Adsorption Techniques. *Macromol. Chem. Phys.* **2014**, *215*, 1818–1828. [CrossRef]
27. Antinucci, G.; Pucciarelli, A.; Vittoria, A.; Zaccaria, F.; Urciuoli, G.; Ehm, C.; Cannavacciuolo, F.D.; Cipullo, R.; Busico, V. Fast Analytics of High-Impact Polypropylene (HIPP). *ACS Appl. Polym. Mater.* **2023**, *5*, 3894–3897. [CrossRef]
28. Polymer Char, S.L. CEF Datasheet: Crystallization Elution Fractionation. Available online: https://polymerchar.com/products/analytical-instruments/cef (accessed on 5 June 2025).
29. De Boor, C. *A Practical Guide to Splines*; Springer: Berlin/Heidelberg, Germany, 1978.
30. Chawla, N.V.; Bowyer, K.W.; Hall, L.O.; Kegelmeyer, W.P. SMOTE: Synthetic Minority over-Sampling Technique. *J. Artif. Intell. Res.* **2002**, *16*, 321–357. [CrossRef]
31. Breiman, L. Random Forests. *Mach. Learn.* **2001**, *45*, 5–32. [CrossRef]
32. Pedregosa, F.; Varoquaux, G.; Gramfort, A.; Michel, V.; Thirion, B.; Grisel, O.; Blondel, M.; Prettenhofer, P.; Weiss, R.; Dubourg, V. Scikit-Learn: Machine Learning in Python. *J. Mach. Learn. Res.* **2011**, *12*, 2825–2830.
33. Tsoumakas, G.; Katakis, I. Multi-Label Classification: An Overview. In *Data Warehousing and Mining: Concepts, Methodologies, Tools, and Applications*; IGI Global Scientific Publishing: Hershey, PA, USA, 2008; Volume 3, pp. 64–74.

polymers

MDPI

Review

Boosting-Based Machine Learning Applications in Polymer Science: A Review

Ivan Malashin [1,*], Vadim Tynchenko [1,*], Andrei Gantimurov [1], Vladimir Nelyub [1,2] and Aleksei Borodulin [1]

1 Artificial Intelligence Technology Scientific and Education Center, Bauman Moscow State Technical University, 105005 Moscow, Russia
2 Scientific Department, Far Eastern Federal University, 690922 Vladivostok, Russia
* Correspondence: ivan.p.malashin@gmail.com (I.M.); vadimond@mail.ru (V.T.); Tel.: +7-926-875-7128 (I.M.)

Abstract: The increasing complexity of polymer systems in both experimental and computational studies has led to an expanding interest in machine learning (ML) methods to aid in data analysis, material design, and predictive modeling. Among the various ML approaches, boosting methods, including AdaBoost, Gradient Boosting, XGBoost, CatBoost and LightGBM, have emerged as powerful tools for tackling high-dimensional and complex problems in polymer science. This paper provides an overview of the applications of boosting methods in polymer science, highlighting their contributions to areas such as structure–property relationships, polymer synthesis, performance prediction, and material characterization. By examining recent case studies on the applications of boosting techniques in polymer science, this review aims to highlight their potential for advancing the design, characterization, and optimization of polymer materials.

Keywords: machine learning; boosting methods; AdaBoost; Gradient Boosting; XGBoost; CatBoost; LightGBM; polymer science

1. Introduction

Polymeric materials are widely used across various industries, including medicine [1–3], automotive engineering [4–6], packaging [7,8], and electronics [9–11], due to their diverse properties and adaptability. Understanding and optimizing polymer properties, as well as their manufacturing processes, is a key area of research. However, the inherent complexity of polymers, stemming from their structure–property relationships and diverse formulations, poses significant challenges for accurate modeling and prediction.

Machine learning (ML) techniques have emerged as valuable tools for addressing such challenges, with boosting algorithms being particularly effective. Methods like AdaBoost [12,13], Gradient Boosting [14,15], LightGBM [16,17], CatBoost [18,19], and XGBoost [20,21] have been used to analyze large and complex datasets, offering robust predictive capabilities and the ability to model non-linear relationships. In the context of polymer science, these algorithms have been applied to tasks such as predicting material properties, optimizing processing parameters, and designing polymer formulations.

Figure 1 shows the publication trends for boosting methods in polymer sciense —GradientBoosting, AdaBoost, CatBoost, LightGBM, and XGBoost—show shifts in popularity from 2018 to 2024. GradientBoosting has seen steady growth, from two papers in 2018 to 97 in 2024, reflecting its increasing use in polymer and other fields. AdaBoost, starting with one paper in 2018, had minimal growth, with a slight rise in 2021 and 2022, indicating limited application in polymer research. CatBoost has gained traction, especially in 2022 (5 papers) and 2023 (11 papers), due to its effectiveness with categorical data. LightGBM shows gradual growth, with a rise from zero papers in 2018–2020 to 10 in 2024, likely due to its scalability with large datasets. XGBoost grew steadily, from one paper in 2018 to 72 in 2024, driven by its versatility and strong predictive performance. In summary, GradientBoosting and XGBoost dominate, with increasing adoption, while AdaBoost and LightGBM show slower growth, and CatBoost is gaining popularity. These trends highlight the growing use of boosting methods in polymer research to improve prediction accuracy and handle complex data relationships.

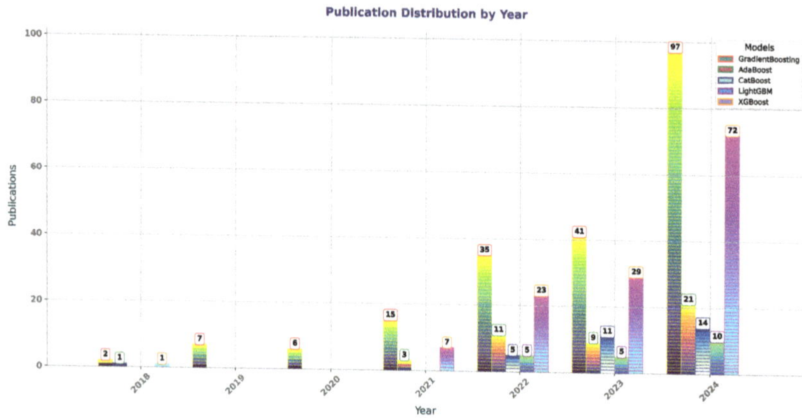

Figure 1. Year-wise distribution of publications featuring 'AdaBoost', 'Gradient Boosting', 'XGBoost', 'CatBoost', 'LightGBM', and 'polymers' in title or abstract.

Figure 2 shows the map with distribution of publications utilizing boosting methods in polymer research across the world. According to the provided data, countries like China (184 publications), India (64 publications), and Iran (35 publications) lead in the volume of research. Other contributors include Australia (33 publications), Canada (25 publications), and Pakistan (33 publications). Smaller but notable contributions come from countries such as the United Kingdom, France, and Turkey, each ranging between 10 to 16 publications. This distribution reflects a global interest in leveraging boosting methods for polymers, with a concentration of research efforts in leading industrial and academic hubs.

The Keywords Occurrence Map (Figure 3) highlights the integration of boosting methods, into polymer research. Techniques like gradient boosting, CatBoost, AdaBoost, and XGBoost are widely applied to predict and optimize material properties such as CS, bond strength, and mechanical performance. These methods excel in capturing complex, non-linear data relationships, enhancing accuracy and reliability. Applications include sustainability-focused studies (e.g., geopolymers, fly ash, and asphalt binders), advanced manufacturing (e.g., additive manufacturing and 3D-printing), and structural materials (e.g., FRP and ultra-high-performance concrete). Boosting models are also leveraged alongside interpretability tools like SHAP analysis and sensitivity analysis to improve understanding of key factors in photodegradation, corrosion, and microplastics.

This review aims to analyze existing research on the application of boosting methods in polymer-related tasks. The focus is on several key areas, including the prediction of mechanical, thermal, and chemical properties of polymers, the optimization of manufacturing processes and polymer blend compositions, and the integration of boosting methods with other ML approaches to improve model accuracy and interpretability.

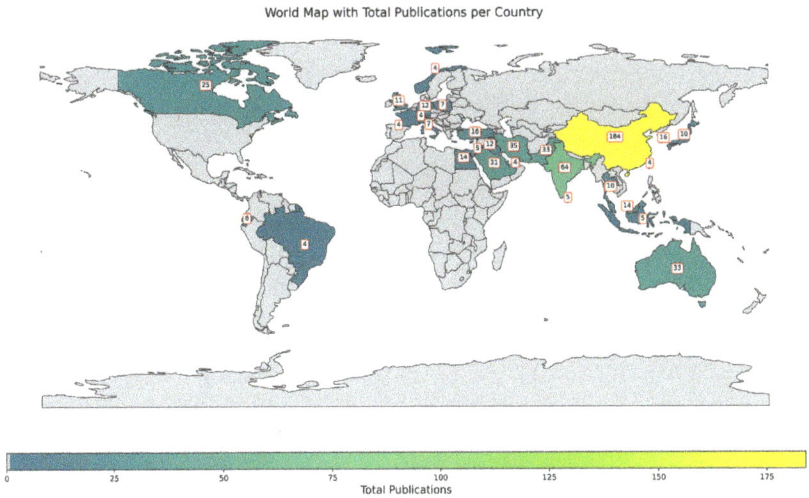

Figure 2. Country-wise distribution of publications featuring 'AdaBoost', 'Gradient Boosting', 'XGBoost', 'CatBoost', 'LightGBM', and 'polymers' in title or abstract.

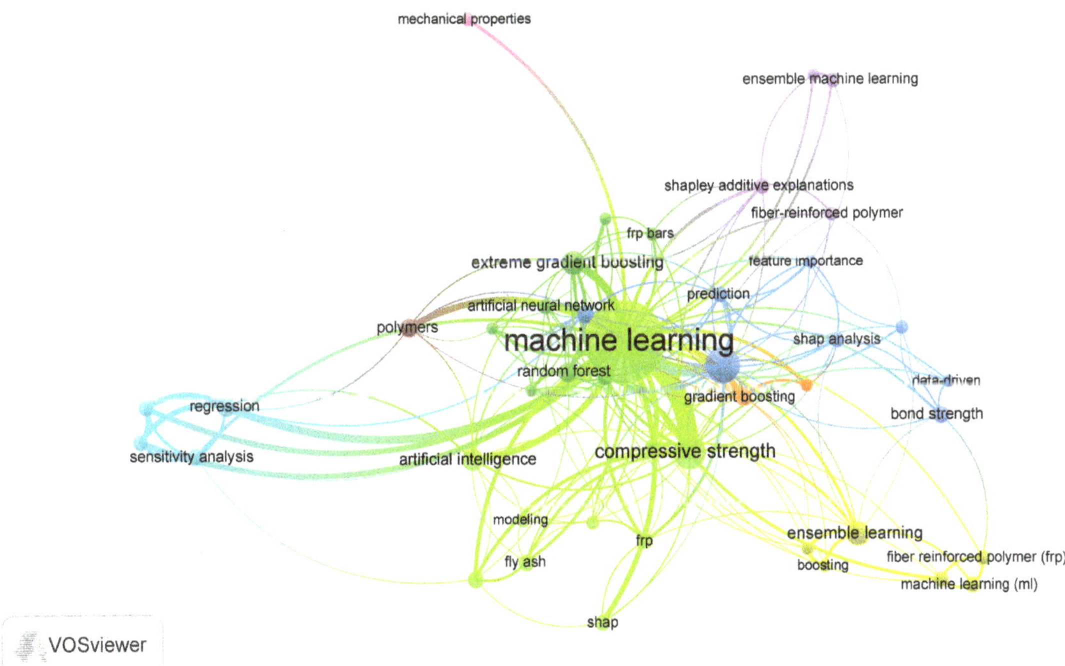

Figure 3. Keyword co-occurrence map based on VOSviewer analysis.

To ensure a systematic and comprehensive review, the PRISMA [22] (Preferred Reporting Items for Systematic Reviews and Meta-Analyses) methodology was employed for identifying, screening, and selecting relevant studies. The search across multiple databases yielded a significant number of records, which were then deduplicated, screened based on titles and abstracts, and assessed for full-text eligibility. After the final selection, studies were categorized into five thematic groups, each chosen based on specific research trends and their relevance to boosting-based ML applications in polymer science.

The first category, concrete and geopolymer composites, was selected due to the increasing interest in using boosting methods to predict and optimize the properties of concrete and geopolymer composites, which play a crucial role in modern construction materials. The second category, FRP and reinforced concrete systems, includes studies focused on FRP composites and reinforced concrete structures, where ML models contribute to improving material performance, durability, and structural behavior.

The third category, material properties prediction, encompasses studies dedicated to predicting the physical, chemical, and mechanical properties of polymer-based materials, which is a key aspect of optimizing material design and applications. The fourth category, advanced manufacturing and processing, highlights research on innovative manufacturing techniques and processing methods where boosting algorithms enhance efficiency, process optimization, and defect detection. Finally, the sustainability, environmental, and structural performance category includes studies assessing the environmental impact, sustainability, and structural efficiency of polymer-based materials, addressing critical challenges related to material life cycle assessment and eco-friendly alternatives.

Figure 4 is a PRISMA flowchart illustrating the study selection process for this systematic review. The diagram outlines the identification, screening, and eligibility assessment stages, leading to the final inclusion of studies.

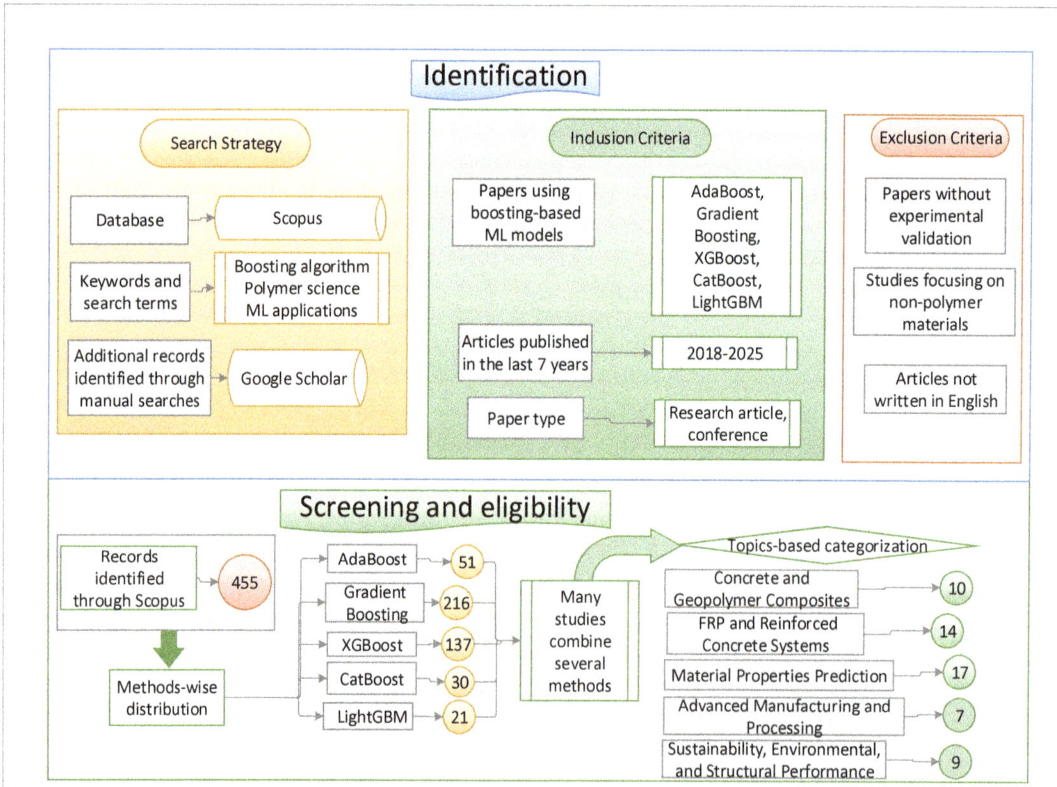

Figure 4. PRISMA flowchart outlining the study selection process for the systematic review.

The objective is to organize the current body of work, examine the achievements and limitations of boosting methods in this field, and suggest directions for future research. This analysis underscores the growing role of ML in materials science and explores the potential of these technologies to drive innovation in the development of advanced polymeric materials.

2. Theoretical Background of Boosting Methods

2.1. Gradient Boosting (GB)

Gradient Boosting (GB) is an ensemble method that builds a strong predictive model by combining several weak models (often decision trees) [23]. Each subsequent model is trained to predict the residual errors made by the previous models. This method can be mathematically described as follows:

Let $f_0(x)$ be the initial model. Typically, $f_0(x)$ is the constant value that minimizes the loss function. In the case of regression, this is often the mean of the target variable, $\hat{y}_i$, over all data points:

$$f_0(x) = \arg\min_{\gamma} \sum_{i=1}^{N} (y_i - \gamma)^2$$

where $\gamma = \frac{1}{N}\sum_{i=1}^{N} y_i$ is the mean of the target values.

At the t-th iteration, the model $f_t(x)$ is constructed by adding a new tree $h_t(x)$ to the previous model $f_{t-1}(x)$ in such a way that the loss function is minimized:

$$L_t = \sum_{i=1}^{N} [y_i - f_{t-1}(x_i) - \eta h_t(x_i)]^2$$

where:

- y_i is the actual target value,
- $f_{t-1}(x_i)$ is the predicted value from the previous model,
- $h_t(x_i)$ is the new decision tree that is fitted to the residuals (errors) from $f_{t-1}(x_i)$,
- η is the learning rate, which controls the contribution of each tree.

The new model is then updated iteratively by adding each new tree's contribution [24]:

$$f(x) = f_0(x) + \sum_{t=1}^{T} \eta h_t(x)$$

where T is the total number of iterations (trees).

In GB, the model is trained by optimizing the following objective function:

$$\mathcal{L}(\theta) = \sum_{i=1}^{N} \ell(y_i, f(x_i)) + \sum_{t=1}^{T} \Omega(h_t)$$

where:

- $\ell(y_i, f(x_i))$ is the loss function that measures the error of the model's prediction,
- $\Omega(h_t)$ is the regularization term that penalizes overly complex trees, often given by:

$$\Omega(h_t) = \gamma T + \frac{1}{2}\lambda \sum_{j=1}^{T} (w_j)^2$$

where T is the number of leaves in the tree, w_j are the leaf weights, and γ and λ are regularization parameters.

The final model $f(x)$ is an additive combination of all trees, which can be written as:

$$f(x) = f_0(x) + \sum_{t=1}^{T} \eta h_t(x)$$

This additive model helps correct the errors made by the previous trees.

To minimize the loss function, the gradient is computed of the loss with respect to the current prediction $f_{t-1}(x_i)$ and a new tree $h_t(x)$ fitted to the residuals [25]:

$$g_i = \nabla_{f_{t-1}(x_i)} \mathcal{L}(y_i, f_{t-1}(x_i))$$

This gradient g_i represents the residuals (the errors) from the current model. The next tree $h_t(x)$ is then trained to fit these residuals.

The process is similar to performing a gradient descent optimization, where each subsequent tree steps in the direction of minimizing the loss.

GB can be applied to various polymer research problems, such as predicting polymer material properties (strength, viscosity, thermal stability, etc.) based on experimental data. For instance:

$$\hat{y}_i = f(x_i) = f_0(x_i) + \sum_{t=1}^{T} \eta h_t(x_i)$$

where x_i represents the chemical composition and structural features of the polymer, and $\hat{y}_i$ is the predicted property (e.g., tensile strength).

GB is useful when the relationship between the polymer features and the material properties is non-linear and complex, which is often the case in polymer science [26]. By iteratively fitting trees to the residuals of previous models, GB can provide highly accurate predictions even for complex datasets. For instance Park et al. [27] proposed a boosting-based probabilistic model (NGBoost) [28] to predict the physical properties of polypropylene composites, addressing data imbalance and uncertainty.

2.2. *AdaBoost*

AdaBoost (Adaptive Boosting) is a popular ensemble learning technique that combines multiple weak classifiers to form a strong classifier [29,30]. It operates by focusing on the instances that previous classifiers misclassified, increasing their weight to make the next classifier pay more attention to those examples.

The AdaBoost algorithm works by iteratively adding weak classifiers to the ensemble, with each classifier trained on the weighted data. The general process is as follows:

The first classifier $f_1(x)$ is trained by minimizing the weighted loss function. The initial weight for each data point is the same, and the loss function typically used is the exponential loss function:

$$L(f_1) = \sum_{i=1}^{N} w_i (y_i - f_1(x_i))^2$$

where:

- $f_1(x)$ is the weak classifier,
- w_i is the weight of the i-th instance,
- y_i is the true label for the i-th instance.

The goal is to minimize this weighted loss function to obtain the first weak classifier $f_1(x)$.

After the first classifier is trained, the weights of misclassified instances are increased, and the weights of correctly classified instances are decreased. The weight update for the i-th instance at the t-th iteration is given by:

$$w_i^{(t)} = w_i^{(t-1)} \cdot \exp(-\alpha_t y_i f_t(x_i))$$

where:

- α_t is the weight of the t-th classifier, computed as:

$$\alpha_t = \frac{1}{2}\ln\left(\frac{1-\epsilon_t}{\epsilon_t}\right)$$

and ϵ_t is the weighted error rate of the t-th classifier:

$$\epsilon_t = \sum_{i=1}^{N} w_i^{(t-1)} \cdot ⊮(y_i \neq f_t(x_i))$$

where $⊮(y_i \neq f_t(x_i))$ is an indicator function that takes the value 1 if y_i is misclassified, and 0 otherwise.

The final strong classifier $F(x)$ is the weighted sum of all the weak classifiers $f_t(x)$. Each weak classifier $f_t(x)$ is weighted according to its accuracy:

$$F(x) = \sum_{t=1}^{T} \alpha_t f_t(x)$$

where:

- T is the total number of weak classifiers,
- α_t is the weight (coefficient) assigned to each weak classifier based on its performance.

The final classification decision is typically made by applying a sign function to the output of the ensemble:

$$\hat{y} = \text{sign}(F(x))$$

In polymer research, AdaBoost can be applied to classification tasks, such as categorizing polymers based on their chemical composition, molecular structure, or resistance to external factors like temperature or chemical exposure. The task is to classify polymers into different categories (e.g., strong vs. weak, thermoplastic vs. thermoset), based on various features extracted from experimental data.

Let us consider the task of classifying polymers based on their thermal stability:

$$\hat{y}_i = \text{sign}\left(\sum_{t=1}^{T} \alpha_t f_t(x_i)\right)$$

where $\hat{y}_i$ represents the predicted category (e.g., stable vs. unstable) for the i-th polymer, and x_i represents the feature vector (e.g., chemical composition, molecular weight, etc.) for the polymer.

AdaBoost works by iteratively re-weighting the data and training weak classifiers on the weighted data, giving more importance to misclassified instances. The weak classifiers are then combined to form a final strong classifier. In the context of polymer research, AdaBoost can be used for tasks such as classifying polymers based on their physical properties, chemical characteristics, or other factors. Its ability to focus on hard-to-classify instances and improve model performance makes it a useful tool for applications with complex or imbalanced data.

2.3. CatBoost

CatBoost (Categorical Boosting) is a gradient boosting algorithm that specializes in handling categorical features. Unlike traditional methods that rely on one-hot encoding, CatBoost introduces an innovative approach to minimize overfitting and reduce computational complexity [31]. It builds decision trees optimally using both categorical and continuous features.

Similar to other gradient boosting methods, CatBoost iteratively adds decision trees trained on residuals [32]. It handles categorical variables through ordered target statistics, which compute target values based on previous data points rather than all at once, preventing overfitting. Let X be a dataset and y be the target variable, and x_i represent the i-th observation with categorical feature C_j. The key transformation for a categorical feature is computed as follows:

$$\hat{C}_j = \frac{1}{|S_j|} \sum_{x_i \in S_j} y_i$$

where:

- S_j is the set of observations corresponding to the category C_j,
- $\hat{C}_j$ is the average target value for the categorical feature.

This transformation is performed using the ordered target statistics, where for each data point, the target statistic is computed using previous data points only to avoid data leakage.

After transforming the categorical features, the model proceeds to build decision trees, much like standard gradient boosting. The goal of each decision tree is to minimize the residual error from the previous model:

$$L_t = \sum_{i=1}^{N} [y_i - f_{t-1}(x_i) - \eta h_t(x_i)]^2$$

where:

- $f_{t-1}(x_i)$ is the previous model's prediction,
- η is the learning rate, and
- $h_t(x_i)$ is the decision tree model trained on the residuals.

Each tree aims to minimize this residual error, adjusting the contribution of each tree using the gradient descent approach.

Once the trees are built, the final model $f(x)$ is a weighted sum of all decision trees, where the weight of each tree is controlled by the learning rate η:

$$f(x) = f_0(x) + \sum_{t=1}^{T} \eta h_t(x)$$

where $f_0(x)$ is the initial model, typically, the mean of the target values.

To optimize the CatBoost model, the objective function is similar to that used in other gradient boosting algorithms. The loss function for regression, for example, is the mean squared error (MSE):

$$\mathcal{L}(\theta) = \sum_{i=1}^{N} (y_i - f(x_i))^2$$

The gradient of this loss function is computed to guide the model's optimization. At each iteration, the new tree $h_t(x)$ is trained to fit the negative gradient of the loss function, which corresponds to the residuals from the previous model.

CatBoost also includes regularization techniques to prevent overfitting. Specifically, it uses feature combinations and permutation-driven techniques that improve model generalization by considering various ways to combine the features. The regularization term can be written as:

$$\Omega(h_t) = \gamma T + \frac{1}{2}\lambda \sum_{j=1}^{T} (w_j)^2$$

where:

- T is the number of leaves in the tree,
- w_j is the weight of each leaf node,

- γ and λ are regularization parameters that control the complexity of the trees.

In polymer research, CatBoost is highly beneficial when dealing with complex datasets that contain a mix of categorical and continuous variables. For example, in the study of polymer materials, the polymerization process conditions (e.g., temperature, pressure, catalyst types) can be categorical, while properties such as the tensile strength, viscosity, or melting temperature are continuous. CatBoost can efficiently handle these mixed data types, building robust predictive models without the need for extensive pre-processing or feature engineering.

For instance, in predicting the tensile strength of a polymer based on various processing conditions and chemical additives:

$$\hat{y}_i = f(x_i) = f_0(x_i) + \sum_{t=1}^{T} \eta h_t(x_i)$$

where x_i includes both continuous features (such as temperature) and categorical features (such as the type of chemical additive), and $\hat{y}_i$ is the predicted tensile strength.

Since CatBoost handles categorical features directly, without the need for one-hot encoding, it reduces the computational complexity and improves both the speed and accuracy of the model, making it ideal for polymer science tasks that involve large and complex datasets.

2.4. LightGBM

LightGBM (Light Gradient Boosting Machine) is a gradient boosting framework that uses a histogram-based algorithm to improve both the training speed and memory efficiency [33]. Instead of using all possible feature values, LightGBM groups them into bins, which reduces the complexity of decision trees and speeds up computations. This makes LightGBM especially suitable for handling large datasets, where scalability is a concern.

The basic idea behind LightGBM is to discretize continuous features into bins. The model then makes decisions based on these bins rather than individual feature values, speeding up both the training and prediction processes. The following steps outline the key mathematical aspects of LightGBM

For each continuous feature x_i, LightGBM groups values into bins. The binning process is as follows:

$$\mathrm{Bin}(x_i) = \mathrm{floor}\left(\frac{x_i}{\delta}\right)$$

where:

- x_i is the value of the continuous feature,
- δ is the binning step size (which determines the size of the bins), and
- $\mathrm{Bin}(x_i)$ represents the bin index that the value x_i falls into.

This transformation reduces the number of unique feature values, making the model more computationally efficient.

The algorithm proceeds by building decision trees based on the binned features. The decision tree is built using a gradient-based approach, where each split is made to minimize the residual error from the previous tree. The residual error at the t-th iteration is:

$$L_t = \sum_{i=1}^{N} (y_i - f_{t-1}(x_i) - \eta h_t(x_i))^2$$

where:

- $f_{t-1}(x_i)$ is the prediction from the previous model,
- η is the learning rate,

- $h_t(x_i)$ is the prediction of the new decision tree at the t-th step.

The tree is grown by iterating over the binned feature values and selecting the best split based on the gradient of the loss function. In LightGBM, this is performed using the histogram-based approach, which selects the optimal bin split that minimizes the loss.

In a standard decision tree, splits are found by considering all feature values. LightGBM, however, uses a histogram-based approach. Let the histogram H_j for feature j be defined as:

$$H_j = \{h_1, h_2, \ldots, h_k\}$$

where h_k is the count of instances falling into bin k for feature j. The algorithm then calculates the gradient of the loss function for each bin and chooses the best split based on these gradients.

For a given candidate split on feature j, the gain from the split can be computed as:

$$\text{Gain} = \frac{1}{N} \sum_{i \in L} \left(\frac{\partial L}{\partial f} \right)^2 + \sum_{i \in R} \left(\frac{\partial L}{\partial f} \right)^2$$

where:

- N is the total number of instances,
- L and R represent the left and right child nodes after the split,
- $\frac{\partial L}{\partial f}$ is the gradient of the loss with respect to the feature values.

The optimal split is chosen based on the highest gain, which corresponds to the best reduction in the residual error after the split.

To prevent overfitting, LightGBM applies regularization techniques such as L2 regularization and leaf-wise pruning. The regularization term for the decision tree can be written as:

$$\Omega(h_t) = \gamma T + \lambda \sum_{i=1}^{T} (w_j)^2$$

where:

- T is the number of leaves in the tree,
- w_j is the weight of the j-th leaf node,
- γ and λ are regularization parameters.

The regularization helps to control the complexity of the trees and ensures that the model generalizes well to unseen data.

The final model in LightGBM is a weighted sum of all the decision trees built during the boosting process:

$$f(x) = f_0(x) + \sum_{t=1}^{T} \eta h_t(x)$$

where:

- $f_0(x)$ is the initial model, typically the mean of the target values,
- η is the learning rate,
- $h_t(x)$ is the t-th decision tree model.

LightGBM's ability to efficiently handle large datasets makes it useful for predicting polymer properties when scalability is a concern. In polymer science, the data used for prediction often come from a variety of sources, including experimental measurements of physical properties (e.g., tensile strength, elasticity, viscosity) and process conditions (e.g., temperature, pressure, and type of catalyst).

For instance, LightGBM can be applied to predict the thermal conductivity of a polymer based on a variety of experimental factors. The feature set x_i might include both continuous features (e.g., temperature) and categorical features (e.g., type of polymer or catalyst), which can be binned and used to create an efficient model.

The final prediction for the polymer property $\hat{y}_i$ is given by:

$$\hat{y}_i = f(x_i) = f_0(x_i) + \sum_{t=1}^{T} \eta h_t(x_i)$$

where x_i is the feature vector containing both categorical and continuous features, and $\hat{y}_i$ is the predicted polymer property.

2.5. XGBoost

XGBoost (XGBoost) is an optimized version of the gradient boosting algorithm, designed to improve training efficiency, prediction accuracy, and model interpretability. It incorporates regularization techniques to prevent overfitting, handles sparse data, and uses more efficient training methods, such as parallelization and hardware optimizations [34].

XGBoost builds models by sequentially adding decision trees, where each tree is fitted to the residuals (errors) from the previous model. It aims to minimize the loss function with an additional regularization term to prevent overfitting.

At the t-th iteration, the model is updated as follows:

$$L_t = \sum_{i=1}^{N} (y_i - f_{t-1}(x_i) - \eta h_t(x_i))^2$$

where:

- y_i is the true label for instance i,
- $f_{t-1}(x_i)$ is the model's prediction for instance i at the previous step,
- $h_t(x_i)$ is the prediction of the new decision tree at the t-th step,
- η is the learning rate.

This formulation minimizes the residual error from the previous model, and the final model prediction is the sum of all tree predictions.

XGBoost introduces regularization to penalize complexity and reduce overfitting [35]. The regularization term $\Omega(f)$ is added to the objective function to control the complexity of the model:

$$\Omega(f) = \gamma T + \frac{1}{2}\lambda \sum_{j=1}^{T} (w_j)^2$$

where:

- T is the number of leaves in the tree,
- w_j is the weight of the j-th leaf,
- γ is a regularization parameter controlling the number of leaves in the tree,
- λ is a regularization parameter controlling the size of the weights.

The regularization term discourages overly complex trees with many leaves and large weights, which helps prevent overfitting [36].

The final objective function to be minimized in XGBoost combines the loss function (residuals) and the regularization term:

$$\mathcal{L}(f) = \sum_{i=1}^{N} L(y_i, f(x_i)) + \Omega(f)$$

where:

- $L(y_i, f(x_i))$ is the loss function that measures the difference between the true label y_i and the predicted value $f(x_i)$,
- $\Omega(f)$ is the regularization term as defined earlier.

The goal of XGBoost is to find the function $f(x)$ that minimizes this objective.

XGBoost builds trees using a greedy algorithm, which chooses the best split at each node based on the objective function. The best split q for a given node can be found by maximizing the gain:

$$\text{Gain}(q) = \frac{1}{2}\left(\frac{(\sum_{i\in L} g_i)^2}{h_L + \lambda} + \frac{(\sum_{i\in R} g_i)^2}{h_R + \lambda} - \frac{(\sum_{i\in S} g_i)^2}{h_S + \lambda}\right)$$

where:

- g_i and h_i are the gradient and Hessian of the loss function (first and second derivatives),
- L, R, and S are the left, right, and split node, respectively.

The gain represents the reduction in the loss after the split, and the optimal split maximizes this gain [37].

Once the trees are constructed, the final prediction is obtained by summing the predictions of all trees:

$$f(x) = f_0(x) + \sum_{t=1}^{T} \eta h_t(x)$$

where:

- $f_0(x)$ is the initial model (often the mean value of the target),
- η is the learning rate,
- $h_t(x)$ is the prediction from the t-th tree.

The prediction for a new instance is computed by summing the predictions of all trees, weighted by the learning rate [38]. XGBoost also incorporates early stopping to prevent overfitting [39]. During training, the model evaluates performance on a validation set after each boosting round. If the performance on the validation set does not improve for a specified number of rounds (called the early stopping round), the training stops.

XGBoost is useful for predicting polymer properties when both numerical and categorical features are present. In polymer research, there are often large datasets consisting of experimental measurements (e.g., tensile strength, viscosity, thermal properties) as well as categorical features (e.g., type of polymer, catalyst used, reaction conditions). XGBoost's ability to handle mixed types of data (numerical and categorical) and its efficient training process make it ideal for these types of predictions. For example, Ueki et al. [40] employed ML to predict grafting yields in radiation-induced graft polymerization of methacrylate ester monomers onto polyethylene-coated polypropylene fabric. XGBoost demonstrated the highest prediction accuracy, identifying monomer polarizability and O_2 NMR shift as key factors influencing grafting efficiency.

For example, XGBoost can be used to predict the stability of a polymer under various conditions, such as exposure to UV radiation or chemical reactions, based on features like molecular structure, temperature, and type of additives used.

The final prediction for polymer property $\hat{y}_i$ can be written as:

$$\hat{y}_i = f(x_i) = f_0(x_i) + \sum_{t=1}^{T} \eta h_t(x_i)$$

where x_i is the feature vector containing both numerical and categorical features, and $\hat{y}_i$ is the predicted polymer property.

3. Case Studies

3.1. Concrete and Geopolymer Composites

Zhao et al. [41] developed a high-generalizability ensemble ML (EML) framework to predict the homogenized mechanical properties of short FRP composites. Using a stacking algorithm, the EML combines Extra Trees (ET), XGBoost, and LightGBM as base models. The framework incorporates a micromechanical model employing a two-step homogenization algorithm verified for its accuracy in modeling composites with randomly distributed fibers, integrating finite element simulations for robust datasets. The results obtained show the EML achieves high accuracy (R^2 values of 0.988 and 0.952 for train and test datasets) and efficient generalization on experimental data, outperforming computationally intensive high-fidelity models. SHAP analysis reveals the Young's modulus of the matrix [42,43], fiber, and fiber content as key factors influencing the homogenized properties, with anisotropy dominated by fiber orientation. This framework reduces computational costs significantly while maintaining precision and interpretability, showcasing its applicability for advanced composite material design.

A two-level stacking algorithm framework (Stacking-CRRL) combining Catboost, RF, Ridge Regression (RR), and LASSO regression is proposed by Zhang et al. [44] for predicting the axial compression load capacity of steel-reinforced concrete columns (SRCCs) clad in CFRP. Sparse data were balanced using SMOTE, and 12 predictive features were selected after eliminating redundancy via Spearman correlation analysis. Catboost, RFR, and RR were the chosen base learners, with LASSO as the meta-learner. The results obtained indicate superior predictive performance of the Stacking-CRRL model compared to individual models, traditional ML methods, and simulation techniques. SHAP analysis further elucidated feature impacts on SRCC load capacity.

Ultra-high-performance geopolymer concrete (UHPGC) [45] offers a sustainable and economical alternative to ultra-high-performance concrete (UHPC), delivering comparable mechanical performance. Despite its potential, the absence of a robust mix design methodology limits its broader adoption. Katlav et al. [46] employed ensemble ML models, including RF, XGBoost, LightGBM, and AdaBoost, to predict the CS of UHPGC using a dataset of 181 test results with 13 input features. XGBoost emerged as the top-performing model, achieving an R^2 of 0.918 and low error metrics. Feature importance and SHAP analyses identified age, fiber, silica fume, Na_2SiO_3, and water content as critical factors influencing CS. A user-friendly graphical user interface (GUI) was developed for practical CS predictions, reducing reliance on experimental tests. While promising, the model's reliability could improve with expanded datasets and exploration of advanced AI techniques like deep learning.

Geopolymers, made from waste materials rich in aluminosilicate, present a promising alternative to traditional Portland cement. Research into GPCs is advancing, but laboratory testing remains time-intensive and costly. ML offers a faster, cost-effective method to predict the CS of these materials. Wang et al. [47] utilized a decision tree (DT) model and two ensemble methods—AdaBoost and Random Forest (RF)—were employed to estimate CS. Results showed that ensemble models outperformed the DT model, with R^2 values of 0.90 for AdaBoost and RF, compared to 0.83 for DT. Additionally, lower errors, such as MAE and RMSE, confirmed the ensemble models' higher accuracy. The findings emphasize ML's potential in accelerating material property analysis for the construction industry.

Geopolymers, made from aluminosilicate-rich waste, are a promising alternative to traditional cement. Khan et al. [48] investigated predicting the CS of GPCs using ML techniques. Three models were used: support vector machine (SVM), GB, and XGBoost. Ensemble methods like GB and XGB outperformed SVM, with XGB achieving the highest R^2 of 0.98. SHAP analysis revealed ground granulated blast-furnace slag (GGBS) as a significant positive factor for CS. Other factors like NaOH molarity and fly ash had mixed effects. These findings demonstrate ML's potential to create fast, cost-effective solutions for eco-friendly construction materials.

GeoPC is a sustainable alternative to traditional concrete, offering environmental benefits and reliable strength performance. Zhou et al. [49] employed three ensemble ML models—Gradient Boosting Regressor (GBR), AdaBoost, and XGBoost to predict GPC's CS and split-tensile strength (STS). Among these, XGBoost achieved the highest accuracy, with R^2 values exceeding 0.90 and lower error metrics, including MAE and RMSE. Sensitivity analysis revealed that blast furnace slag (BFS), curing duration (CD), and fine aggregate quantity were critical factors influencing GPC's mechanical properties. K-fold analysis confirmed XGBoost's superior performance compared to GB and AdaBoost. These results demonstrate the potential of ensemble ML models for precise GPC property prediction, enabling improved quality control and on-site adaptability in sustainable construction practices.

Amin et al. [50] explored the application of ML to predict the CS of geopolymer concrete (GeoPC), using a dataset of 481 mixes and nine input variables. Four ML models — SVM, multi-layer perceptron neural network (MLPNN) [51], AdaBoost regressor (AR), and RF—were compared to identify the most accurate predictor. Ensemble methods (AR and RF) outperformed individual techniques (SVM and MLPNN), with RF achieving the highest accuracy, yielding an R^2 of 0.95. Statistical analysis and k-fold evaluation validated the superior performance of RF, which exhibited lower error metrics (MAE, RMSE) and closer agreement between predicted and experimental results. Sensitivity analysis revealed curing time, curing temperature, and specimen age as the most significant factors influencing GeoPC's CS, contributing 22.5%, 20.1%, and 18.5%, respectively. These findings highlight the efficiency of ensemble ML models in reducing experimental effort while promoting sustainable and cost-effective construction practices through enhanced GeoPC adoption.

Ansari et al. [52] aimed to predict the CS of GeoPC incorporating fly ash (FlA) using various ML models. Three models were tested: linear regression (LR), ANN, and AdaBoost. The AdaBoost model outperformed both LR and ANN, achieving the highest accuracy with a correlation coefficient (R^2) of 0.944, root mean squared error (RMSE) of 2.506, and mean absolute error (MAE) of 1.259 in the training phase. In contrast, the LR model performed poorly with an R^2 of 0.701 and RMSE of 5.805. The study used 154 datasets, with 70% for training and 30% for testing, and evaluated model accuracy through R^2, RMSE, and MAE. The AdaBoost model showed the best predictive performance with minimal error and deviation from experimental results. The findings emphasize the efficiency of ensemble learning models like AdaBoost in improving the prediction of CS in GPC, saving both time and resources compared to traditional experimental methods.

Dodo et al. [53] investigated the use of supervised ML algorithms (MLAs) to predict the mechanical properties of fly ash/slag-based geopolymer concrete (FASBGeoPC) [54]. AdaBoost and Bagging with an ANN ensemble model were employed to predict CS using 156 data points, considering parameters like GGBS, alkaline activator, fly ash, SP dosage, NaOH molarity, aggregate, and temperature. Python programming in Anaconda Navigator with Spyder was used for model development and validation. Statistical evaluation, including MAE, RMSE, and R^2, confirmed that the ensemble methods outperformed individual models, with AdaBoost-ANN achieving the highest R^2 of 0.914. The Shapley analysis identified GGBS, NaOH molarity, and temperature as the most influential parameters in determining CS. Additionally, ensemble methods, such as boosting and bagging, demonstrated more reliable performance, with AdaBoost-ANN showing the least errors and highest accuracy compared to other models. The research indicates that these models, especially the ensemble methods, are effective for predicting FASBGPC properties and can be applied in civil engineering.

Wudil et al. [55] introduced an innovative approach combining ensemble ML and experimental data to predict the carbon dioxide footprint (CO_2-FP) of fly ash GeoPC. Utilizing Adaboost to enhance decision tree regression (DTR) and support vector regression (SVR), the methodology accurately captures complex relationships between material features and CO_2 emissions. Optimal feature combinations (Combo-3) yielded the best predictive performance, with Adaboost-DTR achieving the highest accuracy (CC = 0.9665, NSE = 0.9343). Evaluation through metrics like MAE and RMSE, alongside SHAP analysis, emphasized the critical role of NaOH, curing temperature, and fly ash content in emissions. While dataset limitations and applicability to broader concrete types are challenges, the findings support material optimization for sustainable construction and integration with IoT systems for real-time CO_2 monitoring. Future work includes expanding datasets and exploring other concrete formulations to enhance generalizability.

Table 1 summarizes studies that applied ensemble boosting techniques in predicting the properties of concrete and GPCs across a variety of applications. The table includes details on the materials or properties predicted, the boosting methods used, model performance, key influencing factors, and additional techniques, such as feature importance analysis and data balancing approaches.

Table 1. Comparison of results for studies related to concrete and GPCs.

Study	Boosting Technique	Application	Materials/ Properties Predicted	Dataset	Model Performance (R^2, MAE, RMSE)	Key Influencing Factors	Additional Techniques/ Analysis
Zhao et al. (2023) [41]	XGBoost, LightGBM, Extra Trees	Short FRP composites	Homogenized mechanical properties (e.g., Young's modulus)	High fidelity composite datasets, experimental data	R^2 of 0.988 (train), 0.952 (test)	Fiber orientation, fiber content, matrix Young's modulus	SHAP analysis, micromechanical model integration
Zhang et al. (2023) [44]	CatBoost, RF, Ridge, LASSO	Steel-reinforced concrete columns (SRCCs) clad in CFRP	Axial compression load capacity	Sparse data, 12 features	High predictive accuracy, better than individual models	Load capacity factors	SMOTE for data balancing, SHAP analysis
Katlav et al. (2024) [46]	XGBoost, LightGBM, AdaBoost, RF	UHPGC	Compressive strength (CS)	181 test results, 13 input features	R^2 = 0.948	Age, fiber content, water content	SHAP analysis, user interface for practical predictions
Wang et al. (2022) [47]	AdaBoost, RF	GPCs	CS	Experimental datasets	R^2 = 0.90	Fly ash, curing time, NaOH molarity	SHAP analysis
Khan et al. (2022) [48]	XGBoost, GB	GPCs	CS	500+ mixes	R^2 = 0.98	GGBS, NaOH molarity, fly ash	SHAP analysis
Zhou et al. (2024) [49]	XGBoost, AdaBoost, Gradient Boosting	GPC	CS, STS	Experimental data	R^2 > 0.90	Blast furnace slag, curing duration, fine aggregate	K-fold analysis

Table 1. *Cont.*

Study	Boosting Technique	Application	Materials/ Properties Predicted	Dataset	Model Performance (R^2, MAE, RMSE)	Key Influencing Factors	Additional Techniques/ Analysis
Amin et al. (2022) [50]	AdaBoost, RF	GeoPC	CS	481 mixes, 9 variables	$R^2 = 0.95$	Curing time, temperature, specimen age	Sensitivity analysis, k-fold validation
Ansari et al. (2023) [52]	AdaBoost	GPC with fly ash	CS	154 datasets	$R^2 = 0.944$, RMSE = 2.506, MAE = 1.259	Fly ash content, water-to-binder ratio	Evaluation through R^2, MAE, RMSE
Dodo et al. (2024) [53]	AdaBoost, Bagging with ANN	FASBGeoPC	CS	156 data points	$R^2 = 0.914$	GGBS, NaOH molarity, temperature	SHAP analysis, ensemble methods
Wudil et al. (2025) [55]	AdaBoost	Fly ash GeoPC	Carbon dioxide footprint (CO_2-FP)	Experimental data, material features	CC = 0.9665, NSE = 0.9343	NaOH, curing temperature, fly ash content	SHAP analysis, IoT integration

3.2. FRP and Reinforced Concrete Systems

Developments in FRP composites significantly impact civil engineering, especially in strengthening concrete structures. ML models for predicting FRP–concrete bond strength often fall short of optimal performance. Kim et al. [56] employed the CatBoost algorithm to enhance prediction accuracy, utilizing data from 855 single-lap shear tests. CatBoost outperformed other ensemble methods (XGBoost, HGBoost, RF) with metrics like lower RMSE (2.31) and higher R^2 (0.96). It also surpassed ANN-based models, confirming its efficacy with small datasets and categorical features. This highlights CatBoost's robustness and suitability for bond strength prediction tasks in FRP–concrete systems.

FRP have proven to be effective in strengthening reinforced concrete (RC) structures, but accurately assessing their fire resistance remains a challenge due to limited guidance in building codes. Kumarawadu et al. [57] explored the use of ML to predict the fire resistance of FRP-strengthened RC beams, using a dataset of over 21,000 data points from numerical simulations and experimental tests. Twelve ML models, including ensemble methods like XGBoost and CatBoost, were evaluated, with some achieving accuracy rates over 92%. The study also utilized Bayesian optimization for model tuning and SHAP analysis to assess the influence of key features such as loading ratio and insulation depth. The results highlighted that ensemble ML models outperformed traditional methods, showcasing their ability to accurately predict fire resistance. Key factors affecting fire resistance included the loading ratio, area of tensile reinforcement, insulation depth, and concrete cover thickness. The study concludes that ML models, especially ensemble techniques, provide valuable insights for optimizing fire safety in FRP-strengthened RC beams, with further research needed to expand the dataset to cover a wider range of real-world scenarios.

Wang et al. [58] utilized ML to identify the factors affecting the contribution of externally bonded fiber-reinforced polymer (FRP) composites to the shear strength of reinforced concrete (RC) beams. A comprehensive database of 442 FRP-strengthened RC beams was created, and anomaly detection was applied using the isolation forest algorithm. Six ML models were trained, with XGBoost achieving the highest prediction accuracy compared to traditional equations commonly used in design codes. Key influencing factors for the FRP contribution to shear strength were identified, including the effective height of FRP, shear span ratio, and reinforcement method. The trained models revealed that different reinforcement methods, such as U-wrap, full wrapping, and side-bonding, significantly affect the shear contribution of FRP. Further analysis of parameter importance showed that the effective height of FRP had the greatest impact. A new equation for predicting the shear strength contribution of FRP was derived, integrating the ML models and key influencing parameters, such as the shear span ratio and reinforcement method. This combination of ML and traditional models provides a novel, interpretable method for predicting shear strength in RC beams.

GFRP bars are prone to bonding failure due to their low bond strength with concrete. Mahmoudian et al. [59] utilized four tree-based ML models—Decision Tree, RF, AdaBoost, and XGBoost—to predict flexural bond strength and failure modes at the concrete–GFRP interface. Genetic algorithms were employed to optimize model hyperparameters, increasing R^2 scores by up to 4%. The XGBoost classifier achieved perfect accuracy (100%) in predicting failure modes from test data. Additionally, Shapley Value analysis provided a detailed understanding of feature importance, enhancing model interpretability. These findings highlight the potential of ML in advancing GFRP–concrete interface reinforcement methods.

Mahmoudian et al. [60] evaluated the bond strength of FRP rebars in ultra-high-performance concrete (UHPC) using ML models trained on experimental datasets. The variables considered included rebar type, diameter, elastic modulus, tensile strength, concrete CS, embedment length, and test method. Various boosting ML models, including AdaBoost, CatBoost, Gradient Boosting, XGBoost, and Hist Gradient Boosting, were tested, with XGBoost achieving the highest R^2 score of 0.95 and the lowest RMSE of 2.21. Shapley values analysis identified tensile strength, elastic modulus, and embedment length as the most influential factors. Hyperparameter tuning significantly improved model accuracy, with ensemble approaches like Voting Regressor further enhancing prediction reliability. The study also highlighted the advantages of ML models over traditional methods, which often lack adaptability across diverse scenarios. Additionally, a user interface was developed to facilitate the practical application of these models in structural engineering, providing a customizable platform for engineers to predict bond strength in FRP-reinforced UHPC.

Wang et al. [61] introduced a genetic evolutionary deep learning framework to assess the fire resistance of FRP-strengthened reinforced concrete (RC) beams. The approach uses the Light Gradient-Boosting Machine (LightGBM) algorithm, optimized with a Genetic Algorithm, and Genetic Programming (GP) to predict fire resistance performance. A dataset of 20,000 data points from numerical models and experimental studies was used. The LightGBM model achieved high accuracy, with R^2 values of 0.923 for fire resistance time and 0.789 for deflection at failure, while the GP model provided explicit equations but with lower accuracy (R^2 values of 0.642 and 0.643). The study identified that geometric features, such as insulation thickness and reinforcement area, have a significant impact on fire resistance, which traditional models fail to capture. A graphical user interface (GUI) was developed, enabling engineers to use these insights without coding skills. Additionally, model interpretability techniques like SHAP values and trend analysis were employed to enhance the practical application of the model in engineering decisions.

Hu et al. [62] presented an ML-assisted framework for optimizing the stacking sequence and orientation of CFRP/metal composite laminates, aiming to enhance mechanical properties under quasi-static loading. By integrating experimental data with finite element simulations, the study expands ML analysis in composite material design. Nine ML models, including XGBoost and gradient boosting, were evaluated for their ability to predict tensile and bending strengths. XGBoost and gradient boosting excelled in tensile strength predictions, while decision trees, KNN, and RF performed best in bending strength predictions. The study identifies optimal layup sequences, with sequence 2 showing superior mechanical properties. The combination of ML, numerical, and experimental approaches provided deep insights into CFRP/metal composites' performance. Overall, the findings offer valuable design references and highlight the importance of advanced analytical models for composite material optimization.

Aydın et al. [63] examined the wear behavior of multiwall carbon nanotube (MWCNT)-doped non-crimp fabric carbon fiber-reinforced polymer (NCF-CFRP) composites. The results showed that a 1 wt% MWCNT reinforcement reduced wear loss by 48.1% and 61.1% under 10 N and 30 N loads, respectively, over a sliding distance of 1000 m. Various ML models were evaluated for predicting wear loss, including Deep Multi-Layer Perceptron (DMLP), RF Regression (RFR), Gradient Boosting Regression (GBR), linear regression (LR), and polynomial regression (PR). Among these, the DMLP model exhibited the best predictive performance, achieving an R^2 of 0.9726 in testing, and showed effective generalization without overfitting across varying loads. The study also found that maximum wear resistance occurred at 1 wt% MWCNT content, with wear loss increasing as load and sliding distance grew. SEM and EDS analyses revealed matrix delamination and CF fractures at higher loads. The study is the first to use ML to predict the wear behavior of epoxy matrix hybrid nanocomposites.

Li et al. [64] proposed an ML-based method using RF and AdaBoost algorithms to predict the bond strength between basalt fiber-reinforced polymer (BFRP) bars and concrete in corrosive environments. The model was trained on 355 samples, incorporating factors such as corrosion, concrete strength, and BFRP bar properties. The AdaBoost model outperformed RF, achieving an R^2 value of 0.925, RMSE of 0.0769, and MAE of 0.0589, showing high accuracy. The SHAP method was used to analyze the impact of various factors on bond strength, with the corrosion factor being the most influential. The ML models outperformed traditional empirical models, which had a much higher coefficient of variation. This research highlights the potential of ML techniques in predicting bond strength, offering a more reliable and generalizable alternative to traditional methods. These models can be extended to other types of FRP concrete systems, enhancing prediction accuracy in the field.

Khodadadi et al. [65] introduced a novel Particle Swarm Optimization [66,67]-Categorical Boosting (PSO-CatBoost) model for predicting the CS of CFPR Confined-Concrete (CFRP-CC). The model, trained on 916 experimental results from 105 studies (1991–2023), integrates PSO with CatBoost, leveraging advanced feature evaluation methods, such as SHAP and Permutation Feature Importance (PFI). Comparative analysis shows the proposed model achieved the highest R^2 (0.9572 for testing), and lowest MSE, MAE, and RMSE, outperforming six other ML algorithms and traditional empirical models. Key influencing factors identified include the CFRP reinforcement ratio and unconfined concrete CS. The PSO-CatBoost model represents a significant advancement, providing higher predictive accuracy and generalizability. A user-friendly graphical interface further enhances its practical applicability, setting a new standard for predictive modeling in CFRP-CC research. This work underscores the transformative potential of data-driven approaches in engineering domains.

Accurately predicting the compressive behavior of FRP-confined concrete is critical for optimizing structural designs, meeting safety standards, and minimizing costs and environmental impacts. Alizamir et al. [68] examined four ML models—GBRT, RF, ANNMLP, and ANNRBF—using data from 765 circular specimens. GBRT improved predictions of the strength ratio (f'_{cc}/f'_{co}) with an RMSE reduction of up to 69.94% compared to empirical models. ANNMLP excelled in predicting the strain ratio ($\epsilon_{cc}/\epsilon_{co}$), outperforming GBRT, RF, and others by up to 83.74% in RMSE. These findings demonstrate that ML models, particularly GBRT and ANNMLP, outperform empirical methods, offering enhanced precision for FRP-confinement design. Future research could refine these models using advanced algorithms, robust feature selection, and extended datasets to further improve accuracy and generalization capabilities.

FRP rebars are increasingly used in construction due to infrastructure demands and the need for seawater and sea sand concrete. Amin et al. [69] sought to estimate the flexural capacity of FRP-reinforced concrete beams using decision tree (DT) and gradient boosting tree (GBT) models, incorporating six input parameters, such as beam depth and concrete CS. The models were trained on 60% of the dataset and validated on 40% using the correlation coefficient (R), mean absolute error (MAE), and root mean square error (RMSE). The GBT model outperformed DT, achieving higher R values (0.94 during validation) and a regression slope closer to 1 (0.83 for GBT vs. 0.75 for DT). Sensitivity analysis identified beam depth as the most critical factor influencing flexural strength. While the GBT model showed superior accuracy compared to prior gene expression programming (GEP) models, the American Concrete Institute (ACI) equations remain more reliable overall. Combining R with additional error indices like MAE ensures robust AI model evaluation.

Amin et al. [70] developed ML models to predict the interfacial bond strength (IBS) of FRP laminates on concrete prisms with grooves. Three ensemble models—RF regression, XGBoost, and Light Gradient Boosting Machine (LIGHT GBM)—were evaluated. The models were trained using 70% of the dataset, with the remaining 30% used for validation. LIGHT GBM outperformed the other models, achieving an R^2 of 0.942 for the training data and 0.865 for the testing data, demonstrating its superior accuracy. A SHAPASH analysis revealed that the elastic modulus × thickness of FRP and the width of the FRP plate were the most influential factors on IBS. All models showed reliable performance, but LIGHT GBM provided the highest prediction precision, with low RMSE and MAE values. The results highlight the potential of LIGHT GBM as a robust and efficient tool for predicting IBS in FRP-retrofitted concrete structures.

Tian et al. [71] investigated the influence of FRP bar surface types on bond strength to concrete and developed a practical equation for predicting interfacial bond strength. A database of 158 pull-out test results for helically wrapped and ribbed FRP bars was compiled, considering eight influencing factors, including rib spacing (wc), rib width (wf), rib height (rh), and concrete properties. Twelve ML models were trained, with CatBoost achieving the highest accuracy, reducing RMSE by 58.3% compared to the best existing equation. Geometric indices like wf/d, wc/d, and fc were identified as the most critical factors. A new equation derived from the CatBoost model and existing formulas integrated predictive accuracy with physical interpretability. This equation offers a robust tool for practical design applications. Future research should focus on optimizing geometric indices for varying FRP bar diameters.

Table 2 summarizes studies that applied ensemble boosting techniques to predict properties of polymer-based materials, specifically in civil engineering contexts, such as FRP-concrete bond strength, shear strength of FRP-RC beams, and fire resistance of FRP-strengthened RC beams. It outlines the boosting algorithms employed, the materials and properties predicted, the datasets used, model performance metrics (such as R^2, RMSE, and MAE), key influencing factors, and additional techniques or analyses used to improve prediction accuracy.

Table 2. Summary of studies applying ensemble boosting techniques in predicting properties of FRP and reinforced concrete systems.

Study	Boosting Technique	Application	Materials/ Properties Predicted	Dataset	Model Performance (R^2, MAE, RMSE)	Key Influencing Factors	Additional Techniques/ Analysis
Kim et al. [56]	CatBoost	FRP-concrete bond strength	FRP bond strength	855 shear test data	RMSE: 2.31, R^2: 0.96	Small dataset, categorical features	Compared with XGBoost, HGBoost, RF
Kumarawadu et al. [57]	XGBoost, CatBoost	Fire resistance of FRP-strengthened RC beams	Fire resistance	21,000 data points	Accuracy: >92%	Loading ratio, insulation depth, concrete cover	Bayesian optimization, SHAP analysis
Wang et al. [58]	XGBoost	Shear strength of FRP-RC beams	Shear strength	442 RC beam data	High prediction accuracy	Effective height of FRP, shear span ratio	Isolation forest anomaly detection
Mahmoudian et al. [59]	Decision Tree, RF, AdaBoost, XGBoost	Flexural bond strength of GFRP	GFRP-concrete bond	Experimental data	Accuracy: 100%	Concrete type, GFRP bar properties	Hyperparameter tuning, SHAP analysis
Mahmoudian et al. [60]	AdaBoost, XGBoost, CatBoost, GB, Hist GB	Bond strength in FRP-UHPC	FRP-UHPC bond strength	Experimental dataset	R^2: 0.95, RMSE: 2.21	Tensile strength, elastic modulus, embedment length	Shapley values, Voting Regressor
Wang et al. [61]	LightGBM, Genetic Programming	Fire resistance of FRP-strengthened RC beams	Fire resistance, deflection	20,000 data points	R^2: 0.923 (Fire Resistance), 0.789 (Deflection)	Insulation thickness, reinforcement area	Genetic Algorithm, SHAP analysis
Hu et al. [62]	XGBoost, Gradient Boosting	CFRP/metal composite laminates' mechanical properties	Tensile and bending strength	Experimental and simulation data	Best for tensile (XGBoost), bending (RF)	Laminate stacking sequence	Numerical and experimental integration
Aydın et al. [63]	DMLP, RF, GBR, LR, PR	Wear behavior of MWCNT-CFRP composites	Wear loss prediction	Experimental data	R^2: 0.9726	MWCNT content, load, sliding distance	SEM, EDS analysis
Li et al. [64]	RF, AdaBoost	Bond strength of BFRP-concrete in corrosive environments	BFRP-concrete bond strength	355 samples	R^2: 0.925, MAE: 0.0589	Corrosion, concrete strength, BFRP properties	SHAP analysis
Khodadadi et al. [65]	PSO-CatBoost	Compressive strength of CFRP-confined concrete	CFRP-CC compressive strength	916 experimental results	R^2: 0.9572	CFRP reinforcement ratio, unconfined CS	SHAP, PFI, Graphical interface

Table 2. *Cont.*

Study	Boosting Technique	Application	Materials/ Properties Predicted	Dataset	Model Performance (R^2, MAE, RMSE)	Key Influencing Factors	Additional Techniques/ Analysis
Alizamir et al. [68]	GBRT, RF, ANNMLP, ANNRBF	FRP-confinement in concrete strength	Concrete strength ratio	765 specimens	RMSE reduction: 69.94% (GBRT)	Concrete type, specimen geometry	Advanced feature selection
Amin et al. [50]	DT, GBT	Flexural capacity of FRP-RC beams	Flexural strength	60% training, 40% validation	R: 0.94 (GBT)	Beam depth, concrete CS	Sensitivity analysis
Amin et al. [70]	RF, XGBoost, LIGHT GBM	Bond strength of FRP on concrete prisms	Interfacial bond strength (IBS)	70% training, 30% testing	R^2: 0.942 (training), 0.865 (testing)	FRP thickness, elastic modulus	SHAP analysis
Tian et al. [71]	CatBoost	Bond strength of FRP bars to concrete	Bond strength	158 pull-out test results	RMSE reduction: 58.3%	Rib spacing and width, concrete properties	Integration with traditional formulas

3.3. Material Properties Prediction

Cheng et al. [72] introduced an ML-based method for predicting the friction coefficient of polymer–metal pairs by analyzing friction noise across a wide temperature range (−120 °C to 25 °C) and under various working conditions. Three ML algorithms—XGBoost, LightGBM, and CatBoost—were used to establish a relationship between the time-frequency features of the friction noise and the friction coefficient. Among the models, LightGBM provided the highest accuracy for friction coefficient prediction, while XGBoost excelled in predicting aluminum alloy–polymer pairs. The results show that LightGBM achieved average RMSE and R^2 values of 0.0135 and 0.615, respectively. The study demonstrated that ML can effectively predict the friction coefficient for different polymer–metal pairs, providing a basis for real-time, in situ monitoring of tribological properties. Future research will focus on applying this approach to a wider range of polymer materials and contact modes, and improving the algorithm's robustness under more severe environmental conditions.

Fatriansyah et al. [73] investigated the use of Simplified Molecular Input Line Entry System (SMILES) descriptors in ML models to predict the glass transition temperature (Tg) of polymers. Five models—k-nearest neighbors (KNN), support vector regression (SVR), XGBoost, ANN, and recurrent neural network (RNN)—were applied to predict Tg. The research highlights that SMILES descriptors with fewer than 200 characters are insufficient for accurate predictions, while those over 200 characters reduce model performance due to the curse of dimensionality. The ANN model achieved the highest R^2 of 0.79, but the XGBoost model, with an R^2 of 0.774, showed greater stability and faster training times, making it the preferred model. The study also found that the One Hot Encoding (OHE) method outperformed Natural Language Processing (NLP) in terms of training efficiency. Validation of the XGBoost model on new polymer data showed robust performance with an average deviation of 9.76% from actual Tg values. This research underscores the need for optimizing SMILES conversion and model parameters to improve prediction accuracy, with future work aimed at enhancing model generalizability.

Ascencio-Medina et al. [74] explored the dielectric permittivity of polymers, which is influenced by electronic, ionic, and dipolar polarization mechanisms. A dataset of 86 polymers was analyzed to develop two Quantitative Structure–Property Relationship (QSPR) models using the GB. From an initial 1273 descriptors, the most relevant ones were selected using a genetic algorithm. The GBR models showed high R^2 values of 0.938 and 0.822 for training and test datasets, respectively. An Accumulated Local Effect [75] (ALE) analysis was conducted to examine the relationship between the selected descriptors and their impact on permittivity, revealing key descriptors that positively and negatively affect dielectric properties. Compared to other ML models like multiple linear regression (MLR) [76] and partial least squares (PLS) [77], GBR models excelled in handling non-linear relationships and multicollinearity. The results highlight the potential of GBR models in accurately predicting dielectric permittivity, offering insights for the design of polymer materials with desired electrical properties. This approach can accelerate polymer development, reducing the need for extensive experimental testing.

Recently, the Ramprasad group introduced a QSPR model for predicting E_{gap} values of 4209 polymers, achieving an R^2 score of 0.90 and an RMSE of 0.44 at an 80/20 train-test split. Goh et al. [78] proposed an improved model, LGB-Stack, utilizing a two-level stacked generalization with LightGBM. Four molecular fingerprints were calculated from SMILES strings and reduced via recursive feature elimination to enhance input features for training. The model combines weak learners' outputs to form a strong final prediction model. LGB-Stack achieved R^2 and RMSE scores of 0.92 and 0.41 at the 80/20 split and further improved to 0.94 and 0.34 at a 95/5 split, surpassing the benchmark Ramprasad model [79]. This demonstrates LGB-Stack's effectiveness in accurately predicting polymer properties while offering a foundation for future enhancements and potential applications in transfer learning.

Rajaee et al. [80] evaluated the efficacy of decision tree and AdaBoost algorithms in predicting the mechanical and fracture properties of polypropylene nanocomposites reinforced with nanoparticles and toughened with thermoplastic elastomers. AdaBoost outperformed decision tree models in accuracy for predicting tensile strength, Young's modulus, elongation at break, elastic work, and plastic work. AdaBoost achieved an R^2 value of 0.90 for Young's modulus and demonstrated lower mean absolute percentage errors (<4% for some parameters). Sensitivity analysis identified thermoplastic polyolefin (TPO) [81] levels and nanoparticle content as the most influential features, significantly affecting tensile strength and Young's modulus. Results showed that low TPO levels with high nanoparticle content yielded the highest mechanical strength, while increasing TPO influenced other parameters like elongation at break non-linearly. These findings underline the superiority of AdaBoost in handling complex datasets and the pivotal role of material composition in determining mechanical properties.

Abdi et al. [82] explored the effectiveness of various ML models, including CatBoost, LightGBM, XGBoost, AdaBoost, GBDT, ET, DT, and RF, to predict tetracycline (TC) photodegradation from wastewater using metal–organic frameworks [83,84] (MOFs). A dataset of 374 data points was used, with input parameters like catalyst dosage, antibiotic concentration, illumination time, solution pH, and the MOFs' surface area and pore volume. The CatBoost model outperformed other models, achieving the highest accuracy with an AAPRE of 1.19% and an STD of 0.0431. This model accurately predicted TC degradation and followed the expected trends with varying operational parameters. Outlier detection confirmed the reliability of CatBoost, with 85% of predictions having errors below 1%. The results suggest that CatBoost is a reliable and efficient tool for predicting TC degradation in environmental applications like wastewater treatment.

Surface modification with hydrophilic polymer coatings offers a sustainable solution to prevent membrane clogging and reduce replacement frequency in water treatment systems. By combining molecular descriptors from RDKit and time-domain NMR (TD-NMR) data, Okada et al. [85] developed an ML approach for feature selection to predict surface properties. Polyacrylamide coatings were synthesized via UV-initiated copolymerization of ionic and nonionic monomers on PET films, with cross-linkers influencing the polymer chain dynamics. TD-NMR revealed differences in chain mobility linked to structural variations in cross-linkers, while contact angle measurements quantified surface hydrophilicity. Feature selection using Gradient Boosting Machine-Recursive Feature Elimination (GBM-RFE) demonstrated superior accuracy, identifying key molecular and dynamic properties influencing hydrophilicity. The findings highlight the importance of combining molecular descriptors and TD-NMR data to advance the development of hydrophilic polymer coatings and material-specific informatics methodologies.

Salehi et al. [86] investigated the use of ensemble ML (EML) models to predict the rheological properties of recycled plastic modified bitumen [87] (RPMB). Four models—RF, XGBoost, CatBoost, and Light Gradient-Boosting Machine (LightGBM)—were developed to predict complex shear modulus and phase angle under unaged and short-term aged conditions. Among these, the CatBoost model achieved the highest performance, with R^2 values of 0.98 for complex shear modulus and 0.93 for phase angle. SHAP analysis revealed that the penetration of base bitumen and the quantity of recycled plastic, especially HDPE pellets, were crucial factors affecting these properties. The study used various techniques, such as partial dependence plots and individual conditional expectation plots, to analyze feature interactions and validate model predictions. The data-driven models offer a cost-effective and efficient alternative to traditional laboratory testing for RPMB mixtures, providing valuable insights for material and pavement engineers. Future work could enhance model accuracy by incorporating larger datasets and other plastic types.

Polypropylene composites (PPCs) [27,88] are increasingly utilized due to their versatility, with heat deflection temperature (HDT) serving as a critical property indicator. To address the lack of theoretical equations linking material composition to HDT, an ML approach was proposed by Chonghyo et al. [89]. Among three algorithms—MLR, XGBoost, and CatBoost—CatBoost emerged as the most effective model for HDT prediction, achieving the highest R^2 value (0.8965) and lowest RMSE (7.3477) for the entire dataset. When tested on a subset of 59 "same recipes," CatBoost maintained superior accuracy (R^2 = 0.9801, RMSE = 2.6105). Its ordered encoding approach efficiently handled categorical data, outperforming mean encoding in MLR and XGBoost. A novel dimensionless number "A" was introduced to normalize and analyze variations within categorical groups, providing insights into HDT distributions. These results highlight CatBoost's potential in optimizing PPCs by reducing experimental trial and error.

Chepurnenko et al. [90] focused on developing ML models to predict the rheological properties of polymers from experimental stress relaxation curves. The research employed metaheuristic approaches, local search and evolutionary algorithms, to solve combinatorial optimization problems, with a focus on decision tree construction. CatBoost Regressor was used to solve the regression problem, and data normalization and regularization methods were applied to improve model accuracy. The models, developed using generated datasets for the EDT-10 epoxy binder, predict rheological parameters like initial relaxation viscosity and velocity modulus. Performance evaluation showed the models achieved low errors, with the maximum MAPE error of 0.86 and minimum MSE of 0.001, validating their effectiveness. Future work will explore expanding ML tools, including k-nearest neighbors and support vector regression.

Precise control in laser-based powder bed fusion (PBF-LB) of polymers is essential for ensuring the quality of aerospace and automotive components. Hofmann et al. [91] employed ML to predict local solidity using thermal and temporal features extracted from the melt's temperature profile, with infrared thermography data integrated with X-ray micro-computed tomography using LightGBM. Key predictors of porosity include the peak temperature of the melt and adequate reheating of subsurface layers. High prediction accuracy is achieved with a small voxel size and adjacent thermal data. The findings support detecting process defects and optimizing parameters without post-process testing. Future work includes extending datasets to new materials, geometries, and process parameters, ultimately enabling closed-loop feedback control systems to prevent defects and enhance industrial applications.

Gadagi et al. [92] investigated the use of ML techniques, specifically, Gradient Boosting Machine (GBM), AdaBoost, and XGBoost, to predict the surface roughness of jute/basalt epoxy composites in turning processes. The experiments, guided by Taguchi's L27 array, examined the effects of spindle speed, feed rate, and depth of cut on surface roughness. Among the models, XGBoost demonstrated the highest predictive accuracy, with minimal errors in both training and testing datasets. The optimal turning parameters for achieving the minimum surface roughness of 0.773 μm were identified as 1500 RPM spindle speed, 0.05 mm/rev feed rate, and 0.3 mm depth of cut. An analysis of variance (ANOVA) highlighted that feed rate and spindle speed significantly impacted surface roughness, while the depth of cut showed minimal effect. ML insights revealed that the feed rate had the greatest influence on surface roughness, followed by the spindle speed and depth of cut. The findings emphasize XGBoost's superior performance in predicting surface roughness and the effectiveness of ML in optimizing manufacturing processes.

Wang et al. [93] explored the hybridization of the Imperialist Competitive Algorithm (ICA) [94,95] with the Light Gradient Boosting Machine (LightGBM) to predict the CS of geopolymer concrete (CSGCo). The hyperparameters of the LightGBM model were optimized using ICA to enhance its accuracy. The hybrid ICA-LightGBM model was compared to the traditional LightGBM and four ANN topologies, including multi-layer perceptron (MLP), radial basis function (RBF), generalized feed-forward neural network (GFFNN), and Bayesian regularized neural network (BRNN). The evaluation was based on R^2, RMSE, and VAF metrics, with the ICA-LightGBM outperforming all other models in terms of prediction accuracy. Specifically, the ICA-LightGBM achieved an R^2 of 0.9871 (training) and 0.9805 (testing), significantly outperforming the traditional LightGBM and ANN models. The results confirm that ICA is an effective optimizer for improving LightGBM's predictive capabilities. This hybrid model can be used for accurate predictions of CSGCo, contributing to enhanced safety and efficiency in civil and construction applications.

Ahmad et al. [96] explored the use of ML algorithms to predict the CS of high-calcium fly-ash-based GPC. The study compared the performance of ensemble ML techniques—boosting and AdaBoost—against the individual ANN approach. The results show that boosting performed the best, achieving an R^2 of 0.96, while AdaBoost reached 0.93, and ANN lagged behind with an R^2 of 0.87. Boosting also had the lowest error values for MAE, MSE, and RMSE, highlighting its high prediction accuracy. Sensitivity analysis revealed that fly ash contributed significantly (45.3%) to the prediction of CS. The study concludes that ensemble techniques like boosting and AdaBoost are highly effective for predicting the mechanical properties of GPC, with boosting proving to be the most accurate. Additionally, incorporating more input parameters and increasing the dataset could further improve accuracy, making ML techniques a valuable tool in civil engineering.

Asadi et al. [97] explored the prediction of asphalt binder elastic recovery (ER) from Multiple Stress Creep Recovery (MSCR) [98] test results using ensemble learning methods.

The ensemble models tested included tree-based bagging (RF, Extra Trees) and boosting methods (XGBoost, LightGBM, CatBoost). Extra Trees and XGBoost emerged as the most accurate models, demonstrating superior performance with R^2 values of 0.852 and 0.842, respectively. These models surpassed traditional ER-DSR tests in predicting ER from MSCR results, despite differing temperature ranges. Key influential features identified were recovery at stress levels of 0.1 and 3.2 kPa. Clustering analysis revealed challenges in distinguishing patterns within the binders, suggesting potential improvements in MSCR analysis. Overall, the study advocates for the adoption of MSCR specifications over PG-Plus in asphalt binder characterization.

Shen et al. [99] developed ML models to predict the punching shear strength of FRP-reinforced concrete slabs, using a dataset of 121 experimental results. Several ML algorithms, including artificial neural network (ANN), SVM, decision tree (DT), and AdaBoost, were compared. AdaBoost demonstrated the best predicted accuracy with an RMSE of 29.83, MAE of 23.00, and R^2 of 0.99. The empirical models and design codes were also compared, with GB 50010-2010 (2015) showing the best performance among the traditional models. SHAP [100] was used to interpret AdaBoost's predictions, revealing the importance of variables such as the slab's effective depth and the Young's modulus of FRP reinforcement. The study highlighted that input variables, including the slab's depth, significantly influence punching shear strength predictions. Overall, AdaBoost outperformed traditional models, making it a reliable tool for predicting shear strength in FRP-reinforced concrete slabs.

Rahman et al. [101] presented an extensive database and ML models to predict the shear capacity of reinforced concrete (RC) beams strengthened with FRP. The database includes 584 experimental results for rectangular and T-beams with 12 input features covering variations in beam geometry and FRP properties. Ten ML models, including CatBoost (CatB), XGBoost, and RF, were developed and validated using 10-fold cross-validation. CatB and XGB exhibited superior performance, achieving R^2 values close to 0.9 and mean absolute errors below 0.25 kN, outperforming prior empirical models and design guidelines. SHAP identified the height of FRP layers and beam depth as key factors, while the type of fiber had minimal impact. The study emphasizes the need for updated databases to improve ML models and highlights the superior accuracy of ensemble learning techniques in predicting shear strength.

Table 3 presents an overview of studies employing ML techniques to predict polymer material properties and behavior across different applications. Each study utilizes a specific boosting algorithm to model and predict key properties, with an emphasis on model performance, key influencing factors, and any additional analysis methods used. The studies encompass a diverse range of materials, from polymers and composites to concrete and wastewater treatment, demonstrating the versatility and effectiveness of ML models in predicting complex material properties.

Table 3. Summary of studies applying boosting techniques in polymer materials properties prediction.

Study	Boosting Technique	Application	Materials/ Properties Predicted	Dataset	Model Performance (R^2, MAE, RMSE)	Key Influencing Factors	Additional Techniques/ Analysis
Cheng et al. [72]	XGBoost, LightGBM, CatBoost	Friction coefficient of polymer–metal pairs	Friction coefficient, temperature range (−120 °C to 25 °C)	Various working conditions	RMSE: 0.0135, R^2: 0.615	Friction noise, temperature	Time-frequency feature analysis

Table 3. *Cont.*

Study	Boosting Technique	Application	Materials/ Properties Predicted	Dataset	Model Performance (R^2, MAE, RMSE)	Key Influencing Factors	Additional Techniques/ Analysis
Fatriansyah et al. [73]	XGBoost, ANN, RNN, KNN, SVR	Glass transition temperature (Tg) of polymers	Tg of polymers	SMILES descriptors	R^2: 0.774, MAE: 9.76% deviation	SMILES descriptor length	One Hot Encoding vs NLP
Ascencio-Medina et al. [74]	GBR	Dielectric permittivity of polymers	Dielectric permittivity	86 polymers	R^2: 0.938 (train), 0.822 (test)	Electronic, ionic, dipolar polarization	Genetic algorithm, ALE analysis
Goh et al. [78]	LightGBM (LGB-Stack)	Polymer properties prediction	Various polymer properties	4209 polymers	R^2: 0.92, RMSE: 0.41	Molecular fingerprints	Feature reduction, Recursive Feature Elimination
Rajaee et al. [80]	AdaBoost, Decision Tree	Mechanical	Tensile strength, Young's modulus, elongation	Polypropylene nanocomposites	R^2: 0.90 for Young's modulus	TPO levels, nanoparticle content	Sensitivity analysis
Abdi et al. [82]	CatBoost	Photodegradation of tetracycline	TC degradation from wastewater	374 data points	AAPRE: 1.19%, STD: 0.0431	Catalyst dosage, pH, surface area	Outlier detection
Okada et al. [85]	GBM-RFE	Hydrophilicity of polymer coatings	Surface hydrophilicity	Polyacrylamide coatings	High accuracy in feature selection	Polymer chain dynamics	TD-NMR, Recursive Feature Elimination
Salehi et al. [86]	CatBoost, XGBoost, LightGBM, RF	Rheological properties of RPMB	Complex shear modulus, phase angle	Recycled plastic modified bitumen	R^2: 0.98 (shear modulus)	Base bitumen, recycled plastic quantity	SHAP analysis
Chonghyo et al. [89]	CatBoost, XGBoost, MLR	Heat deflection temperature (HDT) of PPCs	Heat deflection temperature	Polypropylene composites	R^2: 0.8965, RMSE: 7.3477	Material composition	Novel dimensionless number "A"
Chepurnenko et al. [90]	CatBoost, Evolutionary algorithms	Rheological properties of polymers	Viscosity, velocity modulus	Epoxy binder	MAPE: 0.86, MSE: 0.001	Stress relaxation	Data normalization, regularization
Hofmann et al. [91]	LightGBM	Local solidity in PBF-LB process	Porosity, solidity	Thermal and temporal features	High prediction accuracy	Peak temperature, reheating	Infrared thermography, X-ray micro-CT
Gadagi et al. [92]	XGBoost, AdaBoost, GBM	Surface roughness of composites	Surface roughness of epoxy composites	Jute/basalt composites	High accuracy in roughness prediction	Spindle speed, feed rate	Taguchi L27 array
Wang et al. [93]	ICA-LightGBM	Geo-polymer concrete CS prediction	Compressive strength (CS) of geo-polymer concrete	Geo-polymer concrete dataset	R^2: 0.9871 (train), 0.9805 (test)	Hyperparameter optimization	Imperialist Competitive Algorithm optimization

Table 3. *Cont.*

Study	Boosting Technique	Application	Materials/ Properties Predicted	Dataset	Model Performance (R^2, MAE, RMSE)	Key Influencing Factors	Additional Techniques/ Analysis
Ahmad et al. [96]	Boosting, AdaBoost	Compressive strength of GPC	Compressive strength of GPC	High calcium fly-ash-based GPC	R^2: 0.96	Fly ash composition	Sensitivity analysis
Asadi et al. [97]	XGBoost, LightGBM, CatBoost, Extra Trees	Asphalt binder elastic recovery (ER) prediction	Elastic recovery (ER) from MSCR test results	Asphalt binders	R^2: 0.852 (Extra Trees), 0.842 (XGBoost)	Stress recovery at 0.1, 3.2 kPa	Clustering analysis
Shen et al. [99]	AdaBoost	Punching shear strength of FRP RC slabs	Punching shear strength of FRP RC slabs	121 experimental results	R^2: 0.99, RMSE: 29.83, MAE: 23.00	Effective depth, Young's modulus of FRP	SHAP analysis
Rahman et al. [101]	CatBoost, XGBoost	Shear capacity of FRP RC beams	Shear capacity of FRP RC beams	584 experimental results	R^2: 0.9, MAE: 0.25 kN	FRP layer height, beam depth	SHAP analysis

3.4. Advanced Manufacturing and Processing

Biruk-Urban et al. [102] investigated the machinability of new GFRP composites, focusing on the impact of drilling parameters on cutting forces and delamination. Four GFRP materials, varying in fiber type (plain or twill woven) and weight fraction (wf) ratio, were tested using a carbide diamond-coated drill. A novel ink penetration method was introduced to assess delamination, proving effective for detecting both push-out and peel-up delamination, as well as fiber pullouts. ML models were used to simulate the relationship between drilling parameters and delamination, with the Gradient Boosting Regressor achieving the highest accuracy. Results showed that feed per tooth significantly influenced delamination and cutting force amplitude, with lower values of feed per tooth reducing both. Twill fiber materials with lower wf ratios exhibited lower cutting forces and delamination factors, highlighting their machinability. The study offers insights for optimizing drilling processes and proposes future research on material properties, drill geometries, and advanced delamination detection techniques.

Jalali et al. [103] investigated the impedance properties of multi-walled carbon nanotube (MWCNT)/polystyrene nanocomposites synthesized via microwave-assisted in situ polymerization, examining the impact of microwave power, exposure time, and frequency. The Taguchi method and ANOVA identified microwave power as the most significant factor influencing impedance. A predictive model with an R^2 of 0.96 was developed, showing high accuracy in predicting impedance values. ML models, including Decision Tree, RF, XGBoost, CatBoost, and LightGBM, were applied to enhance prediction accuracy. RF and CatBoost outperformed the other models, achieving R^2 values of 0.9880 and 0.9811 on testing data, respectively. The results indicate that higher microwave power and extended exposure time increase impedance due to enhanced polystyrene content. This study demonstrates the potential of ML methods for accurately predicting impedance and tailoring the design of MWCNT-based composites for electrical applications.

Ma et al. [104] proposed using the XGBoost ML algorithm to predict the axial compressive capacity of CFRP-confined CFST short columns. The dataset, consisting of 379 data points from literature and experiments, includes factors such as concrete, steel, and CFRP strengths, cross-sectional areas, and section shapes. Eight ML algorithms, including XG-

Boost, were tested, with XGBoost showing the best prediction performance, achieving an R^2 value of 0.9719. Hyperparameter optimization further improved the XGBoost model, increasing R^2 to 0.9850. The study also identifies the importance of features like the cross-sectional area of core concrete and steel tube in determining compressive capacity. The optimized XGBoost model was highly accurate in predicting the axial capacity, outperforming other models, such as RF and Gradient Boosting Decision Trees. These findings suggest that XGBoost is an effective tool for predicting the behavior of CFRP-confined CFST short columns under axial compression.

Lignin plays a vital role in substituting synthetic polymers and reducing energy consumption, but traditional wet chemical methods for determining lignin content are inefficient and environmentally harmful. In the study by Gao et al. [105], the lignin content of Chinese fir was predicted using Raman spectroscopy, similar to a previous method for poplar. The peak at 2895 cm^{-1} was identified as the optimal internal standard, and the XGBoost algorithm demonstrated the highest prediction accuracy. Transfer learning was applied to improve the model's accuracy and robustness, leading to an efficient, environmentally friendly method for predicting lignin content. Comparisons of nine algorithms revealed that advanced Gradient Boosting Machines (GBM) outperformed classic ML algorithms. Although the XGBoost model achieved a high test R^2 and low RMSE, transfer learning was used to overcome challenges related to chemical structure differences. Ultimately, a reliable lignin content prediction model for Chinese fir, achieving a test R^2 of 0.93, was successfully developed using XGBoost or LightGBM. This approach offers significant potential for more accurate and sustainable lignin content analysis.

Donga et al. [106] proposed a novel system for evaluating the hydrophobicity of insulated material surfaces using image processing and decision tree methods. A mixed image segmentation method is introduced to handle challenges like non-controlled illumination and nonstandard surfaces. The system uses four new characteristic parameters to describe the images of each sample, with classification performed using a MultiBoost decision tree, combining AdaBoost and Bagging algorithms. The results show that MultiBoost outperforms AdaBoost in classification accuracy, reducing errors and demonstrating better robustness, especially with k-fold cross-validation. The system uses a Digital Signal Processor (DSP) platform for training and testing, making it suitable for real-time applications. The study also highlights the limitations of traditional segmentation methods and suggests that the proposed approach is more versatile for uneven lighting images, though not universally applicable. Future work will focus on improving segmentation methods and developing adaptive algorithms for better accuracy in diverse conditions.

Kong [107] introduced an intelligent approach using hyperparameter optimization to predict the interfacial bond strength between FRP and concrete. By selecting CatBoost as the primary ML model, it outperformed eight other models, achieving an R^2 of 0.9394 and MAPE of 1.21%. The hyperparameter optimization significantly improved the model's accuracy, reducing dispersion by 90%. The optimized CatBoost model showed better performance than existing models in terms of R^2, root mean square error, and coefficient of variation. The study also demonstrated that ML models, particularly the optimized CatBoost model, outperformed traditional bond strength models by 16.5% in mean accuracy and 14.19% in R^2. Furthermore, grid searching was used to optimize the hyperparameters of models like CBR, MLP, and LightGBM, leading to enhanced prediction performance. The findings highlight the potential of hyperparameter optimization to improve the accuracy of predicting FRP-concrete bond strength, offering a reliable and efficient approach for future applications.

Membrane-based purification of therapeutic agents has gained significant attention as a promising alternative to traditional methods like distillation. Alanazi et al. [108]

introduced a numerical approach employing multiple ML methods to predict solute concentration distributions during membrane-based separations. Key inputs, r and z, and a single target, C, were analyzed using over 8000 data points. Adaboost was applied to three base learners: k-nearest neighbors (KNN), linear regression (LR), and Gaussian process regression (GPR) [109]. The models were further optimized using the Bat Algorithm (BA). Boosted KNN achieved the highest R^2 score of 0.9853, with low MAE and MAPE values, establishing it as the most accurate model. The boosted GPR model followed closely with robust predictive performance. These findings underscore the potential of ML-based strategies for improving membrane separation processes by providing high accuracy and insightful predictions.

Table 4 summarizes several recent studies that demonstrate the use of boosting techniques, focusing on their application in advanced manufacturing and processing settings:

Table 4. Summary related to polymers advanced manufacturing and processing studies.

Study	Boosting Technique	Application	Materials/ Properties Predicted	Dataset	Model Performance (R^2, MAE, RMSE)	Key Influencing Factors	Additional Techniques/ Analysis
Biruk-Urban et al. [102]	GB	GFRP composites machinability	Cutting forces, delamination	Carbide diamond-coated drill data	High accuracy in delamination prediction	Drilling parameters, fiber type, weight fraction	Novel ink penetration method for delamination detection
Jalali et al. [103]	RF, CatBoost	MWCNT-polystyrene nanocomposites impedance	Impedance properties	Microwave-assisted synthesis data	R^2 = 0.9880 (RF)	Microwave power, exposure time, frequency	Taguchi method, ANOVA for feature importance
Ma et al. [104]	XGBoost	CFRP-confined CFST short columns	Axial compressive capacity	379 data points from literature	R^2 = 0.9850 after hyper-parameter optimization	Concrete, steel, CFRP strengths, cross-sectional area	Hyperparameter optimization for improved accuracy
Gao et al. [105]	XGBoost, LightGBM	Lignin content prediction in Chinese fir	Lignin content	Raman spectroscopy data	R^2 = 0.93 (XGBoost)	Raman peaks, chemical structure differences	Transfer learning for model improvement
Donga et al. [106]	MultiBoost (AdaBoost + Bagging)	Hydrophobicity evaluation of insulated materials	Hydrophobicity properties	Image data from surface samples	High classification accuracy with MultiBoost	Illumination and surface irregularities	Image segmentation, DSP platform for real-time training
Kong [107]	CatBoost	FRP-concrete bond strength prediction	Bond strength	Experimental data	R^2 = 0.9394, MAPE = 1.21%	Interfacial bond strength	Hyperparameter optimization, grid search
Alanazi et al. [108]	Adaboost	Membrane separation process in therapeutic agent purification	Solute concentration distribution	Over 8000 data points from experiments	R^2 = 0.9853 (Boosted KNN)	Solute concentration, membrane parameters	Bat Algorithm for model optimization

3.5. Sustainability, Environmental, and Structural Performance

Tahir et al. [110] focused on designing novel polymer donors for organic solar cells using ML. Mordred descriptors were calculated for 271 polymer donors to train four ML models, with the gradient boosting regressor achieving the highest $R^2 = 0.85$. A chemical library of polymer donors was generated using BRICS, and similarity analysis using RDKit revealed clusters with strong structure–performance relationships. The 30 donors with the highest predicted power conversion efficiency (PCE) ranging from 9.13% to 9.44% were identified. Synthetic accessibility scores showed most polymers to be easily synthesizable ($SA < 6$). Structural changes minimally impacted PCE, emphasizing the robustness of the designed materials.

Jiang et al. [111] developed ML models to predict amorphization and chemical stability during the hot-melt extrusion (HME) process for amorphous solid dispersions (ASDs). Using a dataset of 760 formulations, the study found that ECFP-LightGBM and ECFP-XGBoost are the most accurate models for predicting amorphization (92.8% accuracy) and chemical stability (96.0% accuracy), respectively. Key factors such as barrel temperature, drug loading, excipient ratios, and the chemical structure of active pharmaceutical ingredients (APIs) [112] significantly affect the results. SHAP and information gain analyses reveal important API substructures, such as chlorine atoms and nitrogen-containing heterocycles, that influence amorphization and stability. The study's ML models can reduce trial-and-error in ASD development by accurately predicting amorphization and chemical degradation, ultimately streamlining the product development process. Additionally, the findings highlight the critical processing parameters, such as extruder configuration and screw speed, which influence both the amorphization and stability of ASDs.

A method for real-time monitoring of polymer agglomeration in a fluidized bed reactor (FBR) was developed by Pang et al. [113] using voiceprint feature recognition based on acoustic emission detection. Acoustic signals from polymer collisions on the reactor walls are collected, and voiceprint features are extracted using Mel Frequency Cepstrum Coefficients (MFCC) [114] and Linear Prediction Cepstrum Coefficients (LPCC). An improved Adaboost algorithm is proposed to classify these features, incorporating cost factors and the Gini index to better handle unbalanced small samples and improve accuracy. Experimental results from a fluidized bed pilot plant demonstrate the method's effectiveness. The modified Adaboost algorithm outperforms the original in terms of classification accuracy, particularly for detecting micro-agglomeration and severe-agglomeration states. Performance evaluation metrics like the F-score indicate a significant improvement in prediction efficiency and accuracy. The method offers strong potential for industrial applications, particularly in the polyethylene production process, by enhancing agglomeration fault detection.

Fiosina et al. [115] employed advanced ML models to simulate and reverse engineer polymerization processes, addressing the challenges of tailoring polymer properties. Using data from a kinetic Monte Carlo simulator, ML methods (e.g., RF, XGBoost, and CatBoost) predicted key outputs, such as monomer concentration, average molar masses, and molar mass distributions [116] (MMDs) with high accuracy ($R^2 > 0.96$). Reverse engineering models also demonstrated good agreement with targeted MMDs despite a lower R^2 of 0.68. Multi-target regression (MTR) models outperformed single-output approaches by capturing dependencies among outputs. Explainability techniques validated the importance of input variables, aligning with expert expectations. Ensemble-based methods, particularly decision tree models, excelled in accuracy and scalability, reducing training data needs without performance loss. These results enable efficient prediction of polymerization recipes and conditions, advancing ML applications in polymer engineering. Future work will extend these methods to multi-objective optimization and complex polymer microstructures.

Deshpande et al. [117] investigated the prediction of the specific wear rate of glass-filled PTFE composite using ML algorithms, analyzing experimental data from a pin-on-disc wear testing machine. Various operating parameters, such as applied load, sliding velocity, and sliding distance, were varied using an orthogonal array L25 for experimentation. The data were analyzed using linear regression (LR), GB, and RF, with R^2 values of 0.91, 0.97, and 0.94, respectively, showing the highest R^2 value for the GB model, indicating an almost perfect fit. Pearson's correlation analysis revealed that sliding distance and applied load significantly impacted the wear rate, while sliding velocity had a weaker effect. The experimental results showed a minimal wear rate of 3.04186×10^{-5} mm³/Nm at a load of 150 N, sliding velocity of 2 m/s, and sliding distance of 5000 m. The highest recorded wear rate was 4.410698×10^{-5} mm^3/Nm. The study highlights the effectiveness of ML models in predicting wear rates and the importance of optimizing operating parameters for improved material performance.

Huang et al. [118] explored the influence of various factors on the open circuit voltage (Voc) of ternary polymer solar cells (PSCs) with non-fullerene acceptors (NFAs) using ML algorithms, such as XGBoost, k-nearest neighbor (KNN), and RF. The analysis reveals that the doping concentration of the third component has the greatest impact on Voc, with an optimal HOMO and LUMO energy level of the third component around −5.7 eV and −3.6 eV, respectively. The molecular descriptors (MDs) and molecular fingerprints (MFs) of the third component, such as hydrogen bond strength and aromatic ring structure, also significantly affect Voc. XGBoost was found to be the most accurate model for predicting Voc, with a low RMSE of 0.031 and MAE of 0.022. The study also highlights that the third component's composition, including four methyl groups and two carbonyl groups, maximizes Voc. These findings offer valuable insights for designing and optimizing materials to enhance Voc in ternary PSCs, potentially improving their efficiency.

Plain concrete's low tensile strain capacity (TSC) limits its performance, prompting the development of engineered cementitious composites (ECC) with polymer fibers to improve ductility. Inqiad et al. [119] aimed to predict ECC's TSC using ML techniques, including Multi-Expression Programming (MEP), Gene Expression Programming (GEP), AdaBoost, and XGBoost. Among these, XGB achieved the highest accuracy, with a correlation coefficient of 0.986 and the lowest objective function (OF) value of 0.081. Shapley additive analysis revealed that fiber content, age, and water-to-binder ratio significantly impact TSC. While MEP and GEP provided empirical equations, XGB outperformed in precision. The study emphasizes the need for larger datasets and consideration of additional parameters like aggregate fineness and fiber properties to enhance model robustness and utility for predicting other ECC properties. These advancements support faster, cost-effective, and accurate ECC material evaluations.

Nguyen et al. [120] explored the flexural behavior of reinforced concrete beams using experimental tests and advanced ML models. Eight beams, incorporating varying proportions of recycled aggregates, fly ash, silica fume, and CFRP, were tested to analyze structural performance. A comprehensive dataset of 4851 samples enabled the application of ML frameworks, including RF, XGBoost, and LightGBM (LGBM), with hyperparameter tuning via Pareto optimization. Among the models, RFR demonstrated the highest accuracy, achieving the lowest MSE and effectively predicting flexural strength. Sensitivity analysis identified the key factors influencing beam performance, such as aggregate proportions, CS, and CFRP presence. Experimental results revealed notable improvements in CS (up to 53%) and load-bearing capacity (7%) for beams with recycled aggregates and silica fume. This study highlights the synergy of experimental analysis and ML techniques, advancing sustainable construction practices and optimizing structural design.

Table 5 summarizes sustainabilty studies, including targeted variables, the materials or properties predicted, the datasets and performance metrics, as well as the key factors influencing model predictions and additional techniques employed.

Table 5. Summary of polymers studies on sustainability, environmental, and structural performance.

Study	Boosting Technique	Application	Materials/ Properties Predicted	Dataset	Model Performance (R^2, MAE, RMSE)	Key Influencing Factors	Additional Techniques/ Analysis
Gao et al. [105]	XGBoost, LightGBM	Lignin content prediction	Lignin content in Chinese fir	Raman spectroscopy data	Test R^2 = 0.93	Raman peak (2895 cm^{-1}), chemical structure differences	Transfer learning; comparison of 9 algorithms
Tahir et al. [110]	Gradient Boosting Regressor	Design of polymer donors for OSCs	Predicted power conversion efficiency (PCE)	Mordred descriptors for 271 polymer donors	$R^2 = 0.85$	Molecular structure, synthetic accessibility	BRICS-based chemical library; RDKit similarity analysis
Jiang et al. [111]	ECFP-LightGBM, ECFP-XGBoost	Hot-melt extrusion for ASDs	Amorphization and chemical stability	760 formulation data points	Accuracy: 92.8% (amorphization), 96.0% (stability)	Barrel temperature, drug loading, API substructures	SHAP and information gain analyses
Pang et al. [113]	Improved AdaBoost	Real-time monitoring in FBR	Polymer agglomeration states	Acoustic emission signals (MFCC, LPCC)	Improved classification accuracy (F-score elevated)	Acoustic features affected by illumination	Cost factors and Gini index integration; DSP platform
Fiosina et al. [115]	XGBoost, CatBoost	Reverse engineering polymerization	Monomer concentration, molar masses, MMDs	Kinetic Monte Carlo simulator data	$R^2 > 0.96$ for predictions; 0.68 for reverse engineering	Polymerization kinetics input variables	Multi-target regression; explainability techniques
Deshpande et al. [117]	Gradient Boosting (GB)	Wear rate prediction in composites	Specific wear rate of glass-filled PTFE	Pin-on-disc wear test data (L25 array)	$R^2 = 0.97$ (GB model)	Sliding distance, applied load, sliding velocity	Pearson's correlation analysis
Huang et al. [118]	XGBoost	OSC performance optimization	Open circuit voltage (Voc) of ternary PSCs	Data on polymer solar cells with NFAs	RMSE = 0.031, MAE = 0.022	Doping concentration, HOMO/LUMO levels, MDs	Molecular descriptor and fingerprint analysis
Inqiad et al. [119]	XGBoost	ECC TSC prediction	TSC of ECC	Experimental ECC data	Correlation coefficient = 0.986, OF = 0.081	Fiber content, age, water-to-binder ratio	Comparison with MEP and GEP; Shapley additive analysis
Nguyen et al. [120]	XGBoost, LightGBM, RF	Flexural behavior of RC beams	Flexural strength	4851 experimental samples	RF achieved lowest MSE (highest accuracy)	Aggregate proportions, compressive strength, CFRP presence	Pareto optimization for hyperparameter tuning; sensitivity analysis

4. Review Outlook

To provide a structured overview of the challenges, limitations, and potential future directions for the application of boosting methods in polymer science, a diagram (Figure 5) has been constructed. This visual representation organizes the discussed aspects into thematic groups, emphasizing their interrelations and collective impact on the field.

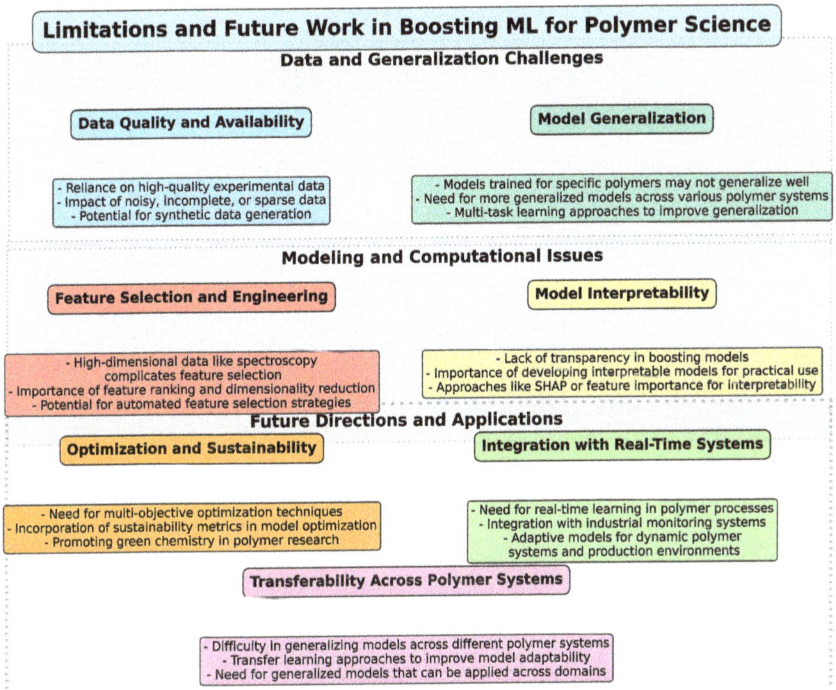

Figure 5. Illustrative diagram summarizing the limitations and future directions in the application of boosting ensemble learning in polymer science.

Boosting methods based on ensemble learning, such as Gradient Boosting and XGBoost, are being applied to capture the complex, non-linear relationships between polymer properties, processing parameters, and performance outcomes. Traditional techniques like linear regression and support vector machines are limited in modeling such interactions, whereas boosting methods iteratively combine multiple weak learners to form a comprehensive model [121]. This approach allows the model to learn intricate patterns within high-dimensional datasets without relying on extensive feature engineering, while reducing bias and variance compared to conventional techniques [122]. In addition, boosting methods integrate diverse input variables—including material characteristics, processing conditions, and environmental factors—into a single predictive framework [123]. Feature importance analysis within these models enables the identification of input variables that contribute most to the prediction, thus providing a clear basis for process optimization and material design in polymer science [124].

4.1. Analysis

In analyzing the boosting-based ML techniques applied to material property prediction in polymer science, the studies can be grouped into several thematic categories based on their application areas and focus.

The first group includes studies focused on predicting the mechanical properties of composites, such as CS, tensile strength, and load-bearing capacities of composites like short FRP composites, geopolymer concrete, and fly ash-based concrete. Boosting algorithms like XGBoost, LightGBM, and AdaBoost have consistently outperformed traditional models (e.g., decision trees or linear regression) in terms of prediction accuracy, as evidenced in studies like those by Zhao et al. [41] and Wang et al. [47]. The inclusion of feature importance analysis (e.g., SHAP) has enhanced the interpretability of results, helping to identify key factors influencing mechanical properties. However, despite high accuracy, some studies noted challenges in improving model reliability with sparse or incomplete datasets. For example, Katlav et al. [46] highlighted that expanding datasets and exploring advanced AI techniques could improve model performance. Additionally, data imbalance (e.g., in experimental datasets) remains an issue, which was addressed in some studies via techniques like SMOTE. Future work could focus on integrating multi-source data (e.g., experimental, simulation, and real-world operational data) to improve dataset diversity. Enhancing the models with deep learning techniques or hybrid models that combine boosting with neural networks might improve predictive accuracy and generalization. Furthermore, increasing the size and variety of experimental datasets could further reduce errors and improve the robustness of predictions.

The second group of studies focuses on sustainability and environmental impact prediction, particularly targeting carbon footprint predictions or the sustainable optimization of concrete materials. Techniques like adaptive boosting (AdaBoost-DTR) demonstrated excellent performance in predicting CO_2 emissions in geopolymer concrete (Wudil et al. [55]). These models help optimize material formulations to reduce carbon emissions while maintaining structural integrity. The primary limitation in this category is the limited generalizability of models, as noted by Wudil et al. [55], due to the dataset's focus on a specific type of geopolymer concrete and its limited size. There are also challenges related to incorporating more complex environmental data, such as lifecycle analysis or real-time IoT data. Broader datasets that incorporate different types of geopolymer concretes and alternative materials could improve model generalization. Additionally, combining boosting models with reinforcement learning or real-time monitoring systems could provide a more dynamic and accurate assessment of environmental impact in real-world applications.

The third group of studies focuses on geopolymer concrete and other sustainable materials. Ensemble methods like AdaBoost and RF Regression (RF) have been successfully applied to predict CS and other properties of geopolymer concrete (e.g., by Amin et al. [50]). These models have proven to be faster and more cost-effective than traditional experimental methods. However, a key limitation in this area is the inconsistent quality and range of datasets used for model training. For example, studies like those by Khan et al. [48] and Amin et al. [50] pointed out that the datasets were limited in terms of geographical diversity and specific material types. Expanding the dataset to include a wider range of geopolymer concrete formulations and real-world conditions (e.g., various curing times, environmental factors) could improve the accuracy and robustness of the predictions. Additionally, applying transfer learning or synthetic data generation techniques could help overcome dataset limitations by simulating material behavior in untested conditions.

XGBoost, LightGBM, and AdaBoost have shown promise in predicting complex material properties with high accuracy. These models are beneficial in the context of composite and sustainable materials, where traditional testing methods are time-consuming and costly. The ability to interpret feature importance (via SHAP or similar methods) is also a major advantage, allowing researchers to identify key factors influencing material performance. However, common limitations across studies include dataset quality (e.g., sparsity, imbalance, and lack of diversity), model overfitting, and generalization issues. Despite the promise of boosting techniques, the accuracy of predictions often depends on the quantity and quality of the data. Furthermore, computational costs and time constraints can be a challenge, especially in high-dimensional datasets or when incorporating complex simulations.

While CatBoost and LightGBM models offer high performance in handling structured data, their suitability varies depending on the specific characteristics of the dataset and prediction requirements. CatBoost is advantageous when dealing with categorical data, as it employs an ordered target-encoding method that prevents data leakage and overfitting [125]. This makes it well suited for scenarios where polymer formulations include categorical variables, such as material types, additive classifications, or process categories. LightGBM, on the other hand, relies on one-hot encoding or label encoding for categorical variables, which can sometimes lead to information loss or increased model complexity [126]. LightGBM is optimized for speed and memory efficiency, making it ideal for large datasets with high-dimensional numerical features [127]. Its histogram-based approach and leaf-wise tree growth strategy allow it to train faster and scale effectively, which is beneficial when modeling large experimental datasets involving high-resolution mechanical testing parameters, such as tensile strength, compression, shear, and fatigue resistance.

For progressive predictions involving dynamic process parameters—such as temperature, pressure, additives, and time variations in polymer molding—LightGBM can be more efficient due to its ability to handle numerical features and complex interactions with reduced computation time [128,129]. However, CatBoost can provide more stable and interpretable predictions when categorical variables significantly influence the mechanical properties of polymer materials [130–132]. CatBoost has built-in Bayesian bootstrapping and an ordered boosting method that improve its robustness to noisy or imbalanced datasets, making it suitable when working with experimental data that may have missing values or skewed distributions [133]. LightGBM, while also capable of handling imbalanced data, may require additional techniques, such as balanced weight adjustments or custom loss functions, to achieve comparable performance [134].

4.2. Limitations

One of the primary limitations is the quality and availability of data. Many studies, such as those predicting lignin content [105] and polymer agglomeration [113], relied on relatively small or unbalanced datasets, which can constrain the reliability and generalizability of ML models. To enhance model performance, future research could focus on expanding and diversifying datasets, potentially through automated data collection methods or collaborative efforts within the research community.

Another challenge is the interpretability of boosting models. While these models often achieve high accuracy, their complexity can make them difficult to understand, which limits their adoption in practical applications. For instance, in studies like [111,119], explainability techniques such as SHAP were employed to gain insights into model behavior, but these techniques need to be further developed to provide clear, actionable insights for stakeholders. Ensuring that models are both accurate and interpretable will be crucial for their widespread acceptance, especially in industries where decision-makers require transparent models to trust automated predictions.

The process of feature selection and engineering also impacts in the success of boosting models. Studies such as [118] demonstrate the importance of molecular descriptors for predicting polymer properties, but automating and improving feature selection processes remains a challenge. Future work could focus on developing more advanced algorithms for feature engineering or domain-specific heuristics that would reduce the burden on researchers and improve model performance.

Moreover, while transfer learning was successfully applied in some studies, such as in the prediction of lignin content [105], generalizing across different polymer systems remains difficult. The need for transferability across polymer systems is essential for building ML models that are universally applicable. Future research could explore the development of generalized frameworks or domain-adaptation techniques that would reduce the dependency on system-specific training data, enhancing the flexibility and scalability of boosting models across a broader range of polymer materials.

Furthermore, although boosting methods have proven effective for predicting outcomes, their integration into experimental workflows has been somewhat limited. For example, the real-time monitoring of polymer agglomeration [113] and the reverse engineering of polymerization processes [115] show great promise, but their validation in industrial settings remains an ongoing challenge. In the future, more robust hybrid frameworks that combine ML models with real-time monitoring systems will be needed to accelerate the adoption of these methods in practical applications.

4.3. Future Work

Another important area for future work is optimization beyond prediction. Most of the studies reviewed emphasized predictive accuracy, but few addressed the optimization of polymer properties, synthesis conditions, or structural designs. A promising direction for future research is the application of multi-objective optimization techniques to balance trade-offs between factors such as performance, cost, and sustainability, as suggested in [115]. This approach could lead to better-designed materials and processes that fulfill the demands of both industry and sustainability.

Sustainability and the role of green chemistry in polymer science are also crucial considerations. Several studies, such as those predicting lignin content [105] and designing polymer donors for organic solar cells [110], emphasized the potential for ML to guide the development of eco-friendly materials. However, future research should expand on this by incorporating life cycle assessments and environmental impact metrics into ML models. This would help ensure that the polymers being designed not only perform well but also minimize their ecological footprint.

Finally, real-time learning systems are an essential future direction. Polymer processes, such as those in fluidized bed reactors [113] or wear testing [117], are highly dynamic, and the ability to adapt to evolving data streams is critical. Future research should focus on developing online learning algorithms capable of maintaining the efficiency of boosting methods in such dynamic environments. These adaptive systems will be crucial for improving accuracy and efficiency in real-time polymer processing and product development.

Moving forward, integrating multiple ML paradigms (such as combining ensemble methods with deep learning) could enhance predictive performance. Additionally, the application of hybrid models that combine the strengths of simulation-based methods with ML could allow for better-informed predictions. For example, Ghasem [135] presented a Computational Fluid Dynamics (CFD) and AI/ML simulation of a polypropylene fluidized bed reactor, aiming to reduce reactor loss and enhance process understanding. By combining CFD with machine learning algorithms, the simulation accurately predicts reactor performance and identifies key operating parameters to optimize polypropylene yield and reactor efficiency. Ensuring that datasets are diverse, well curated, and representative of real-world scenarios will be key to improving model reliability and performance. By addressing the limitations related to data quality, model complexity, and computational efficiency, boosting-based ML models could offer substantial improvements in predicting and optimizing material properties, accelerating the development of advanced polymer and composite materials.

In conclusion, while boosting-based ML methods have shown considerable promise in addressing the challenges of polymer science, future work must tackle the limitations outlined above. By focusing on improving data quality, model interpretability, feature engineering, transferability, and real-time adaptation, as well as integrating optimization and sustainability considerations, boosting methods can become indispensable tools for advancing both fundamental research and industrial applications in polymer science.

Author Contributions: Conceptualization, I.M. and V.T.; data curation, A.G. and V.N.; funding acquisition, A.G., V.N., and A.B.; investigation, V.T. and V.N.; methodology, I.M.; project administration, A.G., V.N., and A.B.; resources, I.M.; software, I.M., V.T., V.N., and A.B.; supervision, A.G., V.N., and A.B.; validation, A.B.; visualization, I.M.; writing—original draft, I.M.; writing—review and editing, V.T. and A.B. All authors have read and agreed to the published version of the manuscript.

Funding: This research received no external funding.

Conflicts of Interest: The authors declare no conflicts of interest.

References

1. Broda, M.; Yelle, D.J.; Serwańska-Leja, K. Biodegradable Polymers in Veterinary Medicine—A Review. *Molecules* **2024**, *29*, 883. [CrossRef] [PubMed]
2. Huang, B.; Yang, M.; Kou, Y.; Jiang, B. Absorbable implants in sport medicine and arthroscopic surgery: A narrative review of recent development. *Bioact. Mater.* **2024**, *31*, 272–283. [CrossRef] [PubMed]
3. Kuperkar, K.; Atanase, L.I.; Bahadur, A.; Crivei, I.C.; Bahadur, P. Degradable polymeric bio (nano) materials and their biomedical applications: A comprehensive overview and recent updates. *Polymers* **2024**, *16*, 206. [CrossRef] [PubMed]
4. Kim, H.C.; Tu, R.; Sodano, H.A. Room temperature 3D printing of high-temperature engineering polymer and its nanocomposites with porosity control for multifunctional structures. *Compos. Part B Eng.* **2024**, *279*, 111444. [CrossRef]
5. Sabet, M. Unveiling advanced self-healing mechanisms in graphene polymer composites for next-generation applications in aerospace, automotive, and electronics. *Polym.-Plast. Technol. Mater.* **2024**, *63*, 2032–2059. [CrossRef]
6. Silva, N.C.; Chevigny, C.; Domenek, S.; Almeida, G.; Assis, O.B.G.; Martelli-Tosi, M. Nanoencapsulation of active compounds in chitosan by ionic gelation: Physicochemical, active properties and application in packaging. *Food Chem.* **2025**, *463*, 141129. [CrossRef]
7. Bharati, S.; Gaikwad, V.L. Biodegradable Polymers in Food Packaging. In *Handbook of Biodegradable Polymers*; Jenny Stanford Publishing: Singapore, 2025; pp. 683–743.
8. Ngasotter, S.; Xavier, K.M.; Sagarnaik, C.; Sasikala, R.; Mohan, C.; Jaganath, B.; Ninan, G. Evaluating the reinforcing potential of steam-exploded chitin nanocrystals in chitosan-based biodegradable nanocomposite films for food packaging applications. *Carbohydr. Polym.* **2025**, *348*, 122841. [CrossRef] [PubMed]
9. Zhang, Z.; Lu, L.; Hong, B.; Ye, Q.; Guo, L.; Yuan, C.; Liu, B.; Cui, B. Starch/polyacrylamide hydrogels with flexibility, conductivity and sensitivity enhanced by two imidazolium-based ionic liquids for wearable electronics: Effect of anion structure. *Carbohydr. Polym.* **2025**, *347*, 122783. [CrossRef] [PubMed]
10. Zhang, Y.; Yu, Y.; Wang, Y.; Xia, C.; He, L.; Xia, Y.; Wang, Z. Self-recovery and self-conducting epoxy-based shape memory polymer microactuator. *Sens. Actuators B Chem.* **2025**, *422*, 136562. [CrossRef]
11. Sun, Y.; Liang, F.; Chen, J.; Tang, H.; Yuan, W.; Zhang, S.; Tang, Y.; Chua, K.J. Ultrathin flexible heat pipes with heat transfer performance and flexibility optimization for flexible electronic devices. *Renew. Sustain. Energy Rev.* **2025**, *208*, 115064. [CrossRef]
12. Tao, W.; Sun, Z.; Yang, Z.; Liang, B.; Wang, G.; Xiao, S. Transformer fault diagnosis technology based on AdaBoost enhanced transferred convolutional neural network. *Expert Syst. Appl.* **2025**, *264*, 125972. [CrossRef]
13. Schapire, R.E. Explaining adaboost. In *Empirical Inference: Festschrift in Honor of Vladimir N. Vapnik*; Springer: Berlin/Heidelberg, Germany, 2013; pp. 37–52.
14. Bentéjac, C.; Csörgő, A.; Martínez-Muñoz, G. A comparative analysis of gradient boosting algorithms. *Artif. Intell. Rev.* **2021**, *54*, 1937–1967. [CrossRef]
15. Chen, H.; Cheng, Y.; Du, T.; Wu, X.; Cao, Y.; Liu, Y. Enhancing the performance of recycled aggregate green concrete via a Bayesian optimization light gradient boosting machine and the nondominated sorting genetic algorithm-III. *Constr. Build. Mater.* **2025**, *458*, 139527. [CrossRef]
16. Meng, S.; Shi, Z.; Xia, C.; Zhou, C.; Zhao, Y. Exploring LightGBM-SHAP: Interpretable predictive modeling for concrete strength under high temperature conditions. In *Structures*; Elsevier: Amsterdam, The Netherlands, 2025; Volume 71, p. 108134.

17. Ke, G.; Meng, Q.; Finley, T.; Wang, T.; Chen, W.; Ma, W.; Ye, Q.; Liu, T.Y. Lightgbm: A highly efficient gradient boosting decision tree. *Adv. Neural Inf. Process. Syst.* **2017**, *30* , 52.
18. Hancock, J.T.; Khoshgoftaar, T.M. CatBoost for big data: An interdisciplinary review. *J. Big Data* **2020**, *7*, 94. [CrossRef] [PubMed]
19. Wei, C.; Li, Z.; Zhu, D.; Xu, T.; Liang, Z.; Liu, Y.; Zhao, N. Regulation of the physicochemical properties of nutrient solution in hydroponic system based on the CatBoost model. *Comput. Electron. Agric.* **2025**, *229*, 109729. [CrossRef]
20. Wang, R.; Zhang, M.; Gong, F.; Wang, S.; Yan, R. Improving port state control through a transfer learning-enhanced XGBoost model. *Reliab. Eng. Syst. Saf.* **2025**, *253*, 110558. [CrossRef]
21. Chen, T.; Guestrin, C. Xgboost: A scalable tree boosting system. In Proceedings of the 22nd ACM Sigkdd International Conference on Knowledge Discovery and Data Mining, San Francisco, CA, USA, 13–17 August 2016; pp. 785–794.
22. Sarkis-Onofre, R.; Catalá-López, F.; Aromataris, E.; Lockwood, C. How to properly use the PRISMA Statement. *Syst. Rev.* **2021**, *10*, 1–3. [CrossRef]
23. Demir, S.; Sahin, E.K. An investigation of feature selection methods for soil liquefaction prediction based on tree-based ensemble algorithms using AdaBoost, gradient boosting, and XGBoost. *Neural Comput. Appl.* **2023**, *35*, 3173–3190. [CrossRef]
24. Zhang, C.; Zhang, Y.; Shi, X.; Almpanidis, G.; Fan, G.; Shen, X. On incremental learning for gradient boosting decision trees. *Neural Process. Lett.* **2019**, *50*, 957–987. [CrossRef]
25. Sobolewski, R.A.; Tchakorom, M.; Couturier, R. Gradient boosting-based approach for short-and medium-term wind turbine output power prediction. *Renew. Energy* **2023**, *203*, 142–160. [CrossRef]
26. Phankokkruad, M.; Wacharawichanant, S. Prediction of mechanical properties of polymer materials using extreme gradient boosting on high molecular weight polymers. In *Complex, Intelligent, and Software Intensive Systems, Proceedings of the 12th International Conference on Complex, Intelligent, and Software Intensive Systems (CISIS-2018), Matsue, Japan, 4–6 July 2018*; Springer: Berlin/Heidelberg, Germany, 2019; pp. 375–385.
27. Park, H.; Joo, C.; Lim, J.; Kim, J. Novel natural gradient boosting-based probabilistic prediction of physical properties for polypropylene-based composite data. *Eng. Appl. Artif. Intell.* **2024**, *135*, 108864. [CrossRef]
28. Kavzoglu, T.; Teke, A. Predictive Performances of ensemble machine learning algorithms in landslide susceptibility mapping using random forest, extreme gradient boosting (XGBoost) and natural gradient boosting (NGBoost). *Arab. J. Sci. Eng.* **2022**, *47*, 7367–7385. [CrossRef]
29. Shahraki, A.; Abbasi, M.; Haugen, Ø. Boosting algorithms for network intrusion detection: A comparative evaluation of Real AdaBoost, Gentle AdaBoost and Modest AdaBoost. *Eng. Appl. Artif. Intell.* **2020**, *94*, 103770. [CrossRef]
30. Hussain, S.S.; Zaidi, S.S.H. AdaBoost Ensemble Approach with Weak Classifiers for Gear Fault Diagnosis and Prognosis in DC Motors. *Appl. Sci.* **2024**, *14*, 3105. [CrossRef]
31. Zhang, L.; Jánošík, D. Enhanced short-term load forecasting with hybrid machine learning models: CatBoost and XGBoost approaches. *Expert Syst. Appl.* **2024**, *241*, 122686. [CrossRef]
32. Mesghali, H.; Akhlaghi, B.; Gozalpour, N.; Mohammadpour, J.; Salehi, F.; Abbassi, R. Predicting maximum pitting corrosion depth in buried transmission pipelines: Insights from tree-based machine learning and identification of influential factors. *Process Saf. Environ. Prot.* **2024**, *187*, 1269–1285. [CrossRef]
33. Osman, M.; He, J.; Mokbal, F.M.M.; Zhu, N.; Qureshi, S. Ml-lgbm: A machine learning model based on light gradient boosting machine for the detection of version number attacks in rpl-based networks. *IEEE Access* **2021**, *9*, 83654–83665. [CrossRef]
34. Dhaliwal, S.S.; Nahid, A.A.; Abbas, R. Effective intrusion detection system using XGBoost. *Information* **2018**, *9*, 149. [CrossRef]
35. Thongsuwan, S.; Jaiyen, S.; Padcharoen, A.; Agarwal, P. ConvXGB: A new deep learning model for classification problems based on CNN and XGBoost. *Nucl. Eng. Technol.* **2021**, *53*, 522–531. [CrossRef]
36. Zabin, R.; Haque, K.F.; Abdelgawad, A. PredXGBR: A Machine Learning Framework for Short-Term Electrical Load Prediction. *Electronics* **2024**, *13*, 4521. [CrossRef]
37. Tian, J.; Tsai, P.W.; Zhang, K.; Cai, X.; Xiao, H.; Yu, K.; Zhao, W.; Chen, J. Synergetic focal loss for imbalanced classification in federated xgboost. *IEEE Trans. Artif. Intell.* **2023**, *5*, 647–660. [CrossRef]
38. Alsulamy, S. Predicting construction delay risks in Saudi Arabian projects: A comparative analysis of CatBoost, XGBoost, and LGBM. *Expert Syst. Appl.* **2025**, *268*, 126268. [CrossRef]
39. Zhuo, H.; Li, T.; Lu, W.; Zhang, Q.; Ji, L.; Li, J. Prediction model for spontaneous combustion temperature of coal based on PSO-XGBoost algorithm. *Sci. Rep.* **2025**, *15*, 2752. [CrossRef] [PubMed]
40. Ueki, Y.; Seko, N.; Maekawa, Y. Machine learning approach for prediction of the grafting yield in radiation-induced graft polymerization. *Appl. Mater. Today* **2021**, *25*, 101158. [CrossRef]
41. Zhao, Y.; Chen, Z.; Jian, X. A High-Generalizability Machine Learning Framework for Analyzing the Homogenized Properties of Short Fiber-Reinforced Polymer Composites. *Polymers* **2023**, *15*, 3962. [CrossRef]
42. Akinpelu, S.; Abolade, S.; Okafor, E.; Obada, D.; Ukpong, A.; Healy, J.; Akande, A. Interpretable machine learning methods to predict the mechanical properties of ABX3 perovskites. *Results Phys.* **2024**, *65*, 107978. [CrossRef]

43. Liu, Z.; Wang, T.; Jin, L.; Zeng, J.; Dong, S.; Wang, F.; Wang, F.; Dong, J. Towards high stiffness and ductility-The Mg-Al-Y alloy design through machine learning. *J. Mater. Sci. Technol.* **2024**, *221*, 194–203. [CrossRef]
44. Zhang, J.G.; Yang, G.C.; Ma, Z.H.; Zhao, G.L.; Song, H.Y. A stacking-CRRL fusion model for predicting the bearing capacity of a steel-reinforced concrete column constrained by carbon fiber-reinforced polymer. In *Structures*; Elsevier: Amsterdam, The Netherlands, 2023; Volume 55, pp. 1793–1804.
45. Zhao, J.; Wang, A.; Zhu, Y.; Dai, J.G.; Xu, Q.; Liu, K.; Hao, F.; Sun, D. Manufacturing ultra-high performance geopolymer concrete (UHPGC) with activated coal gangue for both binder and aggregate. *Compos. Part B Eng.* **2024**, *284*, 111723. [CrossRef]
46. Katlav, M.; Ergen, F.; Donmez, I. AI-driven design for the compressive strength of ultra-high performance geopolymer concrete (UHPGC): From explainable ensemble models to the graphical user interface. *Mater. Today Commun.* **2024**, *40*, 109915. [CrossRef]
47. Wang, Q.; Ahmad, W.; Ahmad, A.; Aslam, F.; Mohamed, A.; Vatin, N.I. Application of Soft Computing Techniques to Predict the Strength of Geopolymer Composites. *Polymers* **2022**, *14*, 74. [CrossRef]
48. Khan, K.; Ahmad, W.; Amin, M.N.; Ahmad, A.; Nazar, S.; Al-Faiad, M.A. Assessment of Artificial Intelligence Strategies to Estimate the Strength of Geopolymer Composites and Influence of Input Parameters. *Polymers* **2022**, *14*, 2509. [CrossRef]
49. Zhou, J.; Tian, Q.; Ahmad, A.; Huang, J. Compressive and tensile strength estimation of sustainable geopolymer concrete using contemporary boosting ensemble techniques. *Rev. Adv. Mater. Sci.* **2024**, *63*, 20240014. [CrossRef]
50. Amin, M.N.; Iqbal, M.; Khan, K.; Qadir, M.G.; Shalabi, F.I.; Jamal, A. Ensemble Tree-Based Approach towards Flexural Strength Prediction of FRP Reinforced Concrete Beams. *Polymers* **2022**, *14*, 1303. [CrossRef] [PubMed]
51. Zadkarami, M.; Shahbazian, M.; Salahshoor, K. Pipeline leakage detection and isolation: An integrated approach of statistical and wavelet feature extraction with multi-layer perceptron neural network (MLPNN). *J. Loss Prev. Process Ind.* **2016**, *43*, 479–487. [CrossRef]
52. Shamim Ansari, S.; Muhammad Ibrahim, S.; Danish Hasan, S. Conventional and Ensemble Machine Learning Models to Predict the Compressive Strength of Fly Ash Based Geopolymer Concrete. *Mater. Today Proc.* **2023** . [CrossRef]
53. Dodo, Y.; Arif, K.; Alyami, M.; Ali, M.; Najeh, T.; Gamil, Y. Estimation of compressive strength of waste concrete utilizing fly ash/slag in concrete with interpretable approaches: Optimization and graphical user interface (GUI). *Sci. Rep.* **2024**, *14*, 4598. [CrossRef]
54. Sidhu, J.; Kumar, P. Experimental investigation on the effect of integral hydrophobic modification on the properties of fly ash-slag based geopolymer concrete. *Constr. Build. Mater.* **2024**, *452*, 138818. [CrossRef]
55. Wudil, Y.S., Al-Fakih, A.; Al-Osta, M.A.; Gondal, M. Effective carbon footprint assessment strategy in fly ash geopolymer concrete based on adaptive boosting learning techniques. *Environ. Res.* **2025**, *266*, 120570120570. [CrossRef] [PubMed]
56. Kim, B.; Lee, D.E.; Hu, G.; Natarajan, Y.; Preethaa, S.; Rathinakumar, A.P. Ensemble Machine Learning-Based Approach for Predicting of FRP–Concrete Interfacial Bonding. *Mathematics* **2022**, *10*, 231. [CrossRef]
57. Kumarawadu, H.; Weerasinghe, P.; Perera, J.S. Evaluating the Performance of Ensemble Machine Learning Algorithms over Traditional Machine Learning Algorithms for Predicting Fire Resistance in FRP Strengthened Concrete Beams. *Electron. J. Struct. Eng.* **2024**, *24*, 47–53. [CrossRef]
58. Wang, C.; Zou, X.; Sneed, L.H.; Zhang, F.; Zheng, K.; Xu, H.; Li, G. Shear strength prediction of FRP-strengthened concrete beams using interpretable machine learning. *Constr. Build. Mater.* **2023**, *407*, 133553. [CrossRef]
59. Mahmoudian, A.; Tajik, N.; Taleshi, M.M.; Shakiba, M.; Yekrangnia, M. Ensemble machine learning-based approach with genetic algorithm optimization for predicting bond strength and failure mode in concrete-GFRP mat anchorage interface. *Structures* **2023**, *57*, 105173. [CrossRef]
60. Mahmoudian, A.; Bypour, M.; Kioumarsi, M. Explainable Boosting Machine Learning for Predicting Bond Strength of FRP Rebars in Ultra High-Performance Concrete. *Computation* **2024**, *12*, 202. [CrossRef]
61. Wang, S.; Fu, Y.; Ban, S.; Duan, Z.; Su, J. Genetic evolutionary deep learning for fire resistance analysis in fibre-reinforced polymers strengthened reinforced concrete beams. *Eng. Fail. Anal.* **2024**, *169*, 109149. [CrossRef]
62. Hu, H.; Wei, Q.; Wang, T.; Ma, Q.; Jin, P.; Pan, S.; Li, F.; Wang, S.; Yang, Y.; Li, Y. Experimental and Numerical Investigation Integrated with Machine Learning (ML) for the Prediction Strategy of DP590/CFRP Composite Laminates. *Polymers* **2024**, *16*, 1589. [CrossRef] [PubMed]
63. Aydın, F.; Karaoğlan, K.M.; Pektürk, H.Y.; Demir, B.; Karakurt, V.; Ahlatçı, H. The comparative evaluation of the wear behavior of epoxy matrix hybrid nano-composites via experiments and machine learning models. *Tribol. Int.* **2025**, *204*, 110451. [CrossRef]
64. Li, B.; Zhang, J.; Qu, Y.; Chen, D.; Chen, F. Data-driven predicting of bond strength in corroded BFRP concrete structures. *Case Stud. Constr. Mater.* **2024**, *21*, e03638. [CrossRef]
65. Khodadadi, N.; Roghani, H.; De Caso, F.; El-kenawy, E.S.M.; Yesha, Y.; Nanni, A. Data-driven PSO-CatBoost machine learning model to predict the compressive strength of CFRP- confined circular concrete specimens. *Thin-Walled Struct.* **2024**, *198*, 111763. [CrossRef]
66. Luo, T.; Xie, J.; Zhang, B.; Zhang, Y.; Li, C.; Zhou, J. An improved levy chaotic particle swarm optimization algorithm for energy-efficient cluster routing scheme in industrial wireless sensor networks. *Expert Syst. Appl.* **2024**, *241*, 122780. [CrossRef]

67. Gong, C.; Zhou, N.; Xia, S.; Huang, S. Quantum particle swarm optimization algorithm based on diversity migration strategy. *Future Gener. Comput. Syst.* **2024**, *157*, 445–458. [CrossRef]
68. Alizamir, M.; Gholampour, A.; Kim, S.; Keshtegar, B.; Jung, W.T. Designing a reliable machine learning system for accurately estimating the ultimate condition of FRP-confined concrete. *Sci. Rep.* **2024**, *14*, 20466. [CrossRef] [PubMed]
69. Khan, K.; Iqbal, M.; Salami, B.A.; Amin, M.N.; Ahamd, I.; Alabdullah, A.A.; Arab, A.M.A.; Jalal, F.E. Estimating flexural strength of FRP reinforced beam using artificial neural network and random forest prediction models. *Polymers* **2022**, *14*, 2270. [CrossRef]
70. Amin, M.N.; Salami, B.A.; Zahid, M.; Iqbal, M.; Khan, K.; Abu-Arab, A.M.; Alabdullah, A.A.; Jalal, F.E. Investigating the bond strength of FRP laminates with concrete using LIGHT GBM and SHAPASH analysis. *Polymers* **2022**, *14*, 4717. [CrossRef] [PubMed]
71. Tian, L.; Wang, L.; Xian, G. Machine learning prediction of interfacial bond strength of FRP bars with different surface characteristics to concrete. *Case Stud. Constr. Mater.* **2024**, *21*, e03984. [CrossRef]
72. Cheng, G.; Xiang, C.; Guo, F.; Wen, X.; Jia, X. Prediction of the tribological properties of a polymer surface in a wide temperature range using machine learning algorithm based on friction noise. *Tribol. Int.* **2023**, *180*, 108213. [CrossRef]
73. Fatriansyah, J.F.; Linuwih, B.D.P.; Andreano, Y.; Sari, I.S.; Federico, A.; Anis, M.; Surip, S.N.; Jaafar, M. Prediction of Glass Transition Temperature of Polymers Using Simple Machine Learning. *Polymers* **2024**, *16*, 2464. [CrossRef] [PubMed]
74. Ascencio-Medina, E.; He, S.; Daghighi, A.; Iduoku, K.; Casanola-Martin, G.M.; Arrasate, S.; González-Díaz, H.; Rasulev, B. Prediction of Dielectric Constant in Series of Polymers by Quantitative Structure-Property Relationship (QSPR). *Polymers* **2024**, *16*, 2731. [CrossRef] [PubMed]
75. Danesh, T.; Ouaret, R.; Floquet, P.; Negny, S. Interpretability of neural networks predictions using Accumulated Local Effects as a model-agnostic method. In *Computer Aided Chemical Engineering*; Elsevier: Amsterdam, The Netherlands, 2022; Volume 51, pp. 1501–1506.
76. Katić, D.; Krstić, H.; Otković, I.I.; Juričić, H.B. Comparing multiple linear regression and neural network models for predicting heating energy consumption in school buildings in the Federation of Bosnia and Herzegovina. *J. Build. Eng.* **2024**, *97*, 110728. [CrossRef]
77. Helmer, M.; Warrington, S.; Mohammadi-Nejad, A.R.; Ji, J.L.; Howell, A.; Rosand, B.; Anticevic, A.; Sotiropoulos, S.N.; Murray, J.D. On the stability of canonical correlation analysis and partial least squares with application to brain-behavior associations. *Commun. Biol.* **2024**, *7*, 217. [CrossRef] [PubMed]
78. Goh, K.L.; Goto, A.; Lu, Y. LGB-Stack: Stacked Generalization with LightGBM for Highly Accurate Predictions of Polymer Bandgap. *ACS Omega* **2022**, *7*, 29787–29793. [CrossRef] [PubMed]
79. Amrihesari, M.; Kern, J.; Present, H.; Moreno Briceno, S.; Ramprasad, R.; Brettmann, B. Machine Learning Models for Predicting Polymer Solubility in Solvents across Concentrations and Temperatures. *J. Phys. Chem. B* **2024**, *128*, 12786–12797. [CrossRef] [PubMed]
80. Rajaee, P.; Ghasemi, F.A.; Rabiee, A.H.; Fasihi, M.; Kakeh, B.; Sadeghi, A. Predicting tensile and fracture parameters in polypropylene-based nanocomposites using machine learning with sensitivity analysis and feature impact evaluation. *Compos. Part C Open Access* **2024**, *15*, 100535. [CrossRef]
81. Mishra, J.K.; Hwang, K.J.; Ha, C.S. Preparation, mechanical and rheological properties of a thermoplastic polyolefin (TPO)/organoclay nanocomposite with reference to the effect of maleic anhydride modified polypropylene as a compatibilizer. *Polymer* **2005**, *46*, 1995–2002. [CrossRef]
82. Abdi, J.; Hadipoor, M.; Hadavimoghaddam, F.; Hemmati-Sarapardeh, A. Estimation of tetracycline antibiotic photodegradation from wastewater by heterogeneous metal-organic frameworks photocatalysts. *Chemosphere* **2022**, *287*, 132135. [CrossRef]
83. Abánades Lázaro, I.; Chen, X.; Ding, M.; Eskandari, A.; Fairen-Jimenez, D.; Giménez-Marqués, M.; Gref, R.; Lin, W.; Luo, T.; Forgan, R.S. Metal–organic frameworks for biological applications. *Nat. Rev. Methods Prim.* **2024**, *4*, 42. [CrossRef]
84. Chen, D.; Zheng, Y.T.; Huang, N.Y.; Xu, Q. Metal-organic framework composites for photocatalysis. *EnergyChem* **2024**, *6*, 100115. [CrossRef]
85. Okada, M.; Amamoto, Y.; Kikuchi, J. Designing Sustainable Hydrophilic Interfaces via Feature Selection from Molecular Descriptors and Time-Domain Nuclear Magnetic Resonance Relaxation Curves. *Polymers* **2024**, *16*, 824. [CrossRef] [PubMed]
86. Salehi, S.; Arashpour, M.; Golafshani, E.M.; Kodikara, J. Prediction of rheological properties and ageing performance of recycled plastic modified bitumen using Machine learning models. *Constr. Build. Mater.* **2023**, *401*, 132728. [CrossRef]
87. Nizamuddin, S.; Jamal, M.; Biligiri, K.P.; Giustozzi, F. Effect of various compatibilizers on the storage stability, thermochemical and rheological properties of recycled plastic-modified bitumen. *Int. J. Pavement Res. Technol.* **2024**, *17*, 854–867. [CrossRef]
88. Gairola, S.; Sinha, S.; Singh, I. Improvement of flame retardancy and anti-dripping properties of polypropylene composites via ecofriendly borax cross-linked lignocellulosic fiber. *Compos. Struct.* **2024**, *354*, 118822. [CrossRef]
89. Chonghyo, J.; Hyundo, P.; Scokyoung, H.; Jongkoo, L.; Insu, H.; Hyungtae, C.; Junghwan, K. Prediction for heat deflection temperature of polypropylene composite with Catboost. In *Computer Aided Chemical Engineering*; Elsevier: Amsterdam, The Netherlands, 2022; Volume 49, pp. 1801–1806.

90. Chepurnenko, A.; Kondratieva, T.; Deberdeev, T.; Akopyan, V.; Avakov, A.; Chepurnenko, V. Prediction of Rheological Parameters of Polymers Using the CatBoost Gradient Boosting Algorithm. *Polym. Sci. Ser. D* **2024**, *17*, 121–128. [CrossRef]
91. Hofmann, J.; Li, Z.; Taphorn, K.; Herzen, J.; Wudy, K. Porosity prediction in laser-based powder bed fusion of polyamide 12 using infrared thermography and machine learning. *Addit. Manuf.* **2024**, *85*, 104176. [CrossRef]
92. Gadagi, A.; Sivaprakash, B.; Adake, C.; Deshannavar, U.; Hegde, P.G.; P., S.; Rajamohan, N.; Osman, A.I. Epoxy composite reinforced with jute/basalt hybrid—Characterisation and performance evaluation using machine learning techniques. *Compos. Part C Open Access* **2024**, *14*, 100453. [CrossRef]
93. Wang, Q.; Qi, J.; Hosseini, S.; Rasekh, H.; Huang, J. ICA-LightGBM Algorithm for Predicting Compressive Strength of Geo-Polymer Concrete. *Buildings* **2023**, *13*, 2278. [CrossRef]
94. Ncir, N.; El Akchioui, N. An advanced intelligent MPPT control strategy based on the imperialist competitive algorithm and artificial neural networks. *Evol. Intell.* **2024**, *17*, 1437–1461. [CrossRef]
95. Abbasi, M.; Sadough, F.; Mahmoudi, A. Solving the fuzzy p-hub center problem using imperialist competitive algorithm. *Int. J. Mach. Learn. Cybern.* **2024**, *15*, 6163–6183. [CrossRef]
96. Ahmad, A.; Ahmad, W.; Chaiyasarn, K.; Ostrowski, K.A.; Aslam, F.; Zajdel, P.; Joyklad, P. Prediction of geopolymer concrete compressive strength using novel machine learning algorithms. *Polymers* **2021**, *13*, 3389. [CrossRef] [PubMed]
97. Asadi, B.; Hajj, R. Prediction of asphalt binder elastic recovery using tree-based ensemble bagging and boosting models. *Constr. Build. Mater.* **2024**, *410*, 134154. [CrossRef]
98. Fares, M.Y.; Marini, S.; Lanotte, M. Multiple Stress Creep Recovery of High-Polymer Modified Binders: Consideration of Temperature and Stress Sensitivity for Quality Assurance/Quality Control Policy Development. *Transp. Res. Rec.* **2024**, 03611981241240765. [CrossRef]
99. Shen, Y.; Sun, J.; Liang, S. Interpretable Machine Learning Models for Punching Shear Strength Estimation of FRP Reinforced Concrete Slabs. *Crystals* **2022**, *12*, 259. [CrossRef]
100. Hamilton, R.I.; Papadopoulos, P.N. Using SHAP values and machine learning to understand trends in the transient stability limit. *IEEE Trans. Power Syst.* **2023**, *39*, 1384–1397. [CrossRef]
101. Rahman, J.; Arafin, P.; Billah, A.M. Machine learning models for predicting concrete beams shear strength externally bonded with FRP. In *Structures*; Elsevier: Amsterdam, The Netherlands, 2023; Volume 53, pp. 514–536.
102. Biruk-Urban, K.; Bere, P.; Józwik, J. Machine Learning Models in Drilling of Different Types of Glass-Fiber-Reinforced Polymer Composites. *Polymers* **2023**, *15*, 4609. [CrossRef] [PubMed]
103. Jalali, S.; Baniadam, M.; Maghrebi, M. Impedance value prediction of carbon nanotube/polystyrene nanocomposites using tree-based machine learning models and the Taguchi technique. *Results Eng.* **2024**, *24*, 103599. [CrossRef]
104. Ma, L.; Zhou, C.; Lee, D.; Zhang, J. Prediction of axial compressive capacity of CFRP-confined concrete-filled steel tubular short columns based on XGBoost algorithm. *Eng. Struct.* **2022**, *260*, 114239. [CrossRef]
105. Gao, W.; Jiang, Q.; Guan, Y.; Huang, H.; Liu, S.; Ling, S.; Zhou, L. Transfer learning improves predictions in lignin content of Chinese fir based on Raman spectra. *Int. J. Biol. Macromol.* **2024**, *269*, 132147. [CrossRef]
106. Dong, Z.; Fang, Y.; Wang, X.; Zhao, Y.; Wang, Q. Hydrophobicity classification of polymeric insulators based on embedded methods. *Mater. Res.* **2015**, *18*, 127–137. [CrossRef]
107. Kong, Q.; He, C.; Liao, L.; Xu, J.; Yuan, C. Hyperparameter optimization for interfacial bond strength prediction between fiber-reinforced polymer and concrete. *Structures* **2023**, *51*, 573–601. [CrossRef]
108. Alanazi, J.; Algahtani, M.M.; Alanazi, M.; Alharby, T.N. Application of different mathematical models based on artificial intelligence technique to predict the concentration distribution of solute through a polymeric membrane. *Ecotoxicol. Environ. Saf.* **2023**, *262*, 115183. [CrossRef]
109. Hai, T.; Basem, A.; Alizadeh, A.a.; Sharma, K.; Jasim, D.J.; Rajab, H.; Ahmed, M.; Kassim, M.; Singh, N.S.S.; Maleki, H. Optimizing Gaussian process regression (GPR) hyperparameters with three metaheuristic algorithms for viscosity prediction of suspensions containing microencapsulated PCMs. *Sci. Rep.* **2024**, *14*, 20271. [CrossRef]
110. Tahir, M.H.; Farrukh, A.; Alqahtany, F.Z.; Badshah, A.; Shaaban, I.A.; Assiri, M.A. Accelerated discovery of polymer donors for organic solar cells through machine learning: From library creation to performance forecasting. *Spectrochim. Acta Part A Mol. Biomol. Spectrosc.* **2025**, *326*, 125298. .: 10.1016/j.saa.2024.125298 [CrossRef]
111. Jiang, J.; Lu, A.; Ma, X.; Ouyang, D.; Williams, R.O. The applications of machine learning to predict the forming of chemically stable amorphous solid dispersions prepared by hot-melt extrusion. *Int. J. Pharm. X* **2023**, *5*, 100164. [CrossRef] [PubMed]
112. Burke, A.J. Asymmetric organocatalysis in drug discovery and development for active pharmaceutical ingredients. *Expert Opin. Drug Discov.* **2023**, *18*, 37–46. [CrossRef] [PubMed]
113. Pang, J.; Zhao, Z. Real-time Monitoring of Fluidized Bed Agglomerating based on Improved Adaboost Algorithm. *J. Physics Conf. Ser.* **2021**, *1924*, 012026. [CrossRef]
114. Chan, R.K.; Wang, B.X. Do long-term acoustic-phonetic features and mel-frequency cepstral coefficients provide complementary speaker-specific information for forensic voice comparison? *Forensic Sci. Int.* **2024**, *363*, 112199. [CrossRef] [PubMed]

115. Fiosina, J.; Sievers, P.; Drache, M.; Beuermann, S. Polymer reaction engineering meets explainable machine learning. *Comput. Chem. Eng.* **2023**, *177*, 108356. [CrossRef]
116. Correia, J.S.; Mirón-Barroso, S.; Hutchings, C.; Ottaviani, S.; Somuncuoğlu, B.; Castellano, L.; Porter, A.E.; Krell, J.; Georgiou, T.K. How does the polymer architecture and position of cationic charges affect cell viability? *Polym. Chem.* **2023**, *14*, 303–317. [CrossRef] [PubMed]
117. Deshpande, A.R.; Kulkarni, A.P.; Wasatkar, N.; Gajalkar, V.; Abdullah, M. Prediction of Wear Rate of Glass-Filled PTFE Composites Based on Machine Learning Approaches. *Polymers* **2024**, *16*, 2666. [CrossRef] [PubMed]
118. Huang, D.; Li, Z.; Wang, K.; Zhou, H.; Zhao, X.; Peng, X.; Zhang, R.; Wu, J.; Liang, J.; Zhao, L. Probing the Effect of Photovoltaic Material on Voc in Ternary Polymer Solar Cells with Non-Fullerene Acceptors by Machine Learning. *Polymers* **2023**, *15*, 2954. [CrossRef]
119. Bin Inqiad, W.; Javed, M.F.; Siddique, M.S.; Khan, N.M.; Alkhattabi, L.; Abuhussain, M.; Alabduljabbar, H. Comparison of boosting and genetic programming techniques for prediction of tensile strain capacity of Engineered Cementitious Composites (ECC). *Mater. Today Commun.* **2024**, *39*, 109222. [CrossRef]
120. Nguyen, T.H.; Vuong, H.T.; Shiau, J.; Nguyen-Thoi, T.; Nguyen, D.H.; Nguyen, T. Optimizing flexural strength of RC beams with recycled aggregates and CFRP using machine learning models. *Sci. Rep.* **2024**, *14*, 28621. [CrossRef] [PubMed]
121. Li, F.; Rana, M.S.; Qurashi, M.A. Advanced machine learning techniques for predicting concrete mechanical properties: A comprehensive review of models and methodologies. *Multiscale Multidiscip. Model. Exp. Des.* **2025**, *8*, 1–41. [CrossRef]
122. Cheng, X. A Comprehensive Study of Feature Selection Techniques in Machine Learning Models. *Insights Comput. Signals Syst.* **2024**, *1*, 10–70088. [CrossRef]
123. Sheng, K.; Jiang, G.; Du, M.; He, Y.; Dong, T.; Yang, L. Interpretable knowledge-guided framework for modeling reservoir water-sensitivity damage based on Light Gradient Boosting Machine using Bayesian optimization and hybrid feature mining. *Eng. Appl. Artif. Intell.* **2024**, *133*, 108511. [CrossRef]
124. Subeshan, B.; Atayo, A.; Asmatulu, E. Machine learning applications for electrospun nanofibers: A review. *J. Mater. Sci.* **2024**, *59*, 14095–14140. [CrossRef]
125. Hussain, S.; Mustafa, M.W.; Jumani, T.A.; Baloch, S.K.; Alotaibi, H.; Khan, I.; Khan, A. A novel feature engineered-CatBoost-based supervised machine learning framework for electricity theft detection. *Energy Rep.* **2021**, *7*, 4425–4436. [CrossRef]
126. Nagassou, M.; Mwangi, R.W.; Nyarige, E. A hybrid ensemble learning approach utilizing light gradient boosting machine and category boosting model for lifestyle-based prediction of type-II diabetes mellitus. *J. Data Anal. Inf. Process.* **2023**, *11*, 480–511. [CrossRef]
127. Yin, L.; Ma, P.; Deng, Z. JLGBMLoc—A novel high-precision indoor localization method based on LightGBM. *Sensors* **2021**, *21*, 2722. [CrossRef] [PubMed]
128. Chen, C.; Zhang, Q.; Ma, Q.; Yu, B. LightGBM-PPI: Predicting protein-protein interactions through LightGBM with multi-information fusion. *Chemom. Intell. Lab. Syst.* **2019**, *191*, 54–64. [CrossRef]
129. Jin, D.; Lu, Y.; Qin, J.; Cheng, Z.; Mao, Z. SwiftIDS: Real-time intrusion detection system based on LightGBM and parallel intrusion detection mechanism. *Comput. Secur.* **2020**, *97*, 101984.
130. Han, R.; Fu, X.; Guo, H. Interpretable machine learning-assisted strategy for predicting the mechanical properties of hydroxyl-terminated polyether binders. *J. Polym. Sci.* **2024**. [CrossRef]
131. Ke, L.; Qiu, M.; Chen, Z.; Zhou, J.; Feng, Z.; Long, J. An interpretable machine learning model for predicting bond strength of CFRP-steel epoxy-bonded interface. *Compos. Struct.* **2023**, *326*, 117639. [CrossRef]
132. Kalladi, A.J.; Ramesan, M.T. In-situ polymerized boehmite/cashew gum/polyvinyl alcohol/polypyrrole blend nanocomposites with tunable structural, electrical, and mechanical properties for enhanced energy storage applications. *J. Mol. Struct.* **2025**, *1322*, 140379. [CrossRef]
133. Nayak, S.; Sharma, Y.K. A modified Bayesian boosting algorithm with weight-guided optimal feature selection for sentiment analysis. *Decis. Anal. J.* **2023**, *8*, 100289. [CrossRef]
134. Zhao, X.; Liu, Y.; Zhao, Q. Improved LightGBM for extremely imbalanced data and application to credit card fraud detection. *IEEE Access* **2024**, *12*, 159316–159335. [CrossRef]
135. Ghasem, N. Combining CFD and AI/ML Modeling to Improve the Performance of Polypropylene Fluidized Bed Reactors. *Fluids* **2024**, *9*, 298. [CrossRef]

polymers

MDPI

Review

Physics-Informed Neural Networks in Polymers: A Review

Ivan Malashin [1,*], Vadim Tynchenko [1,*], Andrei Gantimurov [1], Vladimir Nelyub [1,2] and Aleksei Borodulin [1]

1 Artificial Intelligence Technology Scientific and Education Center, Bauman Moscow State Technical University, 105005 Moscow, Russia
2 Scientific Department, Far Eastern Federal University, 690922 Vladivostok, Russia
* Correspondence: ivan.p.malashin@gmail.com (I.M.); vadimond@mail.ru (V.T.); Tel.: +7-926-875-7128 (I.M.)

Abstract: The modeling and simulation of polymer systems present unique challenges due to their intrinsic complexity and multi-scale behavior. Traditional computational methods, while effective, often struggle to balance accuracy with computational efficiency, especially when bridging the atomistic to macroscopic scales. Recently, physics-informed neural networks (PINNs) have emerged as a promising tool that integrates data-driven learning with the governing physical laws of the system. This review discusses the development and application of PINNs in the context of polymer science. It summarizes the recent advances, outlines the key methodologies, and analyzes the benefits and limitations of using PINNs for polymer property prediction, structural design, and process optimization. Finally, it identifies the current challenges and future research directions to further leverage PINNs for advanced polymer modeling.

Keywords: physics-informed neural networks (PINNs); polymer modeling; multi-scale simulation; ML in materials science; structure–property relationships

1. Introduction

Polymers exhibit complex behaviors due to their hierarchical structures, multi-scale interactions, and dependence on various environmental factors (such as temperature, pressure, and chemical composition). Traditional modeling approaches, such as molecular dynamics (MD) [1–3], Monte Carlo simulations [4–6], and continuum methods [7–9], although successful in capturing specific aspects of polymer behavior, face significant limitations. These approaches often struggle with scalability when dealing with large systems or long time scales as they require immense computational resources. Additionally, they face challenges in efficiently modeling multi-scale phenomena, such as the transition from molecular dynamics to macroscopic behavior.

Furthermore, traditional models often require substantial experimental data to ensure their accuracy and validation, and they may not readily adapt to new, unexplored material systems without extensive recalibration. For example, MD simulations are often confined to small system sizes and short time scales, while continuum models may oversimplify the complexities of polymer microstructures, ignoring molecular interactions.

In light of these challenges, physics-informed neural networks (PINNs) present a promising alternative. By combining machine learning (ML) with physics-based modeling, PINNs allow the incorporation of physical laws into the neural network architecture, ensuring that the model adheres to known scientific principles while learning from data. PINNs are effective in overcoming scalability issues by learning global patterns from data without the need for exhaustive pointwise simulations [10]. Moreover, PINNs can leverage

 https://doi.org/10.3390/polym17081108

optimization techniques that simultaneously minimize both the data-driven error and the residuals of the governing equations, thus offering a more accurate and computationally feasible approach to polymer modeling [11].

This paper aims to address these gaps by proposing a hybrid model that enhances the scalability and efficiency of traditional approaches while maintaining the rigor of physics-based constraints, ultimately leading to more accurate predictions of polymer behaviors across multiple scales.

The number of publications on PINNs applied to polymers has shown a clear upward trend in recent years (Figure 1). From 2020 to 2022, the research activity remained relatively low, with only a few publications each year (two in 2020, ten in 2021, and five in 2022). However, starting in 2023, there was a noticeable increase, with the number of publications reaching seven, followed by a sharp rise to fifteen in 2024. This growth suggests a growing interest in using PINNs for polymer-related problems, likely driven by advancements in ML and increased recognition of the method's potential in modeling complex polymer behaviors. While the data for 2025 are still emerging, with five publications recorded so far, the overall trend suggests a growing use of PINNs in polymer science.

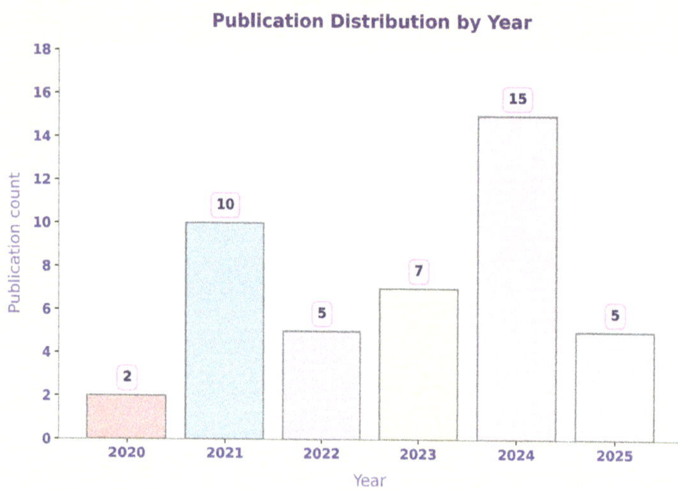

Figure 1. Year-wise distribution of PINN-based publications regarding polymers.

The analysis of the keyword density provides insights into the key research directions and trends in the application of PINNs in polymer science (Figure 2). The VOSviewer-based visualization of keywords reveals a high concentration of terms related to ML, computational mechanics, and polymer property modeling. The most frequently occurring terms can be categorized into several main groups.

First, there are ML- and neural-network-based methods, including PINNs, Artificial Intelligence (AI), Artificial Neural Networks (ANNs), and Deep Learning, which indicate a growing interest in leveraging neural-network-based approaches to model complex physical phenomena in polymers. The presence of transfer learning [12–14] suggests ongoing efforts to generalize PINN applications across different polymeric systems.

Second, the visualization highlights key concepts related to computational mechanics and modeling, such as computational mechanics, constitutive modeling, Elasto-Visco-Plasticity, and Deflection Prediction, confirming the application of PINNs in simulating the mechanical properties of polymers, including their viscoelastic and plastic characteristics [15–17].

The third major theme includes polymer-science- and physics-based models, with terms such as phase separation, Cross-Linked Polymers, Homopolymer Blends, and the Cahn–Hilliard Equation, emphasizing the role of PINNs in modeling microstructure evolution and self-assembly processes in polymeric materials. The high density of the term Cahn–Hilliard Equation suggests the frequent application of PINNs [18,19] in modeling phase separation dynamics within polymer blends.

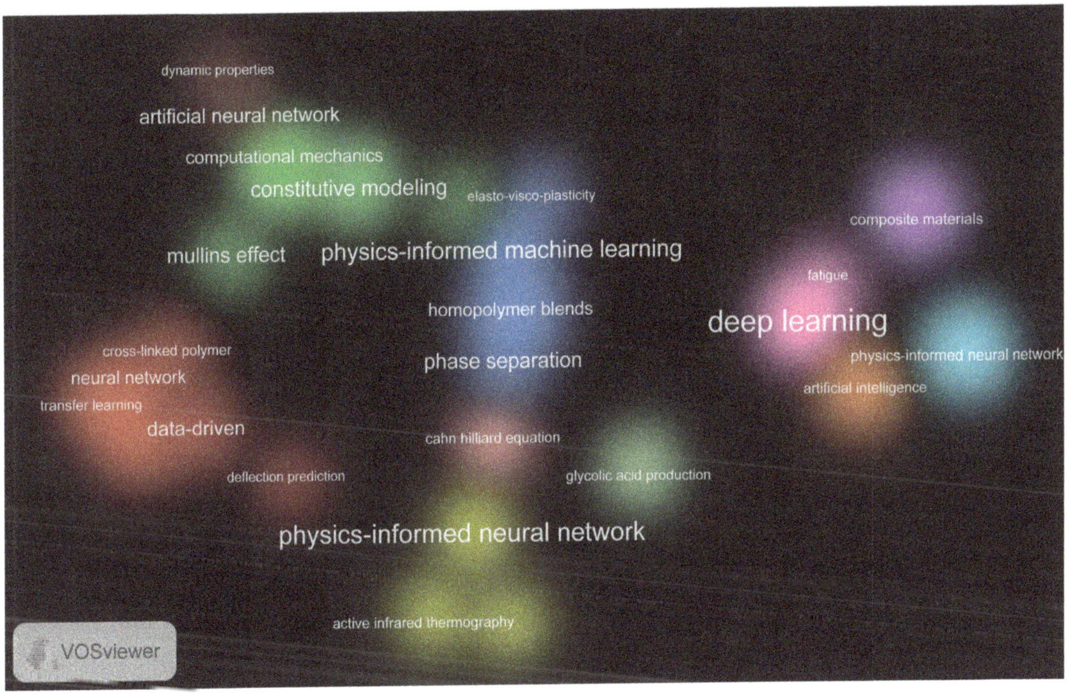

Figure 2. Keyword co-occurrence cluster map based on VOSviewer analysis.

Additionally, experimental techniques and hybrid approaches appear, including Active Infrared Thermography, Dynamic Properties, and Fatigue, highlighting the integration of PINNs with real-world experimental methods to enhance predictive accuracy and model validation [20,21].

The interpretation of keyword density visualization reveals that PINNs are becoming a crucial tool in polymer science by enabling accurate physics-driven modeling while requiring minimal experimental data. The high concentration of physics-related terms, such as constitutive equations and phase behavior, indicates that PINNs are widely used for solving mechanical and thermodynamic problems in polymer systems.

Furthermore, the integration of PINNs with experimental methods like infrared thermography suggests a promising direction toward combining computational predictions with real-world measurements. This trend highlights the transition toward hybrid modeling approaches, where PINNs bridge the gap between data-driven learning and fundamental physics laws to create more accurate and interpretable models for polymer materials.

Future research directions may include expanding the use of PINNs for multi-scale modeling, enhancing model interpretability through uncertainty quantification, and integrating real-time experimental data to refine polymer models adaptively [22].

In addition to the well-established PINNs, several alternative physics-informed learning frameworks have recently emerged, aiming to address limitations in scalability, flexibility, and generalization. One such approach is the physics-informed neural operator (PINO) [23–25], which learns a mapping between entire functions rather than pointwise values. This operator-based formulation allows the PINO to generalize across different boundary conditions and material parameters, offering improved efficiency for high-dimensional or history-dependent systems.

Another notable method is the Physics-Embedded Neural Network (PENN) [26–28], which incorporates governing equations directly into the neural network architecture. Unlike PINNs that rely on residual-based losses, PENNs enforce physical constraints structurally, leading to potentially more stable and interpretable models.

Physics-guided machine learning (PGML) [29–31] represents a broader class of hybrid models where physical principles guide the model selection, feature engineering, or loss formulation. PGML is particularly useful in data-scarce scenarios as it leverages prior knowledge to reduce the solution space.

More recently, the concept of *super-constrained machine learning with L-agents* [32–34] has been introduced to enforce not only physical laws but also logic-based and multi-agent constraints within the learning process. L-agents can be used to represent various types of domain knowledge, including symbolic rules or empirical relations, allowing for highly structured and constrained learning suitable for complex engineering systems.

These emerging approaches complement the capabilities of PINNs and expand the toolkit available for solving inverse problems, multi-scale modeling, and uncertainty quantification in physics-driven domains.

This review examines the applications of PINNs in polymer science, focusing on their role in addressing complex problems that traditional computational methods encounter. The advantages of PINNs are analyzed, including their capacity to integrate physical laws into ML frameworks, leading to improved accuracy, data efficiency, and generalizability across various polymer systems. The methodologies employed in PINN-based modeling are discussed, covering areas such as constitutive modeling, degradation prediction, and process optimization. Some emerging challenges are identified, including computational cost, the necessity of high-quality experimental data, and limitations in capturing highly nonlinear behaviors in polymer systems. Potential future directions are considered, with an emphasis on hybrid modeling approaches, uncertainty quantification, and experimental validation to advance the field.

2. Theoretical Background on PINNs

PINNs [35] provide a framework for solving partial differential equations (PDEs) by embedding physical laws into the loss function of a neural network [36]. This method enables the approximation of solutions to PDEs without the need for extensive labeled data.

Figure 3 illustrates the schematic structure of a PINN designed to solve a problem governed by the Burgers equation. The input to the neural network consists of spatial and temporal coordinates (x, t), which pass through several hidden layers with nonlinear activation functions σ. The network outputs the function u, which is then used to compute temporal and spatial derivatives. These derivatives are substituted into the governing PDE $\frac{\partial u}{\partial t} + u\frac{\partial u}{\partial x} - \nu\frac{\partial^2 u}{\partial x^2} = 0$, where $u = u(x, t)$ represents the solution, and ν is a given parameter controlling the viscosity. A loss function evaluates how well the network's output satisfies the PDE. If the error exceeds a predefined tolerance ε, the neural network parameters are updated through gradient descent. This iterative process continues until the solution meets the specified accuracy criterion.

Let the governing equation for a polymer system be given by a general partial differential equation (PDE):

$$\mathcal{N}(u(\mathbf{x},t)) = f(\mathbf{x},t), \quad \mathbf{x} \in \Omega, t \in [0,T], \tag{1}$$

where $\mathcal{N}$ is a differential operator that encapsulates the underlying physics of the system, such as conservation laws, diffusion equations, or viscoelasticity models. The function $u(\mathbf{x},t)$ represents the unknown solution to be approximated, which may describe properties like stress, strain, or concentration in the polymer system. The variable $\mathbf{x}$ denotes the spatial coordinates within the domain Ω, which can be one-, two-, or three-dimensional, while $t \in [0,T]$ represents the time variable over a given interval. The term $f(\mathbf{x},t)$ accounts for external forcing effects, such as applied loads, heat sources, or other external influences on the system.

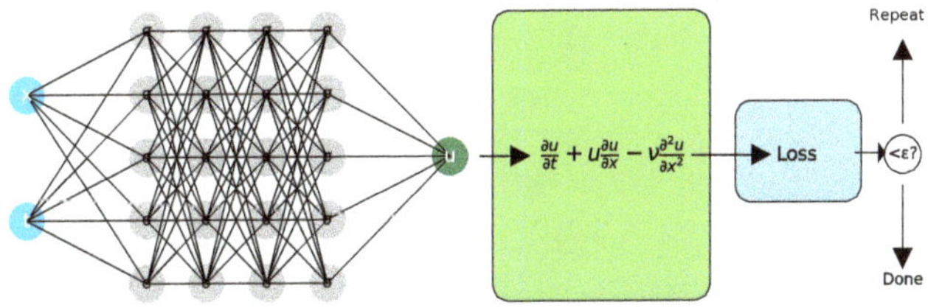

Figure 3. Schematic representation of PINN solving the Burgers equation.

The loss function for PINN consists of multiple components that enforce both data fidelity and physical consistency [37]:

$$\mathcal{L} = \mathcal{L}_{\text{data}} + \lambda\mathcal{L}_{\text{physics}} + \mu\mathcal{L}_{\text{BC}}, \tag{2}$$

where $\mathcal{L}$ is the total loss function minimized during training. The term $\mathcal{L}_{\text{data}}$ represents the data loss, ensuring that the model predictions align with experimental or simulated observations. The term $\mathcal{L}_{\text{physics}}$ enforces the governing PDE constraints, ensuring that the neural network satisfies the differential equation $\mathcal{N}(u(\mathbf{x},t)) = f(\mathbf{x},t)$. The boundary condition loss, $\mathcal{L}_{\text{BC}}$, enforces the required physical constraints at the domain boundaries $\partial\Omega$, such as fixed displacement or zero-flux conditions. The parameters λ and μ are weighting factors that balance the contributions of the physics and boundary loss terms relative to the data loss, allowing for better stability and accuracy in model training.

The data loss is defined as [38]

$$\mathcal{L}_{\text{data}} = \sum_{i=1}^{N_d} |u_{\text{NN}}(\mathbf{x}_i,t_i) - u_{\text{true}}(\mathbf{x}_i,t_i)|^2, \tag{3}$$

where $u_{\text{NN}}(\mathbf{x},t)$ is the neural network approximation of the solution, and $u_{\text{true}}(\mathbf{x},t)$ represents available observational data.

The physics-based loss term ensures the neural network adheres to the PDE constraints [39]:

$$\mathcal{L}_{\text{physics}} = \sum_{j=1}^{N_p} |\mathcal{N}(u_{\text{NN}}(\mathbf{x}_j,t_j)) - f(\mathbf{x}_j,t_j)|^2. \tag{4}$$

This term penalizes deviations from the governing equations at a set of collocation points $\{(\mathbf{x}_j, t_j)\}_{j=1}^{N_p}$.

The boundary and initial conditions are incorporated through an additional constraint:

$$\mathcal{L}_{\mathrm{BC}} = \sum_{k=1}^{N_b} |u_{\mathrm{NN}}(\mathbf{x}_k, 0) - g(\mathbf{x}_k)|^2 + \sum_{m=1}^{N_c} |\mathcal{B}(u_{\mathrm{NN}}(\mathbf{x}_m, t_m)) - h(\mathbf{x}_m, t_m)|^2. \quad (5)$$

Here, $\mathcal{B}$ represents a boundary condition operator, and $h(\mathbf{x}, t)$ is the prescribed boundary value function.

To train the PINN, the total loss $\mathcal{L}$ is minimized using gradient-based optimization methods. Two widely used approaches include the following:

- Adam Optimizer: A first-order gradient-based method that adapts learning rates based on first and second moments of gradients, ensuring stable convergence.
- L-BFGS: A quasi-Newton method that often achieves faster convergence for smooth loss landscapes by leveraging second-order derivative approximations.

The optimization process aims to update the neural network parameters θ by computing the gradient of $\mathcal{L}$ [40]:

$$\theta^{(k+1)} = \theta^{(k)} - \eta \nabla_\theta \mathcal{L}, \quad (6)$$

where η is the learning rate. In the case of L-BFGS, an approximation to the inverse Hessian matrix H_k is used as follows:

$$\theta^{(k+1)} = \theta^{(k)} - H_k \nabla_\theta \mathcal{L}. \quad (7)$$

The gradient $\nabla_\theta \mathcal{L}$ is computed using automatic differentiation (AD), ensuring accurate and efficient backpropagation of errors. Training continues until convergence criteria, such as a tolerance on $\mathcal{L}$ or a maximum number of iterations, are met.

The gradient of $\mathcal{L}$ with respect to the neural network parameters θ is computed via automatic differentiation (AD), which provides an efficient way to obtain exact derivatives for complex nested functions [41]. The training process iteratively updates θ to minimize the total loss, thereby ensuring the learned solution satisfies both the data constraints and the governing physics.

PINNs offer several advantages over traditional numerical methods, including the following:

- Mesh-free formulation, allowing flexibility in handling complex geometries.
- The ability to incorporate sparse and noisy observational data.
- Implicit satisfaction of PDE constraints, reducing the need for explicit discretization.

Figure 4 is a flowchart that outlines the key steps in developing and applying a PINN framework for polymer science, from defining the governing equations to training the model and making predictions. Each step incorporates both computational and physical constraints, ensuring reliable and interpretable results.

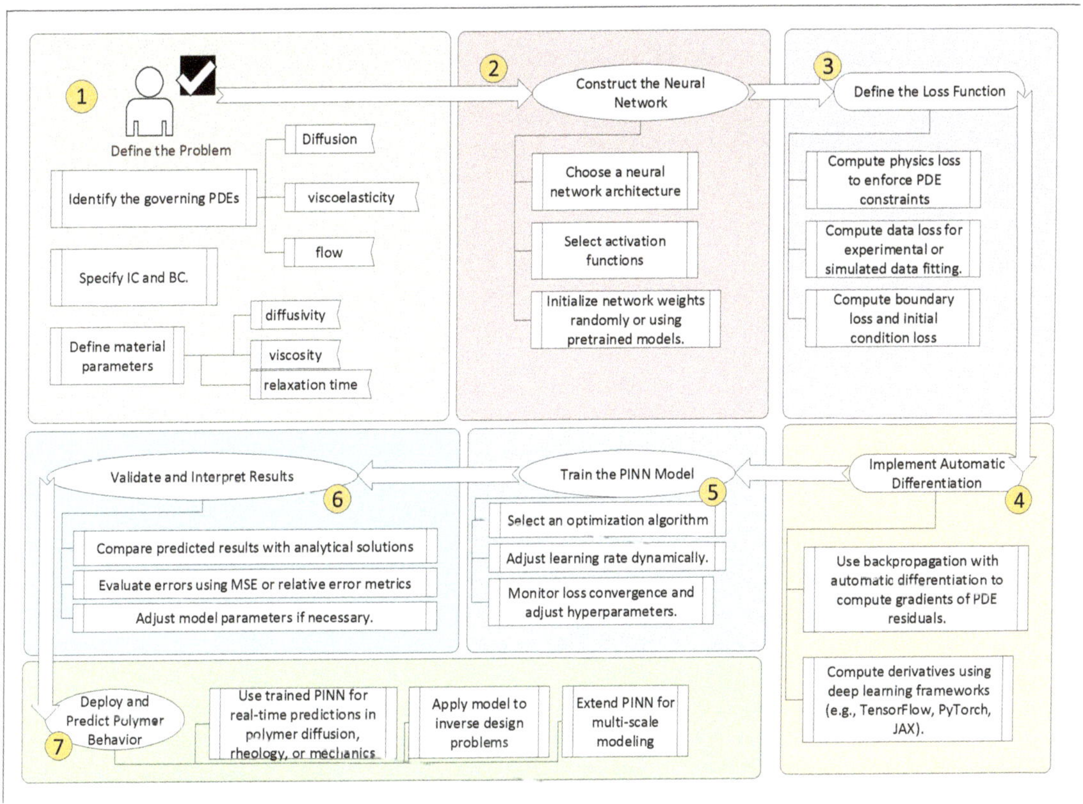

Figure 4. PINN pipeline.

3. Applications of PINNs in Polymer Science

3.1. Temperature

The temperature distribution during the curing process significantly affects the final quality of thermosetting composites. Ensuring temperature histories conform to specifications requires cure optimization, which fundamentally involves solving parametric coupled PDEs with dynamic boundary conditions [42]. Mathematically, this is expressed as solving

$$\frac{\partial T}{\partial t} = k_x \frac{\partial^2 T}{\partial x^2} + k_y \frac{\partial^2 T}{\partial y^2} + k_z \frac{\partial^2 T}{\partial z^2} + \dot{Q} \tag{8}$$

where T represents the temperature of the polymer system, which varies with spatial position and time. The parameters k_x, k_y, k_z denote the thermal conductivities in the x-, y- and z-directions, respectively, capturing the anisotropic heat transfer properties of the material.

The nonlinear exothermic heat source, $\dot{Q}$, accounts for the heat generated within the system due to chemical reactions or phase transformations. It is given by

$$\dot{Q} = v_r \rho_r H_r \frac{\partial \alpha}{\partial t}. \tag{9}$$

In this equation, v_r represents the reaction rate, which governs the speed of the exothermic process. The parameter ρ_r denotes the density of the reacting species, influencing the overall energy release. The term H_r is the reaction enthalpy, quantifying the amount of heat generated per unit mass of reactant. Finally, $\frac{\partial \alpha}{\partial t}$ represents the time derivative of the reaction progress variable α, which describes the extent of the reaction occurring in the system. As α evolves over time, the heat generation rate $\dot{Q}$ dynamically changes, affecting the overall temperature distribution.

Recently, PINNs have emerged as promising solvers for PDEs without labeled data [43]. Conventional PINNs approximate solutions in a pointwise manner, requiring a large number of collocation points, leading to a computational burden of $O(N^3)$ for training. Instead, Meng et el. [23] proposed a physics-informed neural operator (PINO) approach, mapping the entire cure cycle to temperature and degree of cure (DoC) histories as a function-to-function operator for carbon-fiber-reinforced polymers [44] (CFRP) composites. By enforcing global constraints on field outputs, the PINO achieves unsupervised parametric PDE solving while reducing training complexity to $O(N \log N)$ using Fourier neural operators (FNOs).

Compared to traditional PINNs, the PINO introduces a fundamental shift in the learning paradigm. While PINNs aim to approximate the solution $u(x,t)$ by minimizing the residuals of governing PDEs (e.g., $\mathcal{L}[u] = 0$) at discrete collocation points, PINOs learn an operator that maps entire input functions to output solution functions. Specifically, the PINO seeks a functional mapping

$$\mathcal{G} : f(x,t) \mapsto u(x,t),$$

where $f(x,t)$ represents inputs such as boundary/initial conditions or material parameters, and $u(x,t)$ denotes the solution field (e.g., temperature or degree of cure).

In the context of cure modeling for carbon-fiber-reinforced polymers, the PINO maps the full cure cycle to temperature and degree of cure (DoC) histories across space and time. Unlike PINNs, which compute physics residuals pointwise and often encounter scalability issues in high-dimensional problems, the PINO leverages neural operator architectures (e.g., Fourier neural operators) to generalize across varying input conditions with reduced computational cost.

PINNs enforce physical laws explicitly by including PDE residuals in the loss function:

$$\mathcal{L}_{\text{physics}} = \|\mathcal{N}[u_\theta](x,t)\|^2,$$

whereas the PINO captures the physics implicitly through training on solution data that satisfy the underlying equations. This makes the PINO suitable for rapid predictions in parametric or history-dependent systems. To highlight the key methodological differences between PINN and PINO, Table 1 provides a side-by-side comparison of their core features, strengths, and suitable application scenarios.

Table 1. Comparison between PINN and PINO.

Feature	PINN	PINO
Learning Target	Pointwise function approximation	Operator (function-to-function mapping)
Input/Output	Scalar coordinates → scalar solution	Function → function
PDE Enforcement	Explicit via loss function	Implicit through training data or regularization
Scalability	Moderate (can be slow for complex PDEs)	High (efficient once trained)
Generalization	Limited to trained domain	Strong across varying inputs
Suitable Use Case	Low- to moderate-dimensional PDE solutions	High-dimensional, parametric PDE problems

The PINO model reduces training time compared to the fully connected physics-informed neural network (FC-PINN), a standard PINN architecture composed of fully connected layers. In FC-PINN, the network approximates mappings from inputs (e.g., spatial and temporal coordinates) to outputs (e.g., temperature or degree of cure), with physical laws enforced via PDE residuals in the loss function. While FC-PINN serves as a baseline architecture, the PINO achieves much faster convergence, requiring only 84.16 s for a 260-min one-dwell cure cycle, compared to 2370 s for for FC-PINN, while achieving a lower temperature MAE of 0.2 K compared to 1.6 K. For two-dwell and smart cure cases, the PINO maintains high accuracy with temperature MAE values of 0.273 K and 0.257 K, respectively, and DoC MAE values below 0.007. The parametric study shows that, with 50 training samples, the PINO achieves temperature MAE of 0.267 ± 0.068 K for one-dwell and 0.226 ± 0.039 K for two-dwell cases, with relative percentage errors below 0.08%. Training time increases for parametric cases, reaching 3016.61 s for one-dwell and 4880.85 s for two-dwell, yet inference remains highly efficient. The resolution-invariance of the PINO is confirmed as training at a lower resolution ($\triangle t$ = 8 s) still provides accurate predictions at a higher resolution ($\triangle t = 4s$) while reducing computational cost.

While the PINO improves computational efficiency and accuracy in solving parametric PDEs for cure optimization, it still faces notable limitations. The model requires substantial training time for complex parametric cases, reaching 4880.85 s for two-dwell curing, which may limit real-time industrial applications. Despite its improved generalization, the PINO's accuracy depends on the quality and distribution of training samples, making it sensitive to under-represented regions in the parametric space. Additionally, the reliance on FNO may introduce limitations in capturing highly localized temperature gradients, particularly in heterogeneous composites [45]. Finally, while the PINO reduces computational complexity to O(NlogN), achieving further scalability for high-dimensional and multiphysics problems remains an open challenge.

3.2. Viscosity

The production of high-quality polymeric components through additive manufacturing (AM) relies on precise control of melt viscosity (η), which depends on molecular weight (M_w), shear rate ($\dot{\gamma}$), and temperature (T). The viscosity follows a shear-thinning behavior, modeled by

$$\eta(M_w, T, \dot{\gamma}) = \frac{\eta_0(M_w, T)}{1 + \left(\frac{\dot{\gamma}}{\dot{\gamma}_{cr}}\right)^{1-n}} \tag{10}$$

where $\eta_0(M_w, T)$ represents the zero-shear viscosity, $\dot{\gamma}_{cr}$ is the critical shear rate, and n describes shear sensitivity. The temperature dependence is captured by the WLF equation:

$$\eta_0 = \eta_{M_w} \times 10^{-\frac{C_1(T-T_r)}{C_2+(T-T_r)}} \tag{11}$$

where C_1 and C_2 are empirical parameters, and T_r is a reference temperature near the glass transition temperature. Molecular weight dependence follows a piecewise power law:

$$\eta_{M_w} = \begin{cases} k_1 M_w^{\alpha_1}, & M_w < M_{cr} \\ k_2 M_w^{\alpha_2}, & M_w \geq M_{cr} \end{cases} \tag{12}$$

where M_{cr} is the critical molecular weight, with typical values $\alpha_1 \approx 1$ and $\alpha_2 \approx 3.4$. This formulation reflects distinct viscosity behaviors in different molecular weight regimes:

- When $M_w < M_{cr}$: The polymer chains are relatively short, and viscosity follows a weak power law dependence with $\alpha_1 \approx 1$. In this regime, the entanglement between polymer chains is minimal, resulting in a nearly linear increase in viscosity with increasing molecular weight.

- When $M_w > M_{cr}$: The polymer chains exceed the critical entanglement threshold, leading to a significant increase in viscosity characterized by $\alpha_2 \approx 3.4$. This steep increase is attributed to the formation of an entangled polymer network, which restricts molecular motion and enhances resistance to flow.
- When $M_w = M_{cr}$: This represents the transition point where polymer viscosity shifts from the dilute or semi-dilute regime to the entangled regime. At this critical molecular weight, the polymer chains begin to overlap and form entanglements, drastically altering the rheological behavior.

To predict viscosity in unexplored domains, a Physics-Enforced Neural Network (PENN) was developed by Jain et al. [46], enforcing these dependencies while predicting parameters such as n, $\dot{\gamma}_{cr}$, and C_1, C_2. The PENN outperforms physics-unaware ANN and GPR models in extrapolating η for unseen M_w, $\dot{\gamma}$, and T, improving predictive accuracy for novel polymers in AM applications.

The melt viscosity dataset consists of 1903 datapoints, including 1326 homopolymer, 446 co-polymer, and 113 miscible polymer blend samples, spanning 93 unique repeat units with variations in molecular weight (Mw), shear rate ($\dot{\gamma}$), temperature (T), and polydispersity index (PDI). To address under-representation of viscosity (η) at low Mw, 126 additional datapoints were extrapolated using empirical relationships. Model accuracy was assessed using Order of Magnitude Error (OME), with the Physics-Enforced Neural Network (PENN) improving OME by 35.97% on average and achieving up to 79% R^2 for $\dot{\gamma}$ predictions. Compared to Gaussian Process Regression (GPR) and Artificial Neural Network (ANN), the PENN demonstrated lower Kullback–Leibler divergence in empirical parameter estimation, with RMSE values of 0.05 for α_1 and 0.17 for α_2, closely matching theoretical values of 1 and 3.4. The PENN model also outperformed ANN and GPR in capturing shear thinning behavior, with predicted n values between 0.2 and 0.8 and shear rates ($\dot{\gamma}_{cr}$) closely aligned with experimental distributions.

While the PENN model enhances viscosity prediction accuracy, its reliance on extrapolated data for low-molecular-weight regions introduces potential bias and uncertainty. The model's performance is constrained by the availability of high-quality experimental data as errors may propagate when predicting viscosity for novel polymer chemistries with limited training samples [47]. Although the PENN improves OME by 35.97% and achieves up to 79% R^2, its accuracy in predicting extreme viscosity conditions (e.g., highly entangled polymer networks) remains unverified. The use of empirical constraints ensures physically plausible predictions, but it may limit the model's flexibility in capturing unexpected behaviors in complex polymer systems. Finally, while the PENN outperforms ANN and GPR in shear-thinning modeling, further validation across broader AM process conditions is required to confirm its robustness in real-world applications.

3.3. Viscoelasticity

Physics-guided ML (PGML) methods integrate both data and physical knowledge, making them valuable for modeling the constitutive relations of solids. While PGML approaches have successfully modeled elasticity and plasticity, viscoelasticity remains a challenge due to its dependence on time and loading paths [48]. Many existing methods require extensive experimental or simulation data, making accurate modeling difficult in data-scarce scenarios. Qin et al. [49] proposed a physics-guided recurrent neural network (RNN) model combining gated recurrent units (GRUs) and feedforward neural networks (FNNs) to predict the viscoelastic behavior of solids. The model takes time and stretch (or strain) sequences as inputs, allowing stress predictions based on time and loading paths.

A physics-guided initialization strategy using stress–stretch data from the generalized Maxwell model helps to mitigate data scarcity.

Consider a solid B_0 bounded by ∂B_0, which deforms into B_t with surface ∂B_t. The deformation gradient is given by

$$\mathbf{F} = \nabla_R \mathbf{x} \tag{13}$$

where $\mathbf{x} = \chi(\mathbf{X}, t)$ maps an arbitrary material point $\mathbf{X}$ in B_0 to a spatial point in B_t.

The deformation gradient $\mathbf{F}$ describes how an infinitesimal material element in the reference configuration B_0 deforms into the current configuration B_t. The correlation between the deformation gradient and the domain transformations is as follows:

- B_0 (Reference Configuration): This represents the undeformed or initial state of the body, where material points are labeled by their initial coordinates $\mathbf{X}$. The deformation gradient $\mathbf{F}$ is computed relative to this configuration.
- ∂B_0 (Boundary of the Reference Configuration): This is the initial boundary of the material body before deformation. As deformation occurs, boundary points in ∂B_0 are mapped to new positions on ∂B_t, governed by $\mathbf{F}$.
- B_t (Current Configuration): This is the deformed state of the solid at time t. The transformation $\mathbf{x} = \chi(\mathbf{X}, t)$ determines the new position of every material point from B_0 to B_t. The tensor $\mathbf{F}$ quantifies the local stretch and rotation from B_0 to B_t.
- ∂B_t (Boundary of the Current Configuration): The deformed boundary of the material body, which evolves from ∂B_0 under the transformation $\mathbf{x} = \chi(\mathbf{X}, t)$.

The deformation gradient $\mathbf{F}$ relates directly to the changes in surface elements of B_0 and B_t. Specifically, changes in the normal vectors and area elements of ∂B_0 and ∂B_t can be expressed using $\mathbf{F}$ and its determinant, which represents local volume changes. The rate of deformation can be analyzed through its time derivative, often linked to velocity gradients in continuum mechanics. Thus, the deformation gradient provides a link between the reference and current configurations, enabling the analysis of strain, stress, and material behavior under deformation.

Using the generalized Maxwell model, the deformation gradient decomposes as

$$\mathbf{F} = \mathbf{F}_i^e \cdot \mathbf{F}_i^v, \quad i = 1, \ldots, n \tag{14}$$

where $\mathbf{F}_i^e$ and $\mathbf{F}_i^v$ are the elastic and viscous deformation gradients. The Cauchy stress tensor satisfies

$$\sigma \cdot \nabla + \mathbf{b} = 0, \quad \sigma = \sigma^T \tag{15}$$

which, in the reference configuration, takes the form:

$$\mathbf{P} \cdot \nabla_R + \mathbf{b}_R = 0, \quad \mathbf{P} \cdot \mathbf{F}^T = \mathbf{F} \cdot \mathbf{P}^T \tag{16}$$

where $\mathbf{P}$ is the first Piola–Kirchhoff stress tensor. The energy balance equation is

$$\dot{\varepsilon} = \sigma : \nabla \dot{\chi} - \nabla \cdot \mathbf{j}_q + q \tag{17}$$

Introducing the Helmholtz free energy density $\phi_R = \varepsilon - \vartheta\eta$, the entropy inequality is

$$\sigma : \nabla \dot{\chi} - \eta\dot{\vartheta} - \dot{\phi}_R - \frac{1}{\vartheta}\mathbf{j}_q \cdot \nabla\vartheta \geq 0 \tag{18}$$

Assuming the free energy function:

$$\phi_R = \phi_R(C, C_i^e, \vartheta) \tag{19}$$

- 3D-to-1D transition using a microsphere model: The polymer matrix is represented as a network of 1D elements distributed on a unit sphere, where the strain energy is obtained via numerical integration:

$$\Psi_m \approx \sum_{i=1}^{N_d} w_i \Psi_m^{d_i}, \tag{26}$$

where $\Psi_m^{d_i}$ is the energy contribution of an element in direction d_i.
- Network decomposition: The polymer matrix is divided into parallel networks, each describing a specific inelastic effect, leading to a superposition formulation:

$$\Psi_m = \sum_{i=1}^{N_d} \sum_{j=1}^{N_s} w_i \Psi_j^{d_i}. \tag{27}$$

The first Piola–Kirchhoff stress tensor **P** is then derived as

$$\mathbf{P} = \sum_{i=1}^{N_d} \sum_{j=1}^{N_s} w_i \frac{\partial A_j^{d_i}}{\partial \mathbf{F}} - p\mathbf{F}^{-T}, \tag{28}$$

where p is the Lagrange multiplier ensuring incompressibility.

A conditional neural network (CondNN) [55] architecture is used for the L-agents, incorporating both mechanical and environmental damage. The network consists of two branches: one capturing mechanical damage and the other representing environmental effects, combined multiplicatively to model aging effects accurately.

The model introduces constraints to ensure thermodynamic consistency, but these restrictions may limit adaptability to materials with complex aging mechanisms. The microsphere-based 3D-to-1D transition reduces computational cost, yet the accuracy depends on the chosen discretization and weighting scheme. While network decomposition captures multiple inelastic effects, its predictive performance relies on well-calibrated L-agents, which may require extensive training data [56]. The conditional neural network approach accounts for environmental aging, but interactions between mechanical and chemical degradation might not be fully captured. Further validation across diverse elastomer formulations and exposure conditions is necessary to assess its broader applicability.

3.6. Deflection

To characterize the highly nonlinear state response of the Ionic Polymer–Metal Composite (IPMC), coupled with inherent response uncertainties, Zhang et al. [57] reformulated differential equations that encapsulate the highly nonlinear deflection of IPMC, accounting for uncertainties, and proposed a data-driven approach utilizing a physics-informed neural network (PINN) to effectively solve this differential equation and predict the nonlinear deflection of IPMC actuators.

The electric field induces cation migration, and the relationship between the local voltage and ion charge can be described by the Poisson equation:

$$\nabla^2 \phi = -\frac{\rho}{k_e} = -\frac{F(C^+ - C^-)}{k_e} \tag{29}$$

where ϕ is the electric potential, ρ is the charge density, k_e is the effective dielectric constant of the polymer, F is the Faraday constant, and C^+ and C^- are the concentrations of cations and anions, respectively.

The continuity expression relates the ion flux to the cation concentration:

$$\nabla \cdot J = -\frac{\partial C^{+}}{\partial t} \tag{30}$$

where J is the ion flux, which consists of diffusion, migration, and convection components:

$$J = -d\left(\nabla C^{+} + \frac{C^{+}F}{RT}\nabla\phi + \frac{C^{+}\Delta V}{RT}\nabla p\right) + C^{+}v \tag{31}$$

where d is the diffusion coefficient, R is the gas constant, ΔV is the volumetric change, T is the absolute temperature, p is the pressure, and v is the free water velocity field.

Neglecting nonlinear terms, the partial differential equation for charge density is

$$\frac{\partial \rho}{\partial t} - d\frac{\partial^2 \rho}{\partial x^2} + \frac{F^2 dC^{-}}{k_e RT}(1 - C^{-}\Delta V)\rho = 0 \tag{32}$$

The nonlinear control partial differential equation for the IPMC electric field is

$$\frac{\partial^2 E}{\partial t \partial x} = d\left(\frac{\partial^3 E}{\partial x^3} - \frac{F(1 - C^{-}\Delta V)}{RT}\left[\frac{\partial^2 E}{\partial x^2}E + \left(\frac{\partial E}{\partial x}\right)^2\right] - \frac{F^2 C^{-}(1 - C^{-}\Delta V)}{RTk_e}\frac{\partial E}{\partial x}\right) \tag{33}$$

For nonlinear deflection of an IPMC actuator, the geometric deformation relationship is given by

$$\frac{dw}{ds} = \sin\theta, \quad \frac{dx}{ds} = \cos\theta, \quad \frac{dw}{dx} = \tan\theta \tag{34}$$

$$\frac{d\theta}{ds} = \frac{1}{r}, \quad \frac{d^2\theta}{ds^2} + \frac{U}{B}\sin\theta = 0 \tag{35}$$

where U is the input voltage, and B is the bending stiffness of the IPMC.

A PINN integrates information from both measurement results and the governing partial differential equations into the loss function:

$$MSE = MSE_u + MSE_f \tag{36}$$

where

$$MSE_u = \frac{1}{N}\sum_{i=1}^{N_u}\left|u(t_u^i, x_u^i) - u_i\right|^2, \quad MSE_f = \frac{1}{N_f}\sum_{i=1}^{N_f}\left|f(t_f^i, x_f^i)\right|^2 \tag{37}$$

ensuring accurate modeling of the system.

The experimental results show that Kolmogorov–Arnold Neural PINN (KAN-PINN) [58,59] achieves a 27.54% improvement in prediction accuracy compared to Multilayer-Perceptron-based PINN (MLP-PINN) [60], with an average relative error of 0.316% for Pt-IPMC and 0.277% for Ag-IPMC. KAN-PINN also reduces the convergence cycle significantly, reaching convergence at Epoch 500 for Pt-IPMC and Epoch 400 for Ag-IPMC, compared to Epoch 1400 and Epoch 1200 for MLP-PINN, respectively. However, KAN-PINN's per-iteration training time is approximately ten times longer than MLP-PINN, leading to a longer total training time. The total loss of KAN-PINN is 10.3% lower for Pt-IPMC and 25.75% lower for Ag-IPMC, demonstrating superior convergence. Ag-IPMC exhibits higher voltage sensitivity and lower surface resistance, resulting in faster convergence and improved accuracy compared to Pt-IPMC.

The PINN-based approach leverages physical laws to guide learning, but its accuracy depends on the proper formulation of differential equations and boundary conditions. The method reduces data dependence, yet the increased computational cost, especially with KAN-PINN, may limit practical applications. While KAN-PINN achieves lower loss and faster convergence per epoch, its per-iteration training time is significantly higher than MLP-PINN, leading to longer total training times. The model's performance varies between Pt-IPMC and Ag-IPMC, suggesting material-dependent factors influence prediction accuracy and convergence rates. Further validation across diverse IPMC compositions and operating conditions is needed to assess its robustness in real-world applications.

3.7. Polymerization

Multiphysics engineering impacts chemical reactors due to the complex interactions among fluid mechanics, chemical reactions, and transport phenomena, which significantly impact reactor performance. Recently, PINNs have been successfully applied to engineering problems due to their ability to generalize across domains. Ryu et al. [61] introduced a novel application of PINNs for modeling multiphysics in a chemical reactor. They examined the effectiveness of PINNs in reconstructing and extrapolating ethylene conversion in a polymerization reactor. CFD simulations were conducted to generate training data, and a PINN model was built by incorporating conventional neural network loss with the residuals of fundamental physics equations: continuity, Navier–Stokes [62], and species transport.

The governing equations for fluid flow include the Reynolds-averaged Navier–Stokes equation:

$$\frac{d(\rho v)}{dt} = -\nabla \cdot (\rho v v) - \nabla P + \mu \nabla^2 v + \tau_{Re} \tag{38}$$

and the continuity equation:

$$\frac{d\rho}{dt} = -\nabla \cdot (\rho v). \tag{39}$$

The reactor was modeled as incompressible with constant density and viscosity given its operation at low conversion (20%). The rotating motion of the stirrer was approximated using a moving reference frame, leading to the modified Navier–Stokes equation for relative velocity v_r:

$$v_r = v - \Omega \times r \tag{40}$$

$$\frac{d(\rho v_r)}{dt} = -\nabla \cdot (\rho v_r v_r) - \nabla P + \mu \nabla^2 v_r + \tau_{Re} + \Omega \times (\Omega \times r) + 2\Omega \times v_r. \tag{41}$$

Radical polymerization kinetics were described using three primary reactions: initiator decomposition, polymerization, and termination. The kinetic laws are

$$I \to 2\lambda, \quad k_d = 1.54 \times 10^{14} \exp\left(\frac{-15023}{T}\right) \tag{42}$$

$$M + \lambda \to P, \quad k_p = 1.25 \times 10^{8} \exp\left(\frac{-3800}{T}\right) \tag{43}$$

$$\lambda + \lambda \to P, \quad k_t = 1.25 \times 10^{9} \exp\left(\frac{-327}{T}\right). \tag{44}$$

PINN training involved minimizing a loss function composed of empirical loss and physical residuals:

$$L = w_{emp} \frac{1}{N_{emp}} \sum_{i=1}^{N_{emp}} |y_i^{\text{approx}} - y_i^{\text{CFD}}|^2 + \frac{1}{N_{phy}} \sum_{i=1}^{N_{phy}} |L_i^{\text{physical}}|^2. \tag{45}$$

The physical residuals incorporated continuity, Navier–Stokes, and species balance equations:

$$L_i^{\text{physical}} = w_{\text{cont}} L_i^{\text{cont}} + w_{\text{NS}} L_i^{\text{NS}} + w_{\text{species}} L_i^{\text{species}}. \tag{46}$$

The results demonstrated that PINNs accurately predicted ethylene concentration profiles with an 18% lower mean absolute error (0.1028 mol/L) compared to conventional neural networks (0.1267 mol/L). Furthermore, PINNs successfully captured the conversion concaveness effect, a unique feature in radical polymerization, whereas traditional neural networks failed to do so. These findings highlight the potential of PINNs to efficiently model and extrapolate multiphysics in chemical reactors.

The use of PINNs allows the model to incorporate physical laws, reducing reliance on extensive CFD training data. However, the accuracy of PINN predictions depends on the proper weighting of empirical and physics-based loss terms, which can be challenging to tune. While the model outperforms conventional neural networks in capturing ethylene concentration profiles, its effectiveness in highly turbulent or non-ideal reactor conditions remains uncertain. The approach assumes constant density and viscosity, which may limit its applicability to systems with significant property variations. Further studies are needed to assess PINN robustness when applied to different reactor designs and operating conditions.

3.8. Rheology

Time- and rate-dependent material functions in non-Newtonian fluids pose challenges in integrating constitutive models into CFD. The goal is to solve coupled PDEs relating shear stress to deformation, capturing fluid behavior under different conditions. Mahmoudabadbozchelou et al. [63] introduce non-Newtonian physics-informed neural networks (nn-PINNs) to solve these PDEs using automatic differentiation (AD), eliminating the need for mesh generation.

The power law (PL) model, representing shear-thinning or shear-thickening behavior, is

$$\tau_{xy} = \eta\dot{\gamma}^n \tag{47}$$

where η is the consistency index and n is the power law exponent.

The Carreau–Yasuda (CY) model accounts for viscosity variation with shear rate:

$$\tau_{xy} = \eta_\infty + (\eta_0 - \eta_\infty)[1 + (\lambda\dot{\gamma})^a]^{(n-1)/a}\dot{\gamma} \tag{48}$$

where λ, a, and n define viscosity transition characteristics.

Yield stress fluids are modeled using the Bingham plastic model:

$$\tau_{xy} = \tau_y + \eta\dot{\gamma} \tag{49}$$

where τ_y is the yield stress.

The Herschel–Bulkley (HB) model generalizes yield stress fluids:

$$\tau_{xy} = \tau_y + \eta\dot{\gamma}^n \tag{50}$$

The Maxwell model incorporates viscoelastic effects:

$$\tau_{xy} + \frac{\eta}{G}\dot{\tau}_{xy} = \eta\dot{\gamma} \tag{51}$$

where G is the elastic modulus.

For more complex behavior, the TEVP model includes elastic, plastic, and thixotropic effects:

$$\begin{cases} \dot{\tau}_{xy}(t) = G\frac{1}{\eta_s+\eta_p}\left[-\tau_{xy}(t) + \tau_y l(t) + (\eta_s + \eta_p l(t))\dot{\gamma}(t)\right] \\ \dot{l}(t) = k_+(1 - l(t)) - k_- l(t)\dot{\gamma}(t) \end{cases} \tag{52}$$

where k_+ and k_- define structure evolution.

The governing equations for incompressible fluid motion are

$$\nabla \cdot \mathbf{v} = 0 \tag{53}$$

$$\frac{D\mathbf{v}}{Dt} = -\frac{1}{\rho}\nabla p + \frac{1}{\rho}\nabla \cdot \tau \tag{54}$$

where $\mathbf{v}$ is velocity, p is pressure, and ρ is density.

Loss functions in nn-PINN training minimize residuals from PDEs and boundary conditions:

$$\text{MSE} = \text{MSE}_R + w_2\text{MSE}_{ICs} + w_3\text{MSE}_{BCs} \tag{55}$$

where each term represents the squared error in the residuals.

nn-PINN framework successfully predicts the spatiotemporal behavior of NN fluids with a maximum error of 4% for power law fluids and under 2% for generalized Newtonian fluid (GNF) models. The model accurately captures transitions between shear-thinning (n = 0.8), Newtonian (n = 1.0), and shear-thickening (n = 1.2) behaviors while maintaining consistency across different flow protocols. In complex cases, such as viscoelastic and thixotropic models, nn-PINN reconstructs velocity and stress fields with as few as 50 sparse data points. The framework also adapts to unknown boundary conditions, including slip effects, demonstrating its robustness. Overall, nn-PINN generalizes well across a range of constitutive equations, providing an efficient and reliable alternative for solving fluid dynamics problems.

The nn-PINN framework eliminates the need for mesh generation, but its reliance on automatic differentiation can lead to high computational costs for large-scale simulations. While it accurately captures shear-dependent behaviors, its performance for highly nonlinear viscoelastic fluids with strong memory effects remains uncertain. The approach assumes well-defined constitutive models, which may limit its ability to handle fluids with poorly understood or evolving properties [64]. Despite its ability to adapt to unknown boundary conditions, the sensitivity of its predictions to noise in sparse data points needs further investigation. Future work should explore its applicability to real-world industrial flows where multiple nonlinear effects interact simultaneously.

4. Future Perspectives

Various physics-informed ML approaches have been proposed for modeling polymeric and composite materials. Table 2 provide a comparative analysis of the studied methods based on key characteristics, including the mathematical formulation, data requirements, prediction accuracy, computational efficiency, novelty of the approach, and its limitations.

The integration of PIML into polymeric and composite material modeling has demonstrated significant potential in improving predictive accuracy, reducing computational costs, and enhancing generalization capabilities. However, several challenges and opportunities remain for future research.

Table 2. Comparison of physics-informed models across different studies.

Aspect	**Meng et al. [23]**	**Jain et al. [46]**	**Qin et al. [49]**	**Ghaderi et al. [51]**
Key Challenge	Optimizing temperature distribution during curing	Predicting viscosity for AM polymers	Modeling viscoelasticity under time-dependent loads	Overcoming high dimensionality in stress–strain modeling
Model Proposed	PINO (Physics-Informed Neural Operator)	PENN (Physics-Enforced Neural Network)	PGML (Physics-Guided RNN with GRU-FNN)	Super-constrained ML with L-agents
Mathematical Formulation	Solves parametric coupled PDEs with dynamic BCs	Shear-thinning viscosity models (WLF equation)	Generalized Maxwell model for stress–strain prediction	Reduced-order representation with first and second deformation invariants
Data Used	50 training samples for parametric study	1903 viscosity data points (homopolymers, co-polymers, blends)	Stress–stretch data at different strain rates (VHB4905)	Uni-axial, bi-axial, and shear test datasets (Mars, Treloar, Heuillet)
Key Performance Metrics	MAE: 0.2–0.273 K (temperature), 0.007 (DoC)	35.97% improvement in OME, RMSE: 0.05 (α_1), 0.17 (α_2), R^2 up to 79%	RMSE at 313K: 0.81–4.55; RMSE at 333K: 0.24–4.27	Prediction error: 1.12% (Treloar), outperforming WYPiWYG (5.26%)
Computational Efficiency	Training time: 84.16s (1-dwell), 3016.61s (parametric)	More efficient than ANN and GPR	Uses Backpropagation Through Time (BPTT) for efficiency	Order reduction improves efficiency in high-dimensional problems
Novelty and Advantages	Function-to-function mapping reduces training complexity to $O(N \log N)$	Captures viscosity trends with physics-aware constraints	Combines data-driven and physics-based learning for better generalization	Reduces ML dependency on extensive datasets while ensuring thermodynamic consistency
Limitations	Increased training time for parametric cases	Requires extrapolation for under-represented viscosity regions	Noise sensitivity affects prediction accuracy	Limited confidence intervals in some deformation states
Aspect	**Ghaderi et al. [54]**	**Zhang et al. [57]**	**Ryu et al. [61]**	**Mahmoudabadbozchelou et al. [63]**
Key Challenge	Predicting mechanical performance loss in aging elastomers	High nonlinearities and response uncertainties in IPMC bending	Coupling of fluid mechanics, chemical reactions, and transport phenomena	Complex constitutive equations, varying flow conditions
Model Proposed	Multi-agent constitutive model with neural network learning agents (L-agents)	Physics-Informed Neural Network (PINN) for solving nonlinear PDEs	PINN-based ethylene conversion model for radical polymerization reactor	nn-PINNs
Mathematical Formulation	Strain-energy-based formulation, microsphere model, network decomposition	Poisson equation, charge transport PDEs, nonlinear beam deflection equations	Navier–Stokes equations, continuity equation, radical polymerization kinetics	Power law, Carreau–Yasuda, Herschel–Bulkley, Maxwell, and TEVP models
Data Used	Simulated aging dataset	Experimental IPMC deflection data	CFD-simulated reactor data	Sparse experimental and simulated data (50 sparse points)
Key Performance Metrics	Captures Mullins effect and permanent set	27.54% improved accuracy over MLP-PINN, lower error rates (0.316%, 0.277%)	18% lower mean absolute error compared to conventional NN (0.1028 vs. 0.1267 mol/L)	Maximum error 4% (power law), under 2% for generalized Newtonian fluids
Computational Efficiency	3D stress–strain mapping reduced to constrained 1D problems	Faster convergence, but higher per-iteration training time	Efficiently models multiphysics interactions, capturing conversion concaveness	Adapts to unknown boundary conditions, eliminates need for meshing
Novelty and Advantages	Ensures thermodynamic consistency via constrained ML models	Captures electromechanical coupling and improves generalization of PINN models	Successfully reconstructs and extrapolates polymerization profiles where traditional ML fails	Generalizes across diverse constitutive models, effective in sparse-data regimes
Limitations	Requires extensive hyperparameter tuning for stability	Sensitive to parameter initialization and requires extensive labeled data	Computationally expensive for highly nonlinear coupled systems	Struggles with extreme flow conditions and requires careful scaling for different regimes

One major direction is the development of interpretable models that can account for complex physical interactions while maintaining computational efficiency. The current PIML models often require extensive hyperparameter tuning and struggle with extrapolation beyond training data. Future advancements in PIML may focus on adaptive learning techniques and hybrid models that adjust based on real-time experimental feedback [65]. Adaptive learning methods can integrate data such as temperature, pressure, or molecular weight distributions, allowing models to update predictions during processes like polymerization or extrusion [66]. Active learning may help to select informative data points when data are limited, reducing the need for large datasets. Online learning techniques enable models to adjust parameters as new data are received, supporting real-time process control in polymer processing [67]. Hybrid models combine physics-based models with data-driven approaches by incorporating physical laws and neural network methods to capture nonlinear relationships in polymer behavior [68]. Merging equations for heat transfer, fluid flow, and rheology with data-driven techniques provides predictions of polymer properties under varying conditions. The integration of adaptive learning with hybrid models allows PIML methods to update predictions as new data become available while managing uncertainty [69]. This approach may enhance the design, processing, and performance prediction of polymers in various industries.

Another area is data scarcity and uncertainty quantification [69,70]. Many polymeric and composite systems suffer from limited experimental datasets, making it challenging to train data-driven models effectively. Techniques such as transfer learning, active learning, and Bayesian inference can be leveraged to improve model reliability under sparse-data conditions. Furthermore, incorporating physics-based uncertainty quantification methods will be essential for increasing confidence in model predictions, particularly in safety-critical applications such as aerospace and biomedical engineering.

Advancements in multiphysics and multi-scale modeling present another promising research avenue. The current approaches primarily focus on either macroscale mechanical properties or microscale material behavior, but future research should aim to bridge these scales seamlessly. Coupling PIML with molecular dynamics (MD), density functional theory (DFT), and continuum mechanics could enable more comprehensive material characterization.

Additionally, real-time and edge computing applications for physics-informed models [71,72] could revolutionize industrial processes. Deploying lightweight PIML models on embedded systems and IoT devices could enable in situ monitoring of material behavior during manufacturing, leading to enhanced quality control and predictive maintenance strategies.

Finally, benchmarking and standardization of PIML methodologies are methods for broader adoption. Establishing open-source datasets, evaluation metrics, and standardized training protocols will help to compare different approaches objectively and accelerate their integration into industry workflows.

By addressing these challenges and opportunities, future research can unlock the full potential of physics-informed ML, paving the way for more accurate, efficient, and generalizable models in polymer and composite material science.

5. Further Applications of PINNs in Polymers

Physics-informed neural networks (PINNs) have shown great promise in a wide range of applications beyond those discussed in this paper.

PINNs can be employed to optimize polymer processing techniques, such as extrusion, injection molding, and 3D printing. By embedding the physical laws governing heat transfer, fluid flow, and mechanical behavior into the neural network, PINNs can predict the temperature distribution, viscosity, and stress fields in real time, improving process control and product quality [73]. Additionally, PINNs can be used to optimize the design of molds and predict the behavior of polymers during the curing or cooling stages. For example, in recent works, PINNs have been applied to improve the manufacturing processes of polymer-based bone scaffolds, optimizing heat distribution and processing parameters in 3D printing [74].

PINNs can be used to model complex rheological behaviors in polymer melts and solutions, particularly in scenarios involving non-Newtonian fluids or shear-thinning/thickening behavior. By incorporating constitutive models like the Carreau–Yasuda or Bingham models into PINNs, researchers can predict viscosity and flow stress under varying shear rates, temperatures, and pressure conditions [75]. This is useful in simulating flow in confined spaces or through complex geometries, such as during polymer extrusion or in microfluidic devices. As demonstrated in recent studies, PINNs have been applied to simulate polymer rheology and improve predictions of composite material properties [76,77].

The degradation of polymers over time, due to environmental factors such as temperature, UV radiation, and mechanical stress, can also be modeled using PINNs. PINNs can be trained to capture the physical chemistry of polymer degradation, such as the breakage of chemical bonds, chain scission, and the formation of new cross-links [78]. This application is valuable for predicting the long-term performance and lifetime of polymers in various industries, including the packaging, automotive, and biomedical sectors. In one recent study, PINNs were used to predict the accelerated creep behavior of polymer-alloy geocell sheets, simulating aging effects under different conditions [79].

PINNs can be applied to the modeling of polymer blends and composites, where multiple phases and interactions between different polymer components exist. By embedding the governing equations for phase separation, diffusion, and material property changes in the neural network, PINNs can predict the properties of these complex materials under different processing conditions [80]. This could be particularly beneficial in optimizing the properties of high-performance materials for aerospace, automotive, or electronics applications. Recent work on polymer nanocomposites has utilized PINNs for more accurate predictions of material properties [81,82].

The behavior of polymer nanocomposites, which involve the incorporation of nanoparticles to enhance mechanical, thermal, and electrical properties, can also be predicted using PINNs. These materials often exhibit complex interactions between the polymer matrix and the nanoparticles, and PINNs can be used to model and predict how these interactions influence macroscopic properties such as stiffness, strength, and electrical conductivity [83]. Additionally, PINNs can be applied to the study of smart polymers that respond to external stimuli like temperature, pH, or electric fields, enabling the development of adaptive and responsive materials. Recent studies have demonstrated the potential of PINNs to predict the behavior of such advanced materials [84].

PINNs can be used to predict the molecular weight distribution (MWD) of polymer systems during polymerization processes. This is crucial for controlling the material properties of the polymer, such as tensile strength, elasticity, and melting temperature. By embedding the kinetic equations for polymerization reactions and incorporating experimental data, PINNs can provide insights into how changes in reaction parameters [85] (e.g., temperature, pressure, and initiator concentration) influence the MWD and, consequently, the material properties. Recent advancements in this area include the use of PINNs to predict polymer molecular weight and rheological behavior [81,86].

In future research, it could be beneficial to combine this study with other recent advancements in related fields, such as optical neural ODEs [87], deep neural networks for phase hologram generation, recurrent neural networks in laser modeling, and photonic neural network acceleration. These approaches may provide further insights and broaden the scope of PIML applications

The application of holographic flow equations and neural ODEs to polymers, in strongly coupled systems such as viscoelastic fluids, can be formulated as the shear response of a polymer in a strongly coupled system being governed by a modified flow equation [88]:

$$\frac{\partial}{\partial z}\chi + i\omega\frac{f(z)}{z^2}\left(\chi^2 - \frac{1}{z^2}\right) - \frac{1}{i\omega}\frac{m^2}{z^4} = 0, \tag{56}$$

where χ represents the polymer shear response, $f(z)$ is a function related to the polymer network's elasticity, and m^2 is associated with the interactions between polymer chains.

The real part of the shear viscosity η_{re}, which quantifies the polymer's response to shear stress [89], is given by

$$\eta_{re} = \chi_{re}\big|_{z \to 0}. \quad (57)$$

This relationship reflects how the polymer's internal structure, such as entanglements and elasticity, influences its macroscopic flow behavior.

The dynamics of polymer flow are described by an ODE, which can be solved numerically using a neural ODE framework. The shear response of the polymer is governed by the following equations:

$$\frac{d\chi_{re}}{dz} = 2\omega z^2 f \chi_{re}\chi_{im}, \quad (58)$$

$$\frac{d\chi_{im}}{dz} = \frac{\omega}{f}\left((z\chi_{im})^2 - (z\chi_{re})^2 + \frac{1}{z^2} - \frac{m^2}{\omega z^4}\right). \quad (59)$$

In these equations, χ_{re} and χ_{im} represent the real and imaginary components of the shear response, and $f(z)$ describes the polymer's elasticity. The mass term m^2 represents internal entropic forces within the polymer.

The function $f(z)$ (which governs the polymer's elasticity) or the mass term $m^2(z)$ can be parameterized by a neural network with trainable parameters θ:

$$\frac{dx(t)}{dt} = y(x(t), t, \theta). \quad (60)$$

This network learns how the shear response of the polymer evolves over time or distance, enabling accurate modeling of the polymer's viscoelastic behavior.

After training, the shear viscosity of the polymer at high frequencies (in the UV limit) can be extracted as follows:

$$\eta_{re} = \chi_{re}(z = 0). \quad (61)$$

It provides a quantitative understanding of the polymer's flow properties, making it possible to predict the material's response to various conditions.

These applications highlight the potential of PINNs in polymer science and engineering. By integrating governing physical laws with experimental data, PINNs offer a useful approach for addressing problems where conventional modeling methods may face limitations, particularly in the context of complex multiphysics interactions and sparse data. As the field advances, PINNs have the potential to contribute to the design, processing, and performance prediction of polymers across various industries.

6. Conclusions

PINNs present a transformative approach to polymer modeling by embedding governing equations directly into the learning process. This review has discussed their applications, advantages, and challenges. While hurdles remain, advances in computational strategies and model architectures are expected to further enhance their impact in polymer science. The main findings can be summarized as follows:

- PIML bridges the gap between data-driven and physics-based modeling, enabling more accurate and interpretable predictions in polymeric and composite materials.
- The integration of domain knowledge enhances model reliability and generalization, ensuring thermodynamic consistency and reducing dependency on large datasets.
- Computational efficiency remains a critical trade-off, with advanced models achieving high accuracy but requiring significant training time and computational resources.
- Multiphysics and multi-scale modeling are key to capturing complex material behaviors, allowing for better predictions in nonlinear, time-dependent, and high-dimensional problems.

- Extrapolation and uncertainty quantification remain challenges as some models struggle with under-represented data regimes and noise sensitivity.
- Standardization and benchmarking across studies are necessary to establish best practices and facilitate industrial adoption of PIML approaches.
- Future advancements should focus on hybrid models, adaptive learning strategies, and real-time deployment, paving the way for predictive material design and intelligent manufacturing.
- Computational efficiency varies across models, with the training times ranging from 84.16s for simple cases to 3016.61 s for parametric studies, demonstrating trade-offs between accuracy and complexity.
- Data requirements remain a challenge, with models trained on datasets ranging from 50 sparse experimental points to 1903 viscosity data points, highlighting the need for improved data efficiency.
- Extrapolation and noise sensitivity limit generalizability, with some models achieving up to 79% R^2 accuracy, but others show prediction errors of up to 5.26% in challenging deformation states.
- Multiphysics and multi-scale approaches enhance performance, with PINN-based models reducing mean absolute error (MAE) by 18% compared to conventional neural networks in polymerization modeling.
- Thermodynamic consistency and physics constraints improve reliability, with constrained ML approaches reducing prediction error to as low as 1.12%, outperforming purely data-driven models.
- Standardization and benchmarking are essential as variability in performance metrics (RMSE ranging from 0.24 to 4.55) complicates direct comparisons across studies.
- Industrial adoption remains an ongoing challenge, but real-time deployment could enable in situ monitoring and predictive maintenance, leading to improved material processing and manufacturing efficiency.

Funding: This research received no external funding.

Conflicts of Interest: The authors declare no conflicts of interest.

References

1. Hollingsworth, S.A.; Dror, R.O. Molecular dynamics simulation for all. *Neuron* **2018**, *99*, 1129–1143. [CrossRef] [PubMed]
2. Tang, Y.; Fu, Z.; Raos, G.; Ma, F.; Zhao, P.; Hou, Y. Molecular dynamics simulation of adhesion at the asphalt-aggregate interface: A review. *Surf. Interfaces* **2024**, *44*, 103706. [CrossRef]
3. Li, Y.; Chen, R.; Zhou, B.; Dong, Y.; Liu, D. Rational design of DNA hydrogels based on molecular dynamics of polymers. *Adv. Mater.* **2024**, *36*, 2307129. [CrossRef]
4. Kalateh, F.; Kheiry, M. A review of stochastic analysis of the seepage through earth dams with a focus on the application of monte carlo simulation. *Arch. Comput. Methods Eng.* **2024**, *31*, 47–72. [CrossRef]
5. Schiavo, M. Numerical impact of variable volumes of Monte Carlo simulations of heterogeneous conductivity fields in groundwater flow models. *J. Hydrol.* **2024**, *634*, 131072. [CrossRef]
6. Gawusu, S.; Ahmed, A. Analyzing variability in urban energy poverty: A stochastic modeling and Monte Carlo simulation approach. *Energy* **2024**, *304*, 132194. [CrossRef]
7. Arzovs, A.; Judvaitis, J.; Nesenbergs, K.; Selavo, L. Distributed learning in the iot–edge–cloud continuum. *Mach. Learn. Knowl. Extr.* **2024**, *6*, 283–315. [CrossRef]
8. Sincak, P.J.; Prada, E.; Miková, L.; Mykhailyshyn, R.; Varga, M.; Merva, T.; Virgala, I. Sensing of continuum robots: A review. *Sensors* **2024**, *24*, 1311. [CrossRef]
9. Tu, S.; Li, W.; Zhang, C.; Wang, L.; Jin, Z.; Wang, S. Seepage effect on progressive failure of shield tunnel face in granular soils by coupled continuum-discrete method. *Comput. Geotech.* **2024**, *166*, 106009. [CrossRef]
10. Toscano, J.D.; Oommen, V.; Varghese, A.J.; Zou, Z.; Ahmadi Daryakenari, N.; Wu, C.; Karniadakis, G.E. From pinns to pikans: Recent advances in physics-informed machine learning. *Mach. Learn. Comput. Sci. Eng.* **2025**, *1*, 1–43. [CrossRef]

11. Khalid, S.; Yazdani, M.H.; Azad, M.M.; Elahi, M.U.; Raouf, I.; Kim, H.S. Advancements in Physics-Informed Neural Networks for Laminated Composites: A Comprehensive Review. *Mathematics* **2024**, *13*, 17. [CrossRef]
12. Farea, A.; Yli-Harja, O.; Emmert-Streib, F. Understanding physics-informed neural networks: Techniques, applications, trends, and challenges. *AI* **2024**, *5*, 1534–1557. [CrossRef]
13. Hu, H.; Qi, L.; Chao, X. Physics-informed Neural Networks (PINN) for computational solid mechanics: Numerical frameworks and applications. *Thin-Walled Struct.* **2024**, *205*, 112495. [CrossRef]
14. Donnelly, J.; Daneshkhah, A.; Abolfathi, S. Physics-informed neural networks as surrogate models of hydrodynamic simulators. *Sci. Total Environ.* **2024**, *912*, 168814. [CrossRef]
15. Kapoor, T.; Wang, H.; Núñez, A.; Dollevoet, R. Transfer learning for improved generalizability in causal physics-informed neural networks for beam simulations. *Eng. Appl. Artif. Intell.* **2024**, *133*, 108085. [CrossRef]
16. Jalili, D.; Jadidi, M.; Keshmiri, A.; Chakraborty, B.; Georgoulas, A.; Mahmoudi, Y. Transfer learning through physics-informed neural networks for bubble growth in superheated liquid domains. *Int. J. Heat Mass Transf.* **2024**, *232*, 125940. [CrossRef]
17. Hussain, A.; Sakhaei, A.H.; Shafiee, M. Machine learning-based constitutive modelling for material non-linearity: A review. *Mech. Adv. Mater. Struct.* **2024**, 1–19. [CrossRef]
18. Li, Q.Q.; Xu, Z.D.; Dong, Y.R.; He, Z.H.; Yan, X.; Wang, B.; Guo, Y.Q. Characterization of dynamic mechanical properties of viscoelastic damper based on physics-constrained data-driven approach. *Int. J. Struct. Stab. Dyn.* **2024**, *24*, 2450071. [CrossRef]
19. Bergström, J.S.; Hayman, D. An overview of mechanical properties and material modeling of polylactide (PLA) for medical applications. *Ann. Biomed. Eng.* **2016**, *44*, 330–340. [CrossRef]
20. Guo, J.; Wang, H.; Hou, C. An adaptive energy-based sequential method for training PINNs to solve gradient flow equations. *Appl. Math. Comput.* **2024**, *479*, 128890. [CrossRef]
21. Peng, K.; Li, J. The coupled physical-informed neural networks for the two phase magnetohydrodynamic flows. *Comput. Math. Appl.* **2024**, *166*, 118–128. [CrossRef]
22. Guo, J.; Wang, H.; Gu, S.; Hou, C. TCAS-PINN: Physics-informed neural networks with a novel temporal causality-based adaptive sampling method. *Chin. Phys. B* **2024**, *33*, 050701. [CrossRef]
23. Meng, Q.; Li, Y.; Liu, X.; Chen, G.; Hao, X. A novel physics-informed neural operator for thermochemical curing analysis of carbon-fibre-reinforced thermosetting composites. *Compos. Struct.* **2023**, *321*, 117197. [CrossRef]
24. Jiao, A.; Yan, Q.; Harlim, J.; Lu, L. Solving forward and inverse PDE problems on unknown manifolds via physics-informed neural operators. *arXiv* **2024**, arXiv:2407.05477.
25. Rosofsky, S.G.; Al Majed, H.; Huerta, E. Applications of physics informed neural operators. *Mach. Learn. Sci. Technol.* **2023**, *4*, 025022. [CrossRef]
26. Kim, T.; Lee, H.; Lee, W. Physics embedded neural network vehicle model and applications in risk-aware autonomous driving using latent features. In Proceedings of the 2022 IEEE/RSJ International Conference on Intelligent Robots and Systems (IROS), Kyoto, Japan, 23–27 October 2022; pp. 4182–4189.
27. Zhong, Z.; Ju, Y.; Gu, J. Scalable Physics-Embedded Neural Networks for Real-Time Robotic Control in Embedded Systems. In Proceedings of the 2024 IEEE 67th International Midwest Symposium on Circuits and Systems (MWSCAS), Springfield, MA, USA, 11–14 August 2024; pp. 823–827.
28. Li, P.; Ju, S.; Bai, S.; Zhao, H.; Zhang, H. State of charge estimation for lithium-ion batteries based on physics-embedded neural network. *J. Power Sources* **2025**, *640*, 236785. [CrossRef]
29. Jia, X.; Willard, J.; Karpatne, A.; Read, J.S.; Zwart, J.A.; Steinbach, M.; Kumar, V. Physics-guided machine learning for scientific discovery: An application in simulating lake temperature profiles. *ACM/IMS Trans. Data Sci.* **2021**, *2*, 1–26. [CrossRef]
30. Wang, L.; Zhu, S.P.; Luo, C.; Liao, D.; Wang, Q. Physics-guided machine learning frameworks for fatigue life prediction of AM materials. *Int. J. Fatigue* **2023**, *172*, 107658. [CrossRef]
31. Chen, J.; Chen, Y.; Xu, X.; Zhou, W.; Huang, G. A physics-guided machine learning for multifunctional wave control in active metabeams. *Extreme Mechanics Letters* **2022**, *55*, 101827. [CrossRef]
32. Ghaderi, A.; Dargazany, R. A data-driven model to predict constitutive and failure behavior of elastomers considering the strain rate, temperature, and filler ratio. *J. Appl. Mech.* **2023**, *90*, 051010. [CrossRef]
33. Ghaderi, A.; Ayoub, G.; Dargazany, R. Constitutive behavior and failure prediction of crosslinked polymers exposed to concurrent fatigue and thermal aging: A reduced-order knowledge-driven machine-learned model. *J. Mater. Sci.* **2024**, *59*, 5066–5084. [CrossRef]
34. Ghaderi, A.; Chen, Y.; Dargazany, R. A Physics-Based Data-Driven Approach for Modeling of Environmental Degradation in Elastomers. In Proceedings of the ASME International Mechanical Engineering Congress and Exposition, Columbus, OH, USA, 30 October–3 November 2022; American Society of Mechanical Engineers: Columbus, OH, USA, 2022; Volume 86717, p. V009T12A004.
35. Karniadakis, G.E.; Kevrekidis, I.G.; Lu, L.; Perdikaris, P.; Wang, S.; Yang, L. Physics-informed machine learning. *Nat. Rev. Phys.* **2021**, *3*, 422–440. [CrossRef]

36. Zhang, W.; Ni, P.; Zhao, M.; Du, X. A general method for solving differential equations of motion using physics-informed neural networks. *Appl. Sci.* **2024**, *14*, 7694. [CrossRef]
37. Wu, Y.; Sicard, B.; Gadsden, S.A. Physics-informed machine learning: A comprehensive review on applications in anomaly detection and condition monitoring. *Expert Syst. Appl.* **2024**, *255*, 124678. [CrossRef]
38. Wang, Y.; Yao, Y.; Guo, J.; Gao, Z. A practical PINN framework for multi-scale problems with multi-magnitude loss terms. *J. Comput. Phys.* **2024**, *510*, 113112. [CrossRef]
39. Hashemi, Z.; Gholampour, M.; Wu, M.C.; Liu, T.Y.; Liang, C.Y.; Wang, C.C. A physics-informed neural networks modeling with coupled fluid flow and heat transfer–Revisit of natural convection in cavity. *Int. Commun. Heat Mass Transf.* **2024**, *157*, 107827. [CrossRef]
40. Seo, J. Solving real-world optimization tasks using physics-informed neural computing. *Sci. Rep.* **2024**, *14*, 202. [CrossRef]
41. Jha, N.; Mallik, E. GPINN with neural tangent kernel technique for nonlinear two point boundary value problems. *Neural Process. Lett.* **2024**, *56*, 192. [CrossRef]
42. Onyelowe, K.C.; Kontoni, D.P.N. Numerical modeling of the funnel multiphysical flow of fresh self-compacting concrete considering proportionate heterogeneity of aggregates. *Sci. Rep.* **2024**, *14*, 1601. [CrossRef]
43. Fang, Z.; Wang, S.; Perdikaris, P. Learning only on boundaries: A physics-informed neural operator for solving parametric partial differential equations in complex geometries. *Neural Comput.* **2024**, *36*, 475–498. [CrossRef]
44. Stankovic, D.; Davidson, J.R.; Ott, V.; Bisby, L.A.; Terrasi, G.P. Experimental and numerical investigations on the tensile response of pin-loaded carbon fibre reinforced polymer straps. *Compos. Sci. Technol.* **2024**, *258*, 110915. [CrossRef]
45. Huang, O.; Saha, S.; Guo, J.; Liu, W.K. An introduction to kernel and operator learning methods for homogenization by self-consistent clustering analysis. *Comput. Mech.* **2023**, *72*, 195–219. [CrossRef]
46. Jain, A.; Gurnani, R.; Rajan, A.; Qi, H.J.; Ramprasad, R. A physics-enforced neural network to predict polymer melt viscosity. *npj Comput. Mater.* **2025**, *11*, 42. [CrossRef]
47. Tandia, A.; Onbasli, M.C.; Mauro, J.C. Machine learning for glass modeling. In *Springer Handbook of Glass*; Springer: Cham, Switzerland, 2019; pp. 1157–1192.
48. Haywood-Alexander, M.; Liu, W.; Bacsa, K.; Lai, Z.; Chatzi, E. Discussing the spectrum of physics-enhanced machine learning: A survey on structural mechanics applications. *Data-Centric Eng.* **2024**, *5*, e30. [CrossRef]
49. Qin, B.; Zhong, Z. A Physics-Guided Machine Learning Model for Predicting Viscoelasticity of Solids at Large Deformation. *Polymers* **2024**, *16*, 3222. [CrossRef]
50. Zhang, B. Intelligent Vehicle Lateral and Longitudinal Decoupled Dynamic Modeling and Control System Simulation Based on GRU-FNN. In Proceedings of the 2024 3rd International Conference on Energy and Power Engineering, Control Engineering (EPECE), Chengdu, China, 23–24 February 2024; pp. 153–158.
51. Ghaderi, A.; Morovati, V.; Dargazany, R. A physics-informed assembly of feed-forward neural network engines to predict inelasticity in cross-linked polymers. *Polymers* **2020**, *12*, 2628. [CrossRef]
52. Ghaderi, A. *Physics-Informed Data-Driven Models for Inelastic, Aging, Failure Behavior of Crosslinked Polymers*; Michigan State University: East Lansing, MI, USA, 2023.
53. Torzoni, M.; Rosafalco, L.; Manzoni, A.; Mariani, S.; Corigliano, A. SHM under varying environmental conditions: An approach based on model order reduction and deep learning. *Comput. Struct.* **2022**, *266*, 106790. [CrossRef]
54. Ghaderi, A.; Morovati, V.; Bahrololoumi, A.; Dargazany, R. A physics-informed neural network constitutive model for cross-linked polymers. In Proceedings of the ASME International Mechanical Engineering Congress and Exposition, Online, 16–19 November 2020; American Society of Mechanical Engineers: Columbus, OH, USA, 2020; Volume 84607, p. V012T12A007.
55. Wang, H.; Bocchini, P.; Padgett, J.E. Estimation of wind pressure field on low-rise buildings based on a novel conditional neural network. *J. Wind. Eng. Ind. Aerodyn.* **2024**, *250*, 105752. [CrossRef]
56. Yang, T.; Li, G.; Li, K.; Li, X.; Han, Q. The LPST-Net: A new deep interval health monitoring and prediction framework for bearing-rotor systems under complex operating conditions. *Adv. Eng. Inform.* **2024**, *62*, 102558. [CrossRef]
57. Zhang, L.; Chen, L.; An, F.; Peng, Z.; Yang, Y.; Peng, T.; Song, Y.; Zhao, Y. A physics-informed neural network for nonlinear deflection prediction of Ionic Polymer-Metal Composite based on Kolmogorov-Arnold networks. *Eng. Appl. Artif. Intell.* **2025**, *144*, 110126. [CrossRef]
58. Jiang, Q.; Gou, Z. Solutions to Two-and Three-Dimensional Incompressible Flow Fields Leveraging a Physics-Informed Deep Learning Framework and Kolmogorov–Arnold Networks. *Int. J. Numer. Methods Fluids* **2025**, *97*, 665–673. [CrossRef]
59. Shuai, H.; Li, F. Physics-informed kolmogorov-arnold networks for power system dynamics. *IEEE Open Access J. Power Energy* **2025**, *12*, 46–58. [CrossRef]
60. Zhang, S.; Zhang, C.; Han, X.; Wang, B. MRF-PINN: A multi-receptive-field convolutional physics-informed neural network for solving partial differential equations. *Comput. Mech.* **2025**, *75*, 1137–1163. [CrossRef]
61. Ryu, Y.; Shin, S.; Lee, W.B.; Na, J. Multiphysics generalization in a polymerization reactor using physics-informed neural networks. *Chem. Eng. Sci.* **2024**, *298*, 120385. [CrossRef]

62. Eivazi, H.; Tahani, M.; Schlatter, P.; Vinuesa, R. Physics-informed neural networks for solving Reynolds-averaged Navier–Stokes equations. *Phys. Fluids* **2022**, *34*, 075117. [CrossRef]
63. Mahmoudabadbozchelou, M.; Karniadakis, G.E.; Jamali, S. nn-PINNs: Non-Newtonian physics-informed neural networks for complex fluid modeling. *Soft Matter* **2022**, *18*, 172–185. [CrossRef]
64. Singh, P.; Lalitha, R.; Mondal, S. Saffman-Taylor instability in a radial Hele-Shaw cell for a shear-dependent rheological fluid. *J. Non-Newton. Fluid Mech.* **2021**, *294*, 104579. [CrossRef]
65. Bian, K.; Priyadarshi, R. Machine learning optimization techniques: A Survey, classification, challenges, and Future Research Issues. *Arch. Comput. Methods Eng.* **2024**, *31*, 4209–4233. [CrossRef]
66. Munir, N.; Nugent, M.; Whitaker, D.; McAfee, M. Machine learning for process monitoring and control of hot-melt extrusion: Current state of the art and future directions. *Pharmaceutics* **2021**, *13*, 1432. [CrossRef]
67. Castillo, M.; Monroy, R.; Ahmad, R. A cyber-physical production system for autonomous part quality control in polymer additive manufacturing material extrusion process. *J. Intell. Manuf.* **2024**, *35*, 3655–3679. [CrossRef]
68. Kasilingam, S.; Yang, R.; Singh, S.K.; Farahani, M.A.; Rai, R.; Wuest, T. Physics-based and data-driven hybrid modeling in manufacturing: A review. *Prod. Manuf. Res.* **2024**, *12*, 2305358. [CrossRef]
69. Shi, Y.; Wei, P.; Feng, K.; Feng, D.C.; Beer, M. A survey on machine learning approaches for uncertainty quantification of engineering systems. *Mach. Learn. Comput. Sci. Eng.* **2025**, *1*, 11. [CrossRef]
70. Soibam, J.; Aslanidou, I.; Kyprianidis, K.; Fdhila, R.B. Inverse flow prediction using ensemble PINNs and uncertainty quantification. *Int. J. Heat Mass Transf.* **2024**, *226*, 125480. [CrossRef]
71. Ju, Y.; Xu, G.; Gu, J. 20.4 A 28nm Physics Computing Unit Supporting Emerging Physics-Informed Neural Network and Finite Element Method for Real-Time Scientific Computing on Edge Devices. In Proceedings of the 2024 IEEE International Solid-State Circuits Conference (ISSCC), San Francisco, CA, USA, 18–22 February 2024; Volume 67, pp. 366–368.
72. Kamath, A.K.; Anavatti, S.G.; Feroskhan, M. A Physics-Informed Neural Network Approach to Augmented Dynamics Visual Servoing of Multirotors. *IEEE Trans. Cybern.* **2024**, *54*, 6319–6332. [CrossRef]
73. Farrag, A.; Kataoka, J.; Yoon, S.W.; Won, D.; Jin, Y. SRP-PINN: A physics-informed neural network model for simulating thermal profile of soldering reflow process. *IEEE Trans. Compon. Packag. Manuf. Technol.* **2024**, *14*, 1098–1105. [CrossRef]
74. Xu, Y.; Zhang, F.; Zhai, W.; Cheng, S.; Li, J.; Wang, Y. Unraveling of Advances in 3D-Printed Polymer-Based Bone Scaffolds. *Polymers* **2022**, *14*, 566. [CrossRef]
75. Urraca, R.; Pernía-Espinoza, A.; Diaz, I.; Sanz-Garcia, A. Practical methodology for validating constitutive models for the simulation of rubber compounds in extrusion processes. *Int. J. Adv. Manuf. Technol.* **2017**, *90*, 2377–2387. [CrossRef]
76. Wang, G.; Sun, L.; Zhang, C. The effect of polyvinylpyrrolidone modified nano-polymers on rheological properties of silicon-based shear thickening fluid. *Phys. Fluids* **2024**, *36*, 073108. [CrossRef]
77. Zhu, S.; Wu, S.; Fu, Y.; Guo, S. Prediction of particle-reinforced composite material properties based on an improved Halpin–Tsai model. *AIP Adv.* **2024**, *14*, 045339. [CrossRef]
78. Jicsinszky, L.; Bucciol, F.; Chaji, S.; Cravotto, G. Mechanochemical Degradation of Biopolymers. *Molecules* **2023**, *28*, 8031. [CrossRef]
79. Zhao, Y.; Xiao, H.; Chen, L.; Chen, P.; Lu, Z.; Tang, C.; Yao, H. Application of the non-linear three-component model for simulating accelerated creep behavior of polymer-alloy geocell sheets. *Geotext. Geomembr.* **2025**, *53*, 70–80. [CrossRef]
80. Wu, Z.; Zhang, H.; Ye, H.; Zhang, H.; Zheng, Y.; Guo, X. PINN enhanced extended multiscale finite element method for fast mechanical analysis of heterogeneous materials. *Acta Mech.* **2024**, *235*, 4895–4913. [CrossRef]
81. Liu, D.; Li, Q.; Zhu, Y.; Cheng, R.; Zeng, T.; Yang, H.; Yuan, C. Physics-informed neural networks for phase-field simulation in designing high energy storage performance polymer nanocomposites. *Appl. Phys. Lett.* **2025**, *126*, 052901. [CrossRef]
82. Qian, F.; Jia, R.; Cheng, M.; Chaudhary, A.; Melhi, S.; Mekkey, S.D.; Hu, M. An overview of polylactic acid (PLA) nanocomposites for sensors. *Adv. Compos. Hybrid Mater.* **2024**, *7*, 75. [CrossRef]
83. Talwar, D.N.; Becla, P. Microhardness, Young's and Shear Modulus in Tetrahedrally Bonded Novel II-Oxides and III-Nitrides. *Materials* **2025**, *18*, 494. [CrossRef]
84. Huang, G.; Zhang, L.; Chu, S.; Xie, Y.; Chen, Y. A highly ductile carbon material made of triangle rings: A study of machine learning. *Appl. Phys. Lett.* **2024**, *124*, 043103. [CrossRef]
85. Pateras, J.; Zhang, C.; Majumdar, S.; Pal, A.; Ghosh, P. Physics-informed machine learning for automatic model reduction in chemical reaction networks. *Sci. Rep.* **2025**, *15*, 7980. [CrossRef]
86. Ren, D.; Wang, C.; Wei, X.; Zhang, Y.; Han, S.; Xu, W. Harmonizing physical and deep learning modeling: A computationally efficient and interpretable approach for property prediction. *Scr. Mater.* **2025**, *255*, 116350. [CrossRef]
87. Proppe, A.H.; Lee, K.L.K.; Sun, W.; Krajewska, C.J.; Tye, O.; Bawendi, M.G. Neural Ordinary Differential Equations for Forecasting and Accelerating Photon Correlation Spectroscopy. *J. Phys. Chem. Lett.* **2025**, *16*, 518–524. [CrossRef]

88. Gu, Z.F.; Yan, Y.K.; Wu, S.F. Neural ODEs for holographic transport models without translation symmetry. *Eur. Phys. J. C* **2025**, *85*, 63. [CrossRef]
89. Pyromali, C.; Taghipour, H.; Hawke, L.G. Entangled linear polymers in fast shear: Evaluation of differential tube-based modeling including flow-induced disentanglement and chain tumbling. *Rheol. Acta* **2024**, *63*, 541–572. [CrossRef]

polymers

MDPI

Review

Recent Progress of Artificial Intelligence Application in Polymer Materials

Teng Long [1,2,*], Qianqian Pang [1,2], Yanyan Deng [1,2], Xiteng Pang [3], Yixuan Zhang [1,2,4], Rui Yang [1,2] and Chuanjian Zhou [1,2,*]

1 School of Materials Science & Engineering, Shandong University, Jinan 250061, China
2 State Key Laboratory of Coatings for Advanced Equipment, Jinan 250061, China
3 School of Information Science and Engineering, Zaozhuang University, Zaozhuang 277160, China
4 Institute of Materials Science, Technical University of Darmstadt, 64287 Darmstadt, Germany
* Correspondence: tenglong@sdu.edu.cn (T.L.); zhouchuanjian@sdu.edu.cn (C.Z.)

Abstract: Artificial intelligence (AI) technology has made remarkable progress in polymer materials, which has changed polymer science significantly. However, this community still relies heavily on the traditional research paradigm instead of the data-driven paradigm. This review advocates for a fundamental paradigm shift in polymer research from traditional experience-driven methods to data-driven approaches enabled by AI. While AI has made transformative advances in polymer design, property prediction, and process optimization, the field remains anchored in conventional methodologies. AI's computational advantages against persistent barriers are also evaluated, such as data scarcity, inadequate material descriptors, and algorithmic complexity. Potential solutions, including collaborative data platforms, domain-adapted descriptor frameworks, and active learning strategies, are also discussed. Furthermore, we demonstrate how high-quality data and explainable AI methodologies overcome computational limitations while ensuring result credibility in other areas, which can benefit polymer research. Ultimately, this work provides a roadmap for accelerating the sustainable convergence of data-driven AI innovation with polymer science.

Keywords: artificial intelligence; polymer materials; database; descriptors; algorithm

1. Introduction

Over the past century, polymer materials have undergone rapid development and have become indispensable components in modern society. Their application domains have continuously expanded to encompass key industries such as energy [1], healthcare [2,3], biomedical [4], electronics [5–7], and construction [8]. Although the production volumes of commodity polymers such as polyethylene, polypropylene, and polystyrene have surpassed those of traditional metallic materials like steel [9], these polymers still exhibit limitations in terms of mechanical properties and thermal stability, which constrain their application. To overcome existing performance bottlenecks of polymer materials, the design and fabrication of novel high-performance polymers have emerged as a critical research frontier in materials science [10]. Nevertheless, the development workflow for polymer materials is complex, involving multiple interrelated stages, including molecular structure design, controllable synthetic characterization, and scalable process optimization [11,12]. This journey typically spans more than a decade and requires substantial research and development investment. Conventional research approaches heavily rely on experience-driven trial-and-error methods, which are inefficient when navigating through the high-dimensional and nonlinear chemical space, thereby significantly limiting the speed and efficiency of new material discovery. Consequently, there is an urgent need for more efficient research paradigms to accelerate new polymer materials development [13].

In recent years, the rapid development of artificial intelligence (AI) technologies has significantly prompted the advancement of materials science [14,15]. This subject focuses on the understanding of the relationships among processing, structure, properties, and performance (PSPP) [16] of different materials (Figure 1). Currently, such study is mainly conducted by experiment paradigms, which is relative slow and inefficient. By integrating AI into this framework, these multidimensional relationships can be explored in a more efficient and fast way, thereby accelerating materials research. Related works are now widely applied in energy materials [17–19], mechanical materials [20,21], bionic materials [22–24], medical materials [25], and other fields [26–28]. Their applications involve the composition optimization [29], experiment cost deduction [30], properties prediction (including Young's modulus [31], melting temperature [32,33], thermal stability, and thermal conductivity [34,35]), and characterization analysis (such as powder diffraction (PD), pair distribution function (PDF), small-angle scattering (SAS), inelastic neutron scattering (INS), and X-ray absorption spectroscopy (XAS) [36,37]). These applications not only improve research efficiency, but also provide strong support for the design and development of new materials.

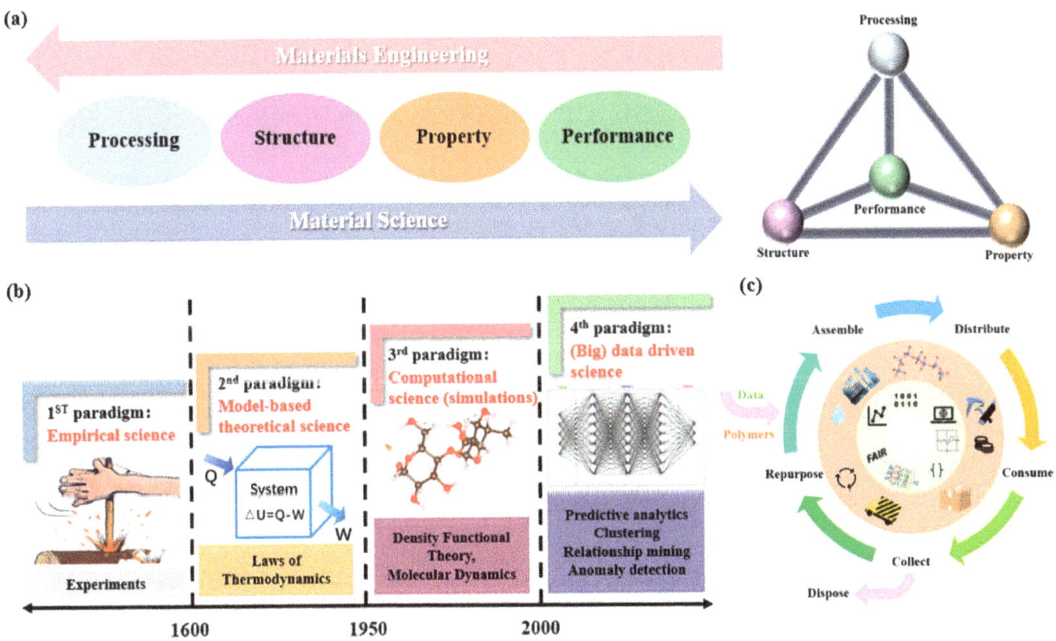

Figure 1. Multi-paradigm development and data-driven process of materials science and engineering. (**a**) Schematic diagram of PSPP relationship; (**b**) four major paradigms of materials development; (**c**) life cycle and data of polymer materials. This figure is inspired by works from Deagen et al. [38].

Compared to inorganic material systems such as metals and ceramics, polymer materials exhibit significant complexity and diversity in their microstructures [39]. Typical polymers exhibit the flexibility of molecular chains, compositional polydispersity, sequence randomness, hierarchical multi-level structures, and strong coupling between processing conditions and final properties. On the one hand, these features endow polymers with excellent mechanical, thermal, electrical, and processing performance [40]; on the other hand, they substantially increase the dimensionality of design variables and the complexity of modeling, rendering traditional "trial-and-error" approaches inadequate for precise design [41]. The emergence of AI, with strong generalization and feature extraction capabilities, has established a new paradigm for the structural design, property prediction, and process optimization of polymer materials [42]. AI can efficiently identify latent relationships within high-dimensional, nonlinear, and multivariable PSPP spaces, thereby enabling coordinated design across multiple structural levels of polymers and significantly enhancing the efficiency and success rate of materials development.

As a result, extensive research efforts have been devoted to study the relationship between materials structure and properties, and the relationship between material performance and processing conditions by AI [43]. For instance, deep neural networks (DNNs) [44] and graph neural networks (GNNs) [45] have been employed to construct predictive models that map molecular structures to properties such as glass transition temperature, modulus, and thermal stability. Reinforcement learning (RL) [46] algorithms have also been applied to automatically optimize polymerization process parameters. These advances collectively demonstrate the unique strengths of AI in uncovering hidden patterns and accelerating design in complex polymer systems. Nevertheless, the successful application of AI in polymer science still faces several critical challenges [47]. First, high-quality and diverse datasets are not available, yet the acquisition of such data is high-cost and low-efficiency [48]. Second, descriptors are not effective due to the multi-scale and multidimensional structural features of polymers [49]. Third, the lack of interpretability of the results limits researchers' ability to understand the underlying scientific relationships [50]. Therefore, reducing dependence on large-scale labeled datasets, establishing effective descriptors, and improving model interpretability represent potential strategies to overcome these challenges.

Thus, this paper will first introduce the current applications of AI in polymer science, with a focus on its practical roles in polymer design, property prediction, and process optimization. Next, we will analyze the key challenges at the intersection of AI and polymer materials—namely, difficulties in data acquisition, insufficient polymer description, and poor interpretability—and the corresponding solutions. Finally, the future of this area will be discussed.

2. The Application of AI in Polymer Materials

AI techniques are revolutionizing polymer materials science by enabling predictions and optimizations. In order to discuss the recent progress, this section will discuss this topic from three different aspects: the development of AI technique, the application of AI in property/structure relationship research, and the application in processing-performance research.

2.1. Foundations and Applications of Machine Learning in Materials Science

By emulating the behavior of human neural networks (NNs), ML identifies and learns patterns within data, thereby establishing complex nonlinear mappings between inputs (e.g., chemical descriptors) and outputs (e.g., material properties) [51]. Ideally, a well-trained ML model can uncover underlying physical or chemical principles, and even approximate certain "effective theories" grounded in quantum mechanics [52–54].

In recent years, ML has found particularly widespread applications in the field of materials science, especially in the prediction of material properties and the design of novel materials. The construction of effective ML models in this domain depends on two fundamental components: high-quality databases and well-designed descriptors. Databases—such as the Materials Project [55], Automatic-FLOW for Materials Discovery (AFLOW) [56], Open Quantum Materials Database (OQMD) [57], the Open Catalyst Project [58], and PolyInfo [59]—contain extensive material data obtained through experiments or simulations, and serve as the foundation for model training and validation. Descriptors, on the other hand, transform complex information related to atomic, molecular, or polymer structures and compositions into numerical features interpretable by ML models. Ideal descriptors should be unique, discriminative, computable, and physically meaningful. Examples include molecular fingerprints [60], structural similarity indices, topological descriptors, molecular orbital energies, atomic electronegativity, molecular weight, and glass transition temperature [61].

Although traditional ML algorithms—such as support vector machines (SVMs) [62,63], random forests (RFs), and kernel ridge regression (KRR)—have demonstrated success in modeling structured data, they often face limitations when dealing with high-dimensional, unstructured data such as images, speech, or text. This challenge has driven the emergence of deep learning (DL) methods, which offer greater expressive power and generalization capabilities. The turning point came in 2011, when Dan Ciresan's team achieved a major breakthrough by training DNNs on GPUs for image classification, drawing widespread attention to DL. Building upon earlier efforts by Geoffrey Hinton, Yoshua Bengio, and Yann LeCun, DL employs multilayer nonlinear architectures [64]—such as convolutional neural networks (CNNs), recurrent neural networks (RNNs), and GNNs—to automatically extract and learn hierarchical feature representations from data. These models have proven particularly effective in capturing the complex dependencies within large datasets. The rapid advancement of hardware, the exponential growth of data, and continuous improvements in training algorithms—such as the Adam optimizer, dropout regularization, and batch normalization—have fueled breakthroughs in various domains, including computer vision, natural language processing (NLP), and speech recognition. These successes have, in turn, catalyzed the integration of DL into materials research.

As a result, a diverse array of ML algorithms has emerged (as illustrated in Table 1), offering researchers new tools to accelerate material design [65–72]. In materials science, ML significantly reduces the time and cost of experimental synthesis and characterization, unlocking new possibilities for high-throughput screening and inverse design. DL, in particular, has been widely adopted for property prediction, synthesis route optimization, and generative modeling of new materials. Although training large models demands substantial computational resources, their capacity to handle complex, large-scale datasets gives them a notable advantage. Furthermore, the applications of AI in polymer science continue to expand, encompassing NLP, computer vision, and laboratory automation. Entering the era of Industry 4.0, AI is increasingly being utilized across various stages of materials research, including design, diagnostics, analysis, predictive maintenance, and early fault detection.

2.2. The Exploration of Property and Structure Relationship for Polymer Materials

Polymer design and property prediction play significant roles in materials science. With the rapid development of AI, researchers can explore new materials with unprecedented efficiency. In this section, discussions will be conducted on the combination of polymer informatics and the Materials Genome Initiative (MGI), the application of structure–property relationships, and the construction of different design methods and prediction models.

Table 1. Classification of machine learning algorithms and their classic and latest developments.

	Class of Algorithm	Classic Important Algorithms	Latest Development Algorithm
Machine learning	Supervised learning	Linear model [73], logistic regression [68], decision tree [74,75], support vector machine [76], random forest [65], gaussian process regression [77,78], multilayer perceptron [79]	XGBoost [80], LightGBM [81], CatBoost [82], TabNet [83], Neural Tangent Kernel [84]
	Unsupervised learning	K-means [85], hierarchical clustering [86], principal component analysis [87]	UMAP [88], Deep clustering [89], Variational autoencoder [90,91], learning contrast (SimCLR [92], MoCo [93])
	Semi-supervised learning	Self-training [94], co-training [95], label propagation [96]	FixMatch [97], MixMatch [98], Pseudo-labeling [99], Snorkel [100], Consistency Regularization [101]
	Deep learning	Convolutional neural network [102], recurrent neural network [103], long short-term memory [104], generative adversarial network [105]	Transformer [106] (BERT, GPT, Vision Transformer), Diffusion model [107], Graph Neural Networks [108]
	Reinforcement learning	Q-learning [109], state-action-reward-state-action [110], deep q-network [111], Monte Carlo method [112]	DreamerV2 [113]
	Meta learning/Auto ML	Grid search [114], random search [114], Bayesian optimization [115]	Neural architecture search [116], Hyperband [117], Meta-Learning [118] (MAML [119], Reptile [120])

2.2.1. Machine Learning Interatomic Potentials

ML interatomic potentials (MLIPs) have substantially expanded the temporal and spatial scales accessible in polymer simulations. Serving as a critical link between primary polymer structures—including composition, conformation, configuration, and monomer sequence—and macroscopic properties, MLIPs enable efficient, accurate modeling across multiple scales. Conventionally, density functional theory (DFT) offers high precision by solving the Schrödinger equation, but its prohibitive computational cost restricts its use to small systems and short timeframes. In contrast, classical molecular dynamics (MD) simulations, which rely on empirical force fields, are significantly more efficient but often fail to capture complex phenomena involving changes in electronic structure. MLIPs offer a compelling solution to this long-standing trade-off by approaching DFT-level accuracy while maintaining MD-level efficiency [121].

The performance of MLIPs relies heavily on the quality of feature descriptors. Early studies predominantly utilized handcrafted features derived from polymer repeat units or chemical heuristics, with researchers manually identifying correlations between structure and performance from large datasets [122]. However, the rapid expansion of polymer databases has rendered manual feature extraction increasingly unsustainable. To address this, Gurnani et al. introduced advanced DL techniques, including GNNs and multi-task learning (MTL), to automatically extract key structural features from polymer repeat units [41]. Compared to traditional handcrafted feature extraction methods, their approach achieved a 1–2 orders of magnitude improvement in feature extraction speed while maintaining comparable prediction accuracy. Further, Zhang et al. proposed the COMFO model, which utilizes three distinct feature extraction methods (the bidirectional encoder transformer model extracts information from simplified molecular input line entry system sequences, the attentive FP network extracts atomic and bond information from molecular graphs, and molecular fingerprints are used to extract substructure information) to accelerate the discovery of polyimide dielectric materials [123]. Compared to traditional feature extraction methods, COMFO achieves a significant improvement in extraction speed while maintaining high prediction accuracy.

The other primary factor to determine the performance of MLIPs is the algorithm. To bridge the gap between quantum-level accuracy and large-scale dynamics, some researchers have developed end-to-end learning frameworks that predict electronic charge densities directly from atomic configurations, therefore bypassing the computationally intensive Kohn–Sham equations. A notable example is the work by del Rio et al. (Figure 2), whose model achieves near-ab initio accuracy with dramatically improved computational efficiency, making it particularly suitable for large-scale polymer systems that lie beyond the practical limits of conventional DFT [124]. Another complementary strategy focuses on enhancing MD simulations by incorporating machine-learned or auxiliary energy terms to improve the accuracy of interatomic interactions. For instance, Parkhill and collaborators developed the TensorMol model, which integrates many-body expansions with NNs to account for long-range physical interactions such as Coulombic and dispersion forces, which are often neglected in standard DFT [125]. TensorMol not only reduces computational cost, but also improves the precision and transferability of molecular simulations across diverse polymer systems. Although these approaches provide strong support for bridging quantum-level accuracy and multi-scale modeling, their generalizability across diverse chemical spaces and stability across different material systems remain open challenges. Currently, they are still in the exploratory and optimization stages; nevertheless, they lay a crucial foundation for the future development of universal models that couple electronic structure with mechanical behavior in polymers (Figure 3).

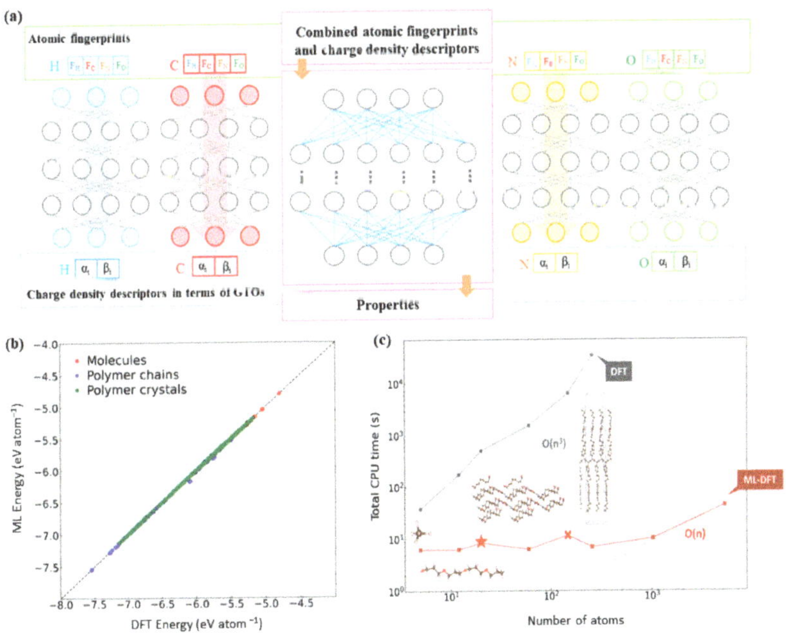

Figure 2. (**a**) Prediction of electron charge density by combining atomic fingerprints with gaussian orbital (GTO) descriptors, and further prediction of other DFT characteristic models using the combined descriptors; (**b**) parity diagram of potential energy for each atom; (**c**) total CPU time of DFT and ML-DFT in electronic structure prediction. This figure is inspired by works from del Rio et al. [124].

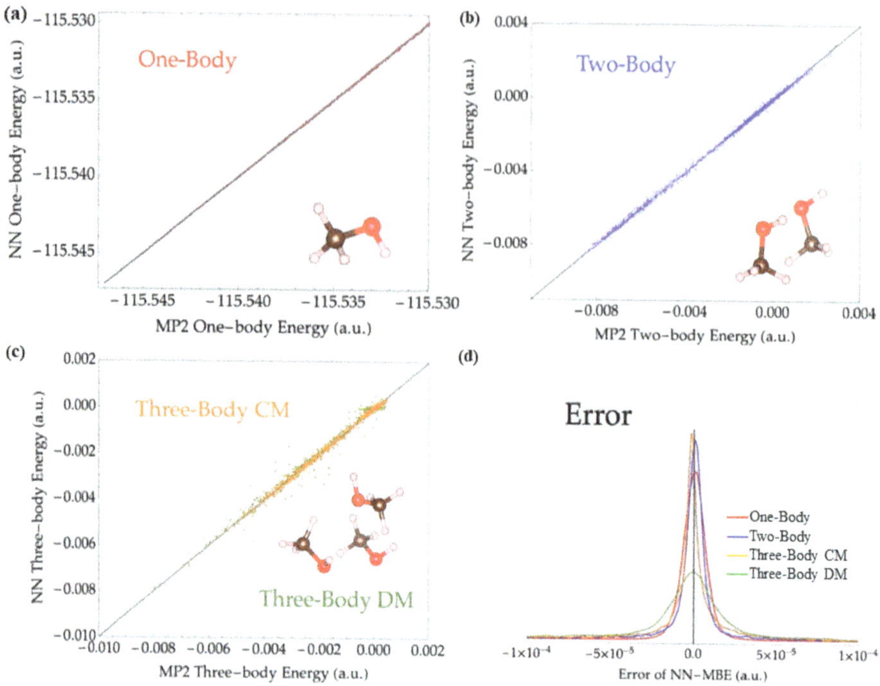

Figure 3. (**a**) From the one-body energy maps calculated by MP2 and neural networks. (**b**) From the two-body energy diagrams calculated by MP2 and neural networks. (**c**) From the three-body energy diagrams calculated by MP2 and neural networks. (**d**) The histogram of the errors of the many-body energy terms predicted by the neural network. This figure is inspired by works from Yao et al. [125].

2.2.2. Structure–Property Modeling and Inverse Design of Polymers

Revealing the coupling mechanisms between structures and properties is a key step toward the design of polymer materials with controllable performance and programmable functions. In recent years, structure–property modeling and property-driven inverse design have emerged as transformative paradigms in polymer research.

A group of studies focuses on predicting local responses based on known structures. For instance, Gautham et al. developed a DNN framework incorporating Behler–Parrinello symmetry functions for modeling the potential of mean force between polymer-grafted nanoparticles [126]. This method uses the radial and angular symmetry functions of the nanoparticle's centroid and grafting points as inputs, and employs a dual-DNN to map the overall interaction potential and local chemical environment potential, effectively capturing the anisotropic interactions induced by the grafted polymers. Although the training data rely on large-scale molecular simulations and have limited generalizability, the method overcomes the limitations of traditional approaches in modeling angle-dependent interactions, significantly improving computational efficiency and laying the foundation for future development of more universal models for interatomic interaction predictions. Shi et al. developed an efficient predictive framework for adhesive free energy from polymer sequences by integrating coarse-grained simulations with support vector regression (SVR) modeling [127]. This framework eliminates the need for complex theoretical assumptions, significantly reducing computational costs. Additionally, by combining GA with the SVR model, the study enables rapid screening of polymer sequences with target properties, providing methodological support for the "inverse engineering" of functional polymers. Although the framework has limitations in generalization capability and adaptation to real-world systems, it is expected to be further extended to diverse interfacial material designs. Yang et al. developed a Fiber Network Deep Learning System that extracts the structural and mechanical information of fiber networks from atomic force microscope images using DL [128]. This method addresses the challenge of low-quality image segmentation, enabling the identification of parameters such as fiber contour length and node distribution. By combining the worm-chain model to calculate mechanical indicators like persistence length, the method was validated in silk fibroin nanofibers. This work lays the foundation for achieving the ultimate goal of "image-driven material design." By extracting chemical-independent features of block polymers from small-angle X-ray scattering (SAXS) data and applying a physics-informed ML algorithm, Fang et al. achieved out-of-sample predictive accuracy of around 95% over the identification of morphologies [129]. Collectively, these efforts provide efficient and deployable structure–property models that make the prediction of complex mechanical and interfacial behaviors possible, significantly improving performance evaluation and experimental planning.

Another line of research takes a further step forward by reversing the modeling direction: from target properties back to optimal material structures—inverse design [130]. Given the vast design space of polymers, such as monomer sequences and component combinations, exhaustive search is intractable. Data-driven inverse design thus offers a new route for performance-oriented structural optimization. For example, Li et al. achieved intelligent regulation of polymer molecular weight distributions through the deep integration of RL and atom transfer radical polymerization simulations, providing efficient and scalable strategies for the synthesis of functional materials [131]. Despite challenges related to model simplifications and limited experimental validation, their demonstrated AI-driven synthesis approach paves a new path for cross-disciplinary research in polymer science and ML, offering a new paradigm for "inverse synthetic design." Kranthiraja et al. demonstrated how to use a random forest model to develop organic photovoltaics with a relatively small number of experimental data points (566) and screen a large number of molecular structures [132]. Although issues regarding descriptors and model interpretability still persist, they presented an experimental-oriented, data-driven research paradigm, offering new ideas for research in this field. Webb et al. achieved targeted polymer sequence design through the deep integration of coarse-grained modeling and ML [133]. This paradigm not only demonstrates prediction accuracy exceeding 95% for target properties and universality in cross-class polymer design, but also enables a substantial reduction in computational cost and ensures target deviation within 10%, proving the significant advantages of data-driven approaches in complex molecular systems. This framework provides critical technological support for overcoming bottlenecks in polymer design and is poised to accelerate the translation of theoretical designs into practical applications for high-performance materials. Hiraide et al. constructed the first DL-based inverse design framework for phase-separated structures in polymer alloys through the collaboration of conditional generative adversarial networks (cGANs) and CNNs, enabling the inference of optimal microstructures from target mechanical stresses [134]. The framework demonstrates excellent performance in both structural generation accuracy (structural factor matching) and mechanical property prediction accuracy ($R^2 > 0.9$), providing an efficient and scalable methodology for the rational design of high-performance materials. This approach holds promise to drive a paradigm shift in materials science from "trial-and-error experiments" to "data-driven design."

Meanwhile, Zhu et al. constructed a "data-driven soft sensor" using a deep belief network (DBN) model, demonstrating excellent performance in nonlinear and data-imbalanced industrial scenarios [135]. This provides an efficient and reliable solution for real-time quality control in polymer production processes, marking a significant application breakthrough of DL in chemical process modeling. Sattari et al. proposed a data-integrated inverse design framework that combines simulation and experimental data, enabling efficient closed-loop optimization from design to manufacturing, offering promising solutions for rapid development of functional polymers (Figure 4) [136].

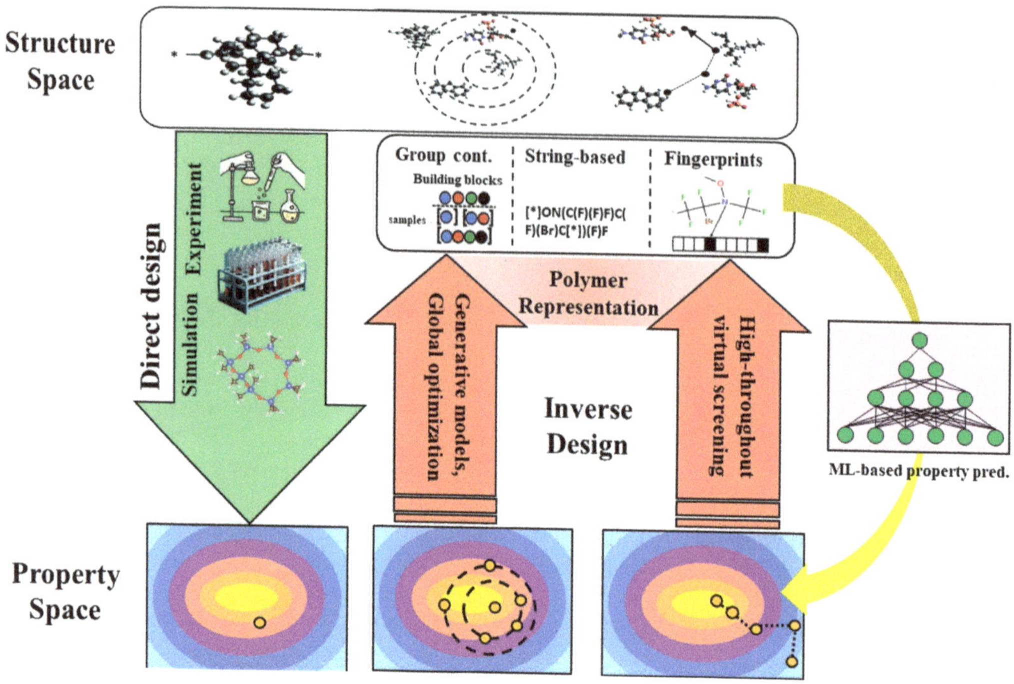

Figure 4. Schematic diagrams of forward and reverse material design. This figure is inspired by works from Sattari et al. [136].

2.2.3. Properties Prediction Model

In the modeling framework of structure–property relationships for polymeric materials, beyond the structure-induced local response behavior, an alternative yet equally critical research direction is the ML-based prediction of intrinsic properties. Intrinsic properties refer to macroscopic physical or chemical characteristics exhibited by polymers under standard conditions, such as glass transition temperature (Tg), electronic bandgap, Young's modulus, dielectric constant, melt flow index, and gas permeability. These properties often determine the applicability, safety, and functional performance of materials in specific engineering contexts. The central objective of intrinsic property prediction is to construct accurate and generalizable mappings from molecular inputs—such as SMILES strings, molecular fingerprints, or graph-based representations—to scalar property outputs [137]. Such models aim to replace time-consuming and costly experimental or computational procedures, thereby enabling high-throughput screening and performance-oriented material design.

A representative early work in this area was reported by Pilania et al., who utilized the RF and KRR models to study polyhydroxyalkanoate (PHA) polymers and established the mapping relationship between their structures and the Tg [138]. This model demonstrated high prediction accuracy. Although this research method has certain limitations in considering the influence of polymer configuration and expanding the application scope, it is superior to traditional methods in terms of prediction accuracy and other aspects, and it is also universal. To further enhance the adaptability and transferability of the model, Shi et al. utilized transfer learning techniques and significantly improved the accuracy of the GNN model pre-trained on shorter oligomers (with errors of excited-state energy that can be less than 200 meV compared to time-dependent density functional theory calculations), even with only a small dataset of longer oligomers, thus solving the problem of insufficient training data in the modeling of optoelectronic properties of long conjugated oligomers and polymers [139].

MTL has also emerged as an effective strategy to improve modeling efficiency and capture inter-property correlations. Kuenneth et al. used a multitask DNN to predict 13 polymer properties, covering thermal, mechanical, and gas permeability aspects [140]. The model leverages data sharing to improve training efficiency and shows good prediction performance. However, there are potential problems in reconciling multi-objective weights. If not properly handled, it will reduce the performance of individual tasks, and future research is needed to explore optimization strategies. This challenge may be mitigated by multi-origin data integration strategies enabled by transfer learning. As demonstrated in pioneering work by Phan et al. [141], language models pre-trained on quantum chemical datasets of small molecules can be effectively fine-tuned using sparse experimental measurements of polymeric macroscopic properties. This approach leverages cross-scale knowledge transfer: the model first captures fundamental physics from computationally tractable systems, then extrapolates to complex polymer behaviors by domain adaptation. Crucially, such frameworks overcome data scarcity by establishing hierarchical feature relationships between theoretically accessible properties (e.g., dipole moments influencing permittivity) and experimentally measured bulk characteristics (e.g., mechanical strength). This paradigm represents a fundamental advancement beyond conventional simulation-reliant approaches. Similarly, Himanshu et al. proposed a method for selecting NN topologies through layer-by-layer expansion and a data selection method based on the latent space, which effectively solved the key problems of DL models in predicting polymer properties [142]. These two methods can be extended to other polymer and material systems and can be further combined to form an efficient workflow, which is of great significance for the feature learning of polymers and other materials and the development of ML models.

Beyond structure-based representations, some studies have explored image-based features for property inference. For instance, Liang et al. utilized convolutional NNs and transfer learning methods to achieve automatic miscibility identification [143]. The accuracy of the model reached as high as 94%, and they also proposed a quantitative criterion for polymer miscibility using this model. Although the performance of this method is limited by the quality and clarity of SEM images, with the increase in the dataset, it is expected that this problem can be solved in the future (Figure 5). Moreover, the success of the MGI has brought new opportunities. Taking advantage of this, Kim et al. developed a fingerprinting method that can capture polymer characteristics from the atomic to the morphological and structural levels [144]. This method, combined with ML models, can rapidly predict polymer properties. The related models have been integrated into the Polymer Genome online platform. At the same time, the research team also plans to expand the chemical and property space to further enhance the platform's support for polymer material research in different fields. By applying evidential neural networks on light scattering and SAXS data of single-chain polymer nanoparticles, Upadhya et al. successfully predicted, synthesized, and characterized 30 novel compact SCNPs, where 58% of them had a Porod exponent ≥ 3.5 [145].

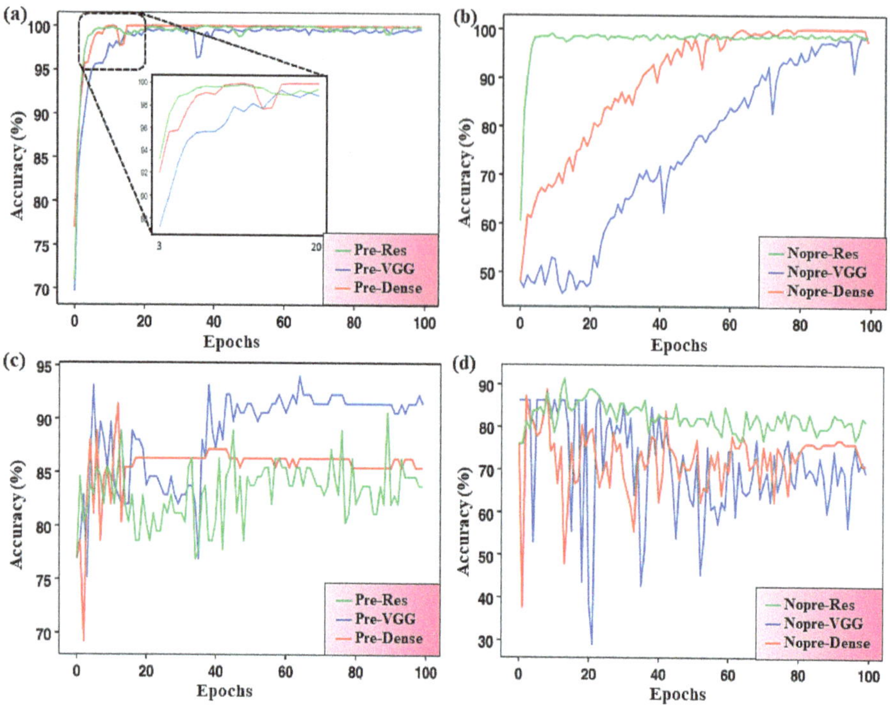

Figure 5. Accuracy of the models on the training set and test set in the training process. (**a**) Accuracy of pretrained models on the training set. (**b**) Accuracy of non-pretrained models on the training set. (**c**) Accuracy of pretrained models on the test set. (**d**) Accuracy of non-pretrained models on the test set. This figure is inspired by works from Liang et al. [143].

In summary, intrinsic property prediction constitutes a "precision modeling" pathway for the application of ML in polymer science [146]. Emphasizing prediction accuracy, structural representation fidelity, and inter-system generalization, these approaches provide a robust foundation for material discovery and targeted optimization.

2.3. Polymer Process Optimization

In the research and development of polymer materials, the high cost of experimental resources, long synthesis cycles, and vast parameter spaces are the core bottlenecks restricting the rapid discovery of high-performance materials. In recent years, AI technology has gradually entered the experimental stage [147]. It is no longer merely used for predicting material properties or evaluating the quality of structures, but is directly integrated into the entire process of experimental design, parameter optimization, and even synthesis decision-making. This has promoted a transformation in the experimental paradigm from "model-assisted" to "system-led". Depending on the optimization objectives and the depth of intervention, existing research can be roughly divided into two categories: parameter optimization with an unchanged process and system-level experimental closed-loop iteration.

2.3.1. Parameter Optimization of the Process

Prior to the widespread adoption of AI, statistical process control and design of experiments (DOE) were the dominant strategies for process optimization in polymer materials. However, these traditional approaches often suffer from long experimental cycles, complex data processing, high costs, limited generalizability, and a lack of ability to capture nonlinear interactions among multiple variables. With the advancement of AI technologies, their integration with polymer science has introduced transformative solutions [147]. Leveraging ML algorithms such as SVMs and neural NNs, researchers have developed data-driven models that enable intelligent optimization of input parameters without altering the underlying experimental or synthetic protocols. These models have significantly improved experimental efficiency and product performance.

ML-based parameter optimization has now become a critical component of intelligent polymer design. Depending on their specific goals, such efforts can be broadly classified into two categories: (1) process-level parameter control, with a strong focus on polymer forming such as additive manufacturing techniques, and (2) structural-level parameter regulation targeting material composition and macroscopic performance.

In the field of additive manufacturing, Khanzadeh et al. employed the self-organizing map method to analyze geometric error patterns in the fused filament fabrication (FFF) process of polymers [148]. By processing point cloud data, this method effectively identified various types of deviations without requiring manual annotation. The study demonstrated that geometric accuracy could be accurately characterized using only 2.4% of the total data points, significantly accelerating scanning and defect detection. Although the research was based on simplified test parts, which limits its generalizability to complex geometries, it laid a solid foundation for the analysis of more intricate shapes and fine features.

In the area of thermal field modeling, the NN model developed by Roy et al. demonstrated exceptionally high computational efficiency [149]. Compared with traditional finite element simulations, the model required only 0.036 s to generate a single temperature curve, with a prediction error of less than 5%. It exhibited strong stability and applicability in FFF processes with fixed structures and processing conditions, providing valuable support for optimization, process planning, and real-time monitoring in closed-loop control during 3D printing.

Beyond these examples, ML has also played a significant role in controlling other processes such as material extrusion, photopolymerization, powder bed fusion, binder jetting, and material jetting. For instance, Zhu et al. used principal component analysis (PCA) and SVM to reduce printing defects [150]; Edlim et al. applied a CNN based on the you only look once (YOLO) architecture for anomaly detection [151], and Lee et al. utilized 3D-CNN and CNN-LSTM models to identify defects in two-photon lithography [152]. In addition, Sassman et al. used SVM to develop a pre-screening method based on powder flowability and surface roughness [153], while Satterlee et al. combined image enhancement with various CNN architectures to detect porosity-related defects [154]. These studies collectively demonstrate the strong adaptability and broad applicability of DL models—including multilayer perceptrons (MLP), RNNs, and CNNs—in polymer forming. Representative architectures and their primary applications are illustrated in Figure 6.

Beyond polymer forming, parameter optimization has also been widely applied to the regulation of material composition and performance. Chen employed an integrated ML strategy combined with the Sure Independence Screening and Sparsifying Operator (SISSO) method to analyze experimental data, optimize the dispersion ratio of graphene oxide–hydrogen-bonded zinc hydroxystannate in a polypropylene (PP) matrix, and developed a predictive model for flame retardancy [156]. This approach significantly improved the flame-retardant properties of the PP composite and shows strong generalizability, with the potential to be extended to other polymer systems.

Li et al. applied Bayesian optimization (BO) to tune the synthesis conditions of short polymer fibers [157]. Their study demonstrated that BO is capable of efficiently searching sparse process spaces and substantially reducing the number of experiments, outperforming traditional grid search methods. However, its applicability to other reaction systems remains to be further explored. Similarly, Takasuka et al. used a flow synthesis system to prepare styrene–methyl methacrylate copolymers and employed BO to adjust processing variables, confirming the method's effectiveness in regulating monomer ratios in copolymers [158]. In the future, multi-objective BO is expected to enable the simultaneous optimization of copolymer composition and physical properties, further advancing the intelligent design of copolymers.

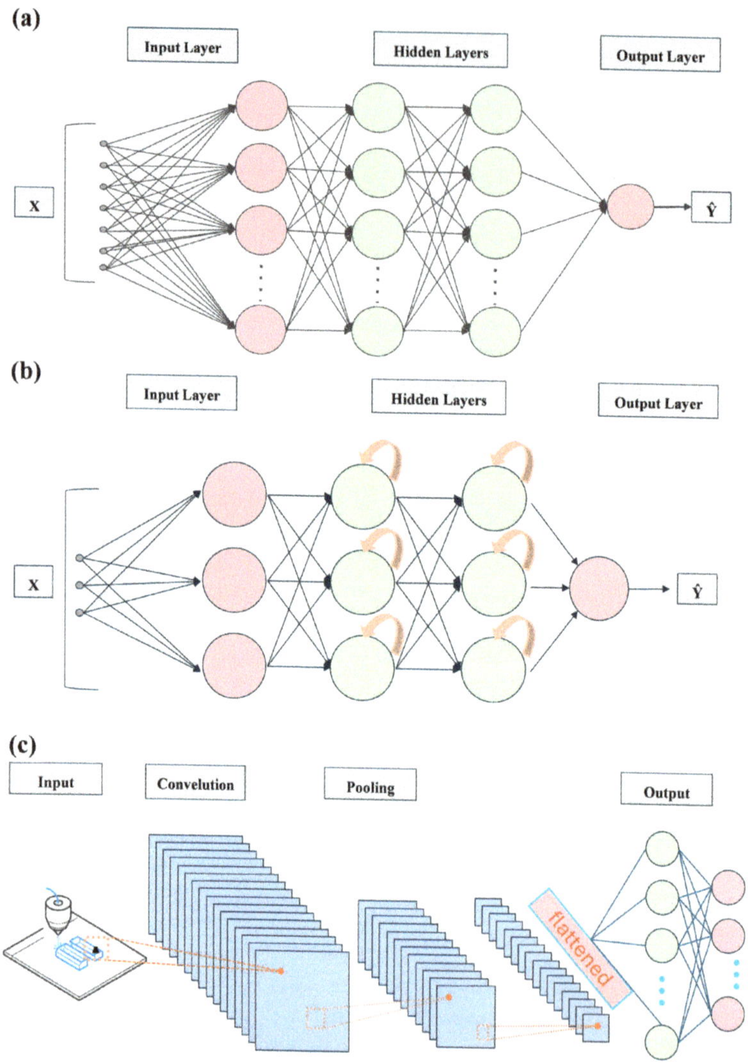

Figure 6. General architectures of different DL models. This figure is inspired by works from Nasrin et al. [155]. (**a**) Multilayer perceptron for predicting polymer AM print properties; (**b**) RNN for quality control in polymer AM prints; (**c**) CNN for defect detection in a polymer AM process.

In summary, parameter optimization strategies have undergone a paradigm shift in polymer research, from early-stage unsupervised clustering and statistical regression to advanced techniques such as BO, MTL, and hierarchical modeling. These methods have demonstrated exceptional performance in structured, parameter-sensitive processes like additive manufacturing, and also provide robust technical support for broader applications in copolymer design, composite formulation, and coarse-grained parameter calibration [159,160].

2.3.2. Process Iteration and Experimental Closed-Loop Systems

Building on previous advances in parameter tuning and process control, recent research has turned to the development of automated, closed-loop systems that enable iterative experimentation, intelligent decision-making, and real-time optimization in polymer materials engineering. In recent years, the continuous advancement of AI and automation technologies has profoundly impacted polymer materials engineering, particularly in enhancing experimental efficiency, precision control, and sustainability [161,162]. The integration of DL algorithms with high-throughput experimental platforms has played a critical role in the construction of process iteration and experimental closed-loop systems, gradually reshaping traditional materials research paradigms.

In the domain of material identification and sorting, Sbrana et al. combined near-infrared hyperspectral imaging technology with the N-BEATS DL model to classify over 4500 plastic samples [163]. The overall classification accuracy of the model reached 79%, with 90% accuracy for colored plastics and 67% for black plastics. This data-driven approach has significantly improved recycling efficiency in material application scenarios, promoting the optimization of raw material procurement and the upstream links of the supply chain.

In polymerization reaction research, Rizkin et al. used a semi-autonomous microfluidic platform combined with in situ thermography and NN to gain a deeper understanding of the catalytic cycle of homogeneous polymerization reactions [164]. The study demonstrated the feasibility of integrating high-throughput microfluidic technology with ML algorithms, reducing chemical waste and energy input. While the reaction system had certain limitations, it significantly improved experimental efficiency and greatly reduced chemical waste in the zirconocene-catalyzed α-olefin polymerization reaction system, opening up new avenues for polymerization reaction research.

Furthermore, process automation has emerged as a key enabler of high-throughput experimentation. Cao et al. developed an ML-guided experimental platform that integrates BO with DOE strategies [165]. The platform, applied to the optimization of complex liquid formulations, conducted hundreds of experiments within 15 working days and identified nine optimal candidate formulations that met all predefined specifications. Compared to traditional trial-and-error workflows, this approach significantly improved both time efficiency and prediction accuracy.

To further enhance synthesis precision and adaptability, Knox et al. successfully developed a platform capable of conducting automated reversible addition-fragmentation chain transfer polymerization under special conditions (Figure 7) [166]. The TSEMO algorithm effectively reduced the number of experiments and achieved closed-loop multi-objective optimization. This technology has broad application prospects in various polymerization techniques with more input variables and objectives, and follow-up research can further clarify the properties of polymers produced under extreme conditions.

In summary, process iteration and experimental closed-loop platforms are emerging as foundational tools in the transition of polymer research from empirical paradigms to data-driven strategies. By tightly integrating experimental design, real-time feedback, algorithmic decision-making, and robotic execution, these platforms address long-standing inefficiencies in materials development [167,168].

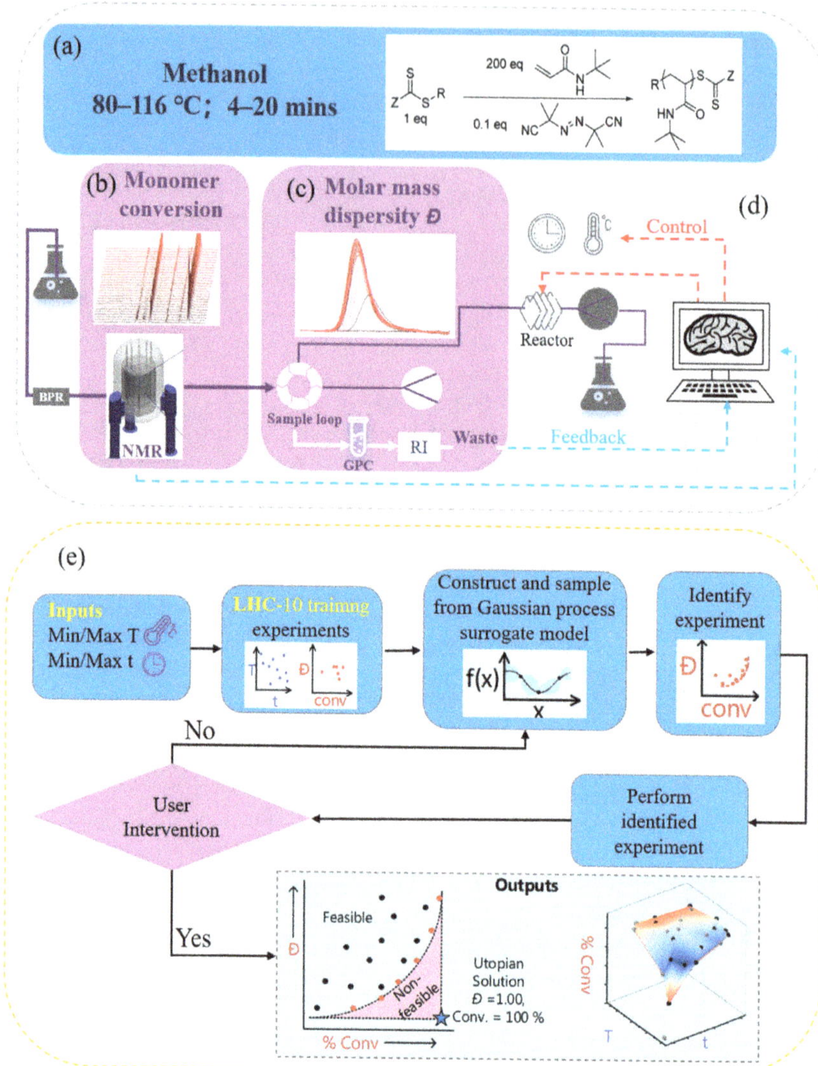

Figure 7. Schematic diagram of a multifunctional fully automatic synthesis platform taking the programmed RAFT polymerization of tert-butylpropanamide as an example. This figure is inspired by works from Knox et al. [166]. (**a**) Generalised scheme for the RAFT synthesis of $P(tBuAm)_{200}$. Example; (**b**) Hydrogen nuclear magnetic resonance (NMR) spectra from the automated continuous flow platform (see ESI for full platform details and all analytical data); (**c**) Gel permeation chromatography hydrogen nuclear magnetic resonance (NMR) spectra from the automated continuous flow platform (see ESI for full platform details and all analytical data); (**d**) Schematic of the automated platform; (**e**) Overview of the machine learning-directed experimental setup using the Thompson-sampling efficient multi-objective optimisation (TSEMO) algorithm.

3. Challenges and Feasible Solutions

AI has demonstrated promising potential in polymer research including molecular structure design, property prediction, process parameter optimization, and the development of intelligent closed-loop laboratories. These advances are gradually driving the transformation of materials research from an experience-driven to a data-driven paradigm.

However, compared to other material systems, the application of AI in polymers remains in its early exploratory stages [169]. The current challenges include: (1) the scarcity of large-scale, high-quality, structured, and representative datasets; (2) the lack of universal descriptor systems capable of effectively representing the multi-scale structures of polymers; and (3) the need for optimization of model architectures and training mechanisms.

3.1. Databases

The construction and utilization of data have become the main bottleneck restricting the further development of this field. Compared with relatively simple systems such as inorganic materials and two-dimensional materials, polymer materials face greater challenges in data acquisition due to their complex molecular structures, diverse synthesis routes, and long experimental cycles. Although the early datasets established by researchers such as Treloar [170] provided valuable references, their limited sample sizes and low dimensions can no longer meet the growing demands of modern large-scale models for data diversity and quantity.

The current challenges mainly focus on two aspects: (1) the high cost of data acquisition makes it difficult to collect large-scale data; and (2) there are significant inconsistencies in formats and measurement standards, increasing the difficulty of data cleaning and integration. In this regard, the successful experiences in fields such as semiconductors, metals, and biomaterials are particularly important [171]. These fields have made remarkable progress in reducing data acquisition costs and unifying formats and measurement standards, providing important references for the polymer materials field. To address the above issues, we can learn from three aspects: data generation, potential data mining, and database construction.

In terms of data generation, the combination of high-throughput experiments and ML has shown great potential. For example, the research team led by Zhang proposed an entropy-based active learning (ET-AL) framework to alleviate the structure–stability bias in material data, especially achieving good results on datasets obtained from DFT calculations [172]. This method uses information entropy as a diversity metric. For the OQMD-8 dataset (containing binary alloy data of elements such as aluminum and titanium) and the J-CFID dataset (containing elastic modulus data), it enhances the diversity of underrepresented regions through AL, significantly improving the accuracy of ML models in predicting material properties (such as bulk modulus and shear modulus) while reducing data bias. This approach provides a new path for data-driven modeling in material discovery. Similar strategies can be applied to the polymer field, combined with high-throughput experiments and ML algorithms, to improve the prediction ability of polymer material properties.

At the same time, multitask learning based on simulated data has become an important supplement to experimental data. For example, the method of Hamiltonian parameterization for perovskite materials proposed by Ma et al. enables a seamless transition from DFT data to large-scale atomic simulations through AL [173]. This research effectively deals with the complex interactions in the complex perovskite system. Polymer systems also have complex interactions such as inter-chain interactions and side-group interactions. The success of this model provides a new way to handle the complex interactions in polymer systems and simulate their structures and behaviors in different environments. In addition, adversarial algorithms have also shown significant potential in enhancing model robustness and enriching data diversity. For example, Yang et al. proposed a two-step data augmentation method based on the GAN model for predicting the hardness of high-entropy alloys, and revealed the feature importance and the reasons for the performance improvement of the two-step method through interpretability analysis [174]. By referring to the two-step data augmentation method, generating polymer data similar to the real data distribution through the generative adversarial network can expand the dataset. When studying the properties of new polymer materials, this method can be used to generate more data, reduce the impact of insufficient data on the model, and improve the prediction accuracy of the model.

In terms of potential data mining, a large number of experimental parameters and performance records in academic literature can be used as important supplementary resources. A large number of experimental parameters and performance records are scattered in scientific publications. Through NLP tools and large language models (LLMs), this information can be efficiently extracted and structured. For example, the ChemMatch framework proposed by Xiuying Chen et al. demonstrates the potential of literature-driven dataset construction, significantly expanding the database capacity while reducing the cost and cycle of experimental data acquisition [175]. This method provides a reference for related research in the polymer field. However, special attention should be paid to the accuracy and semantic integrity when extracting unstructured data such as images and charts to prevent the loss of valuable information during the conversion process. Yan et al. proposed a semi-supervised material information extraction framework based on an automatically generated corpus [176]. They used Snorkel to automatically annotate the corpus and trained the model with an ON-LSTM network, showing good performance in multiple material property extraction tasks, providing a new method for information extraction in the material field. Although there are currently problems such as a small number of datasets and imbalanced positive and negative samples, the idea of automatically annotating corpora has certain reference significance for processing material literature containing images and charts.

In addition, the development of databases has brought new opportunities to solve data-related problems in the polymer field. Although research institutions in various countries have actively established databases related to polymers (as shown in Table 2), there is still a lack of a widely recognized large-scale authoritative platform like The Materials Project in this field. While specialized databases remain essential for application-specific challenges, universal polymer databases would provide complementary foundational value. Such platforms would serve two critical functions: (1) revealing intrinsic structure–property relationships through systematic cross-material comparisons, and (2) enabling transfer learning approaches where large language models pre-trained on universal datasets are fine-tuned using domain-specific experimental data. This hierarchical ecosystem, with universal databases scaffolding specialized repositories, creates pathways for AI acceleration comparable to advances in inorganic materials science. When building such a platform, it is necessary to develop clear data collection standards, metadata specifications, privacy protection measures, and access control systems to improve professionalism and sustainability and enhance data utility. Therefore, promoting interdisciplinary cooperation and establishing a standardized polymer database system is an important direction for future development. In addition, building online resource libraries [177,178] to improve data accessibility and standardization is also an important means. However, attention should be paid to solving problems such as data maintenance, updates, and access rights management. A perfect operation and maintenance mechanism should be developed, and interdisciplinary alliances should be promoted to establish unified data standards and privacy protection guidelines to ensure the long-term operation of the database.

Table 2. Database-related information.

No.	Database	Origin of Data	Description	URL
1	Khazazna	computational	thermoplastic; mechanical, thermal, electrical properties	https://khazana.gatech.edu/dataset/ (accessed on 18 February 2020)
2	PolyInfo	empirical	thermoplastic; mechanical, optical, thermal, rheological properties	https://polymer.nims.go.jp/ (accessed on 22 January 2021)
3	Polymer property predictor and database	empirical	Flory–Huggins parameter, glass transition temperature, binary polymer solution cloud point	https://pppdb.uchicago.edu/ (accessed on 30 March 2016)
4	Material properties database	empirical/computational	thermoplastic, thermoset, rubber; mechanical, thermal, electrical properties	https://www.makeitfrom.com/ (accessed on 16 April 2020)
5	CROW polymer properties database	empirical/computational	thermoplastic, rubber, fiber; physical, thermal properties	https://polymerdatabase.com/ (accessed on 2 March 2019)
6	PI1M	computational	virtual polymers; physical, thermal, electrical properties	https://github.com/RUIMINMA1996/PI1M (accessed on 11 December 2020)
7	Dortmund database	computational	physical properties, phase equilibrium data	https://ddbst.com/ (accessed on 7 October 2020)
8	AI plus Polymers	empirical/computational	thermoset, thermoplastic; physical, mechanical, thermal, electrical properties	https://polymergenome.ecust.edu.cn/ (accessed on 23 March 2019)

In conclusion, to achieve a major breakthrough in the application of AI in polymer materials, it is necessary to address fundamental issues such as limited data availability, insufficient data quality, and inconsistent standards. Only by building an open, standardized, and high-quality data infrastructure can we provide solid support for AI model training and fully unleash the potential of AI in polymer design, property prediction, and process optimization.

3.2 Descriptors

Polymer materials typically exhibit characteristics such as a wide distribution of molecular weights, complex three-dimensional conformations, branching and cross-linking structures, and inter-chain entanglements. These inherent structural complexities make it extremely difficult to accurately capture their microscopic features and effectively map them to macroscopic properties.

The challenges regarding descriptors mainly focus on the following three aspects: (1) given the diversity of polymer structures, it is difficult to construct effective descriptors; (2) high-dimensional descriptors are prone to the "curse of dimensionality"; and (3) there are significant differences in the sources and scales of descriptors, leading to difficulties in integration.

The root cause of these challenges lies in the inherent structural complexity of polymer materials. Compared to crystalline metals or inorganic semiconductors with relatively ordered structures, polymer chains exhibit a wide range of features such as backbone compositions, branching structures, copolymer sequences, and hierarchical features like cross-linking points, crystalline regions, and amorphous regions. This complexity makes traditional low-dimensional or statistically simple features insufficient for performance modeling. For example, in branched polymers, subtle differences in the number, length, or position of side chains can significantly affect mechanical, thermal, or solubility properties. Encoding these subtle yet important features into machine-readable structured descriptors remains a huge challenge. To address these issues, we will elaborate from three aspects, descriptor construction, dimensionality reduction and feature compression, and standardized descriptor systems, drawing on experiences from other fields.

In terms of descriptor construction, research in the field of protein science provides important references for the polymer field. For example, in the case of sequence order, Yu and his colleagues did not directly use the composition of peptide chains as features [179]. Instead, they assigned each peptide chain to a structural category based on its structural class trends and then transformed the protein sequence into a sequence of structural category trends, which improved the prediction accuracy. Similarly, for polymer materials, the monomer sequence in the polymer chain can be used as a similar feature. This helps to reveal the deep-seated relationship between the polymer structure and its properties. Although polymers are more complex than proteins in terms of spatial freedom and interchain interactions, this analogy provides valuable inspiration for the design of polymer-specific descriptor systems.

To improve the expressiveness and adaptability of descriptors, innovation has been attempted from two aspects: multi-scale modeling and time-series analysis. Considering that materials often undergo dynamic evolution during use—such as thermal aging, mechanical fatigue, or environmental responses—the design of descriptors with a time dimension has become a focus of attention. In recent years, sequence models such as long short-term memory (LSTM) and Transformer have been used to capture the time-dependent evolution of materials. These methods enhance the time-resolution ability of descriptors and provide new ideas for predicting non-equilibrium behaviors, service life, and durability. For example, Niu evaluated the performance of four different neural network architectures—MLP, CNN, LSTM, and Transformer—in predicting the Vickers hardness of M50NiL steel [180]. The study found that MLP performed outstandingly in prediction accuracy and iterative efficiency. Meanwhile, LSTM also showed high accuracy in predicting Vickers hardness and had certain advantages in handling data with time-series characteristics or sequential patterns, which may be helpful for analyzing the related properties of complex materials. Inspired by this, Lee's team applied LSTM and feed-forward neural network (FNN) deep-learning models to predict the tensile behavior of polymer–matrix composites (PMC), especially achieving progress in predicting the tensile failure point [181]. However, despite these important advancements in the field, there are still key issues such as data uncertainty, how to handle high-dimensional data, and eliminating redundant features.

To address the problem of dimensionality caused by high-dimensional features, researchers have explored dimensionality reduction and feature compression techniques. Bahoria et al. proposed a model based on a hybrid DL and GA approach for optimizing and predicting the compressive strength of concrete [182]. In this process, PCA was applied to reduce the high-dimensional feature set to 150 principal components, reducing the feature dimension by approximately 65% while retaining 95% of the data variance. This dimensionality reduction method not only significantly reduced the computational burden, but also effectively alleviated the overfitting problem and improved the generalization ability of the model on the training and validation sets. This strategy is helpful for eliminating redundant variables, reducing training costs, and enhancing the interpretability and robustness of the model, which also has reference significance for the polymer field.

Finally, the standardization of descriptor systems remains an important but unresolved issue. Currently, many research teams rely on custom-made or field-specific descriptor frameworks, which leads to difficulties in result comparison, data sharing, and model transfer. To solve this problem, polymer informatics can draw on the standardization efforts in the broader field of materials informatics, such as ISO standards, AFLOWLIB metadata schemas, and MatML protocols, and promote the development of a unified descriptor-encoding system based on physical and chemical properties. Through initiatives by international materials database consortia and professional working groups, a standardized descriptor library and data-sharing platform for various polymer systems can be established. This will significantly enhance the accumulation and transfer of research results and greatly promote the universality, reproducibility, and collaborative potential of AI models in polymer science.

In summary, the key to enhancing the predictive modeling ability of polymer materials lies in a deep understanding of their structural complexity and dynamic characteristics. Constructing hierarchical, multi-scale, and time-resolved structured descriptors—combined with dimensionality reduction and nonlinear modeling techniques—can significantly improve modeling efficiency and expressiveness. At the same time, promoting descriptor standardization will lay a solid foundation for the wide application of AI in polymer design, thus supporting the development of new materials and accelerating technological deployment.

3.3. Task-Orientated Algorithm

At the model level, there are still two core challenges that hinder the further development and practical application of AI in polymer science: (1) the lack of model interpretability; and (2) high computational costs associated with scalability.

Firstly, the "black-box" nature of AI models remains a major obstacle to their credibility and application in polymer material research. Unlike traditional theoretical models based on clear physical principles, although DL models have powerful nonlinear fitting capabilities, they often fail to provide transparent decision-making logic or causal explanations, especially in complex multi-scale polymer systems. To address this issue, the materials science community has widely adopted general interpretable AI methods such as Shapley Additive Explanation (SHAP) and Local Interpretable Model-Agnostic Explanation (LIME). SHAP can conduct global feature importance analysis to help identify the key structural factors determining material properties, while LIME focuses on local interpretability, which is helpful for experimental design and anomaly detection. Tao and his team proposed a model-based knowledge extraction method for the design of multi-principal-element high-temperature alloys (MPESAs) [183]. They constructed two classification models to predict phase composition and used the SHAP method to extract knowledge and transform it into material rules (Figure 8). After verification, the prediction accuracy of these rules exceeded 98%, enabling the rapid design of alloys with an "FCC + $L1_2$" dual-phase structure. Twelve alloys prepared experimentally all exhibited the ideal dual-phase microstructure, verifying the accuracy of the strategy. In the polymer field, similar methods can be borrowed to predict the microstructure and properties of polymers. For example, a multi-scale model of polymers can be constructed, and methods such as SHAP can be used to extract the key structural factors affecting performance. This can not only improve the efficiency of polymer design, but also provide a scientific basis for optimizing the functionality and processability of polymers.

In addition, symbolic regression (SR) has recently attracted attention as a modeling technique, capable of learning explicit mathematical expressions that map structures to properties. For example, Yang et al. used the SR method to find a new mixing-length formula that is generally valid in wall-bounded turbulence [184]. This formula has a physical interpretation and has the correct asymptotic relationships in the viscous sub-layer, buffer layer, logarithmic-law region, and outer region. Its accuracy and generalization were demonstrated through classical cases, showing excellent interpretability and physical expressiveness. Although the application of SR in polymer modeling is still in its infancy, its advantages—such as a transparent expression format and controllable physical consistency—make it have great potential in revealing hidden structure–property mechanisms, especially for generating scientific hypotheses and modeling fundamental principles. Such methods will be crucial for improving the physical credibility and scientific acceptance of AI models in this field.

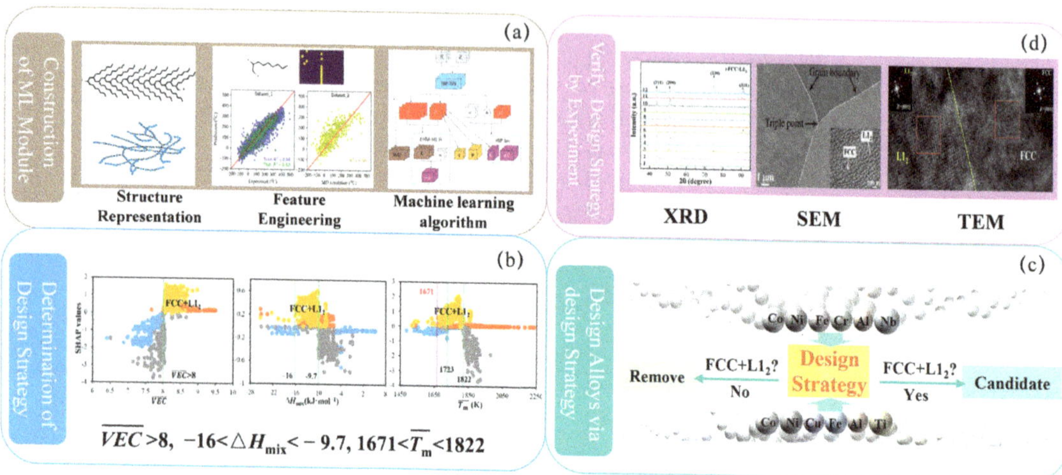

Figure 8. Divided into four parts: (**a**) construction of machine learning models; (**b**) determination of design strategy; (**c**) design alloys via design strategy; (**d**) verify design strategy by experiments. This figure is inspired by works from Tao et al. [183].

Secondly, the high computational requirements of AI models are another key obstacle to their widespread application in polymer research. The inherent structural complexity of polymer chains and their wide conformational space pose a huge challenge to computational resources when training deep models with large-scale high-dimensional data. In many industrial applications, the limitations of GPU capacity and memory availability pose serious challenges to model deployment. To address this issue, researchers are increasingly focusing on lightweight model design strategies that improve computational efficiency at the architectural level. Linear sparse modeling techniques, such as Least Absolute Shrinkage and Selection Operator (LASSO) and SISSO, have shown excellent performance in feature selection and dimensionality reduction. For example, the Ouyang team applied the SISSO method to classify the metal/insulator characteristics in multiple binary systems and achieved an accuracy of 99.0% [185]. This method also showed high accuracy in predicting the enthalpy difference between rock-salt and zinc-blende structures. However, misclassifications were still observed for some compounds. This method significantly reduced the size of the feature space, thereby reducing the computational burden. These techniques effectively reduce model complexity and training costs while maintaining prediction accuracy. Although they may have certain limitations in capturing the highly nonlinear structure–property relationships in polymer systems, they remain valuable tools in feature engineering and model pre-processing.

In summary, AI models in polymer material research still face key challenges such as computational burden and lack of interpretability. Future work should focus on developing lightweight and efficient modeling strategies while establishing an interpretable AI framework for the context of polymer science. These advancements will play a crucial role in achieving more transparent, efficient, and reliable AI-assisted workflows in material discovery and development.

4. Summary and Prospect

This paper outlines the achievements and challenges of AI, polymer research, and proposes feasible solutions. The application of AI in the polymer materials domain has made significant progress, especially in areas such as material design, property prediction, and process optimization, demonstrating immense potential and strongly driving the field's shift from an experience-driven to a data-driven model.

However, despite the strong application potential of AI in the polymer materials field, numerous challenges remain. Specifically, in terms of data, there are high costs for data acquisition in the polymer field, a lack of high-quality datasets, and the absence of mature and comprehensive data platforms. In terms of descriptors, the complex structure of polymers makes the construction of effective descriptors difficult, and there are also issues with dimensionality and inconsistent standards. In terms of models, due to the uniqueness of polymer structures, existing ML models generally suffer from poor interpretability and high computational costs. To address these challenges, strategies like AL, LLM data mining, time-serious model adoption, PCA dimension reduction, descriptor standardization, SHAP and LIME techniques, and SR methods, drawn from experiences in other fields, can be adopted.

Looking forward, the integration of AI with polymer materials will develop in a more efficient, intelligent, and sustainable direction. As AI technology continues to evolve and improve, it will play an increasingly important role in the design, prediction, and optimization of polymer materials. Solving the challenges related to data, descriptors, and models will be the key direction of future research. Moreover, promoting interdisciplinary collaboration will accelerate this process, bringing more innovative solutions to the polymer materials field. In this process, AI will not only be a tool for polymer research, but also the core driving force behind revolutionary changes in the polymer field.

Author Contributions: Conceptualization, T.L., Q.P., and C.Z.; validation, Q.P. and T.L.; formal analysis, T.L., investigation, Q.P., Y.D., and T.L.; resources, T.L. and Y.Z.; data curation, Q.P.; writing—original draft preparation, Q.P.; writing—review and editing, T.L., Y.D., X.P., Y.Z., and R.Y.; visualization, Q.P. and R.Y.; supervision, C.Z.; project administration, T.L.; funding acquisition, T.L. All authors have read and agreed to the published version of the manuscript.

Funding: The authors would like to acknowledge the financial support from the National Natural Science Foundation of China (52301016), research projects from Department of Science and Technology of Shandong Province (2023HWYQ-043, ZR2023QE033, KM20250039), Key Technologies R&D Program of CNBM (2023SJYL05). This work was supported By Young Talent of Lifting engineering for Science and Technology in Shandong, China (SDAST2024QTA010).

Data Availability Statement: No new data were created or analyzed in this study.

Acknowledgments: The authors would like to thank the support of Ji'nan AICC. The authors thank Chenlu Wang for contributing to the design of the figures.

Conflicts of Interest: The authors declare no conflicts of interest.

References

1. Nandi, A.K.; Chatterjee, D.P. Hybrid Polymer Gels for Energy Applications. *J. Mater. Chem. A* **2023**, *11*, 12593–12642. [CrossRef]
2. Alam, A.U.; Qin, Y.; Nambiar, S.; Yeow, J.T.W.; Howlader, M.M.R.; Hu, N.-X.; Deen, M.J. Polymers and Organic Materials-Based pH Sensors for Healthcare Applications. *Prog. Mater. Sci.* **2018**, *96*, 174–216. [CrossRef]
3. Liles, D.; Lin, F. Silicone Elastomeric Particles in Skin Care Applications. *ACS Symp. Ser.* **2010**, *1053*, 207–219. [CrossRef]

4. Sionkowska, A. Current Research on the Blends of Natural and Synthetic Polymers as New Biomaterials: Review. *Prog. Polym. Sci.* **2011**, *36*, 1254–1276. [CrossRef]
5. Pfleger, J.; Cimrová, V. Polymers for Electronics and Photonics: Science for Applications. *Chem. Pap.* **2018**, *72*, 1561–1562. [CrossRef]
6. Huan, T.D.; Boggs, S.; Teyssedre, G.; Laurent, C.; Cakmak, M.; Kumar, S.; Ramprasad, R. Advanced Polymeric Dielectrics for High Energy Density Applications. *Prog. Mater. Sci.* **2016**, *83*, 236–269. [CrossRef]
7. Li, J.; Liu, X.; Feng, Y.; Yin, J. Recent Progress in Polymer/Two-Dimensional Nanosheets Composites with Novel Performances. *Prog. Polym. Sci.* **2022**, *126*, 101505. [CrossRef]
8. Zhang, G.; Cheng, X.; Wang, Y.; Zhang, W. Supramolecular Chiral Polymeric Aggregates: Construction and Applications. *Aggregate* **2023**, *4*, e262. [CrossRef]
9. Geyer, R.; Jambeck, J.R.; Law, K.L. Production, Use, and Fate of All Plastics Ever Made. *Sci. Adv.* **2017**, *3*, e1700782. [CrossRef] [PubMed]
10. Jayaraman, A.; Klok, H.-A. ACS Polymers Au's Grand Challenges in Polymer Science. *ACS Polym. Au* **2023**, *3*, 1–4. [CrossRef]
11. Kumar, S.K.; Krishnamoorti, R. Nanocomposites: Structure, Phase Behavior, and Properties. *Annu. Rev. Chem. Biomol. Eng.* **2010**, *1*, 37–58. [CrossRef]
12. Sarkar, B.; Alexandridis, P. Block Copolymer–Nanoparticle Composites: Structure, Functional Properties, and Processing. *Prog. Polym. Sci.* **2015**, *40*, 33–62. [CrossRef]
13. Correa-Baena, J.-P.; Hippalgaonkar, K.; van Duren, J.; Jaffer, S.; Chandrasekhar, V.R.; Stevanovic, V.; Wadia, C.; Guha, S.; Buonassisi, T. Accelerating Materials Development via Automation, Machine Learning, and High-Performance Computing. *Joule* **2018**, *2*, 1410–1420. [CrossRef]
14. Sha, W.; Guo, Y.; Yuan, Q.; Tang, S.; Zhang, X.; Lu, S.; Guo, X.; Cao, Y.-C.; Cheng, S. Artificial Intelligence to Power the Future of Materials Science and Engineering. *Adv. Intell. Syst.* **2020**, *2*, 1900143. [CrossRef]
15. Papadimitriou, I.; Gialampoukidis, I.; Vrochidis, S.; Kompatsiaris, I. AI Methods in Materials Design, Discovery and Manufacturing: A Review. *Comput. Mater. Sci.* **2024**, *235*, 112793. [CrossRef]
16. Kalidindi, S.R. Feature Engineering of Material Structure for AI-Based Materials Knowledge Systems. *J. Appl. Phys.* **2020**, *128*, 041103. [CrossRef]
17. Hattrick-Simpers, J.R.; Choudhary, K.; Corgnale, C. A Simple Constrained Machine Learning Model for Predicting High-Pressure-Hydrogen-Compressor Materials. *Mol. Syst. Des. Eng.* **2018**, *3*, 509–517. [CrossRef]
18. Choudhary, K.; Bercx, M.; Jiang, J.; Pachter, R.; Lamoen, D.; Tavazza, F. Accelerated Discovery of Efficient Solar Cell Materials Using Quantum and Machine-Learning Methods. *Chem. Mater.* **2019**, *31*, 5900–5908. [CrossRef]
19. Chun, S.; Roy, S.; Nguyen, Y.T.; Choi, J.B.; Udaykumar, H.S.; Baek, S.S. Deep Learning for Synthetic Microstructure Generation in a Materials-by-Design Framework for Heterogeneous Energetic Materials. *Sci. Rep.* **2020**, *10*, 13307. [CrossRef]
20. Jiao, P.; Alavi, A.H. Artificial Intelligence-Enabled Smart Mechanical Metamaterials: Advent and Future Trends. *Int. Mater. Rev.* **2021**, *66*, 365–393. [CrossRef]
21. Guo, K.; Yang, Z.; Yu, C.-H.; Buehler, M.J. Artificial Intelligence and Machine Learning in Design of Mechanical Materials. *Mater. Horiz.* **2021**, *8*, 1153–1172. [CrossRef]
22. Gu, G.X.; Chen, C.-T.; Richmond, D.J.; Buehler, M.J. Bioinspired Hierarchical Composite Design Using Machine Learning: Simulation, Additive Manufacturing, and Experiment. *Mater. Horiz.* **2018**, *5*, 939–945. [CrossRef]
23. Zhang, G.; Ali, Z.H.; Aldlemy, M.S.; Mussa, M.H.; Salih, S.Q.; Hameed, M.M.; Al-Khafaji, Z.S.; Yaseen, Z.M. Reinforced Concrete Deep Beam Shear Strength Capacity Modelling Using an Integrative Bio-Inspired Algorithm with an Artificial Intelligence Model. *Eng. Comput.* **2022**, *38*, 15–28. [CrossRef]
24. Shen, S.C.; Buehler, M.J. Nature-Inspired Architected Materials Using Unsupervised Deep Learning. *Commun. Eng.* **2022**, *1*, 37. [CrossRef]
25. Ishiwatari, Y.; Yokoyama, T.; Kojima, T.; Banno, T.; Arai, N. Machine Learning Prediction of Self-Assembly and Analysis of Molecular Structure Dependence on the Critical Packing Parameter. *Mol. Syst. Des. Eng.* **2024**, *9*, 20–28. [CrossRef]
26. Wei, J.; Chu, X.; Sun, X.-Y.; Xu, K.; Deng, H.-X.; Chen, J.; Wei, Z.; Lei, M. Machine Learning in Materials Science. *InfoMat* **2019**, *1*, 338–358. [CrossRef]
27. Zhou, T.; Song, Z.; Sundmacher, K. Big Data Creates New Opportunities for Materials Research: A Review on Methods and Applications of Machine Learning for Materials Design. *Engineering* **2019**, *5*, 1017–1026. [CrossRef]
28. Goswami, L.; Deka, M.K.; Roy, M. Artificial Intelligence in Material Engineering: A Review on Applications of Artificial Intelligence in Material Engineering. *Adv. Eng. Mater.* **2023**, *25*, 2300104. [CrossRef]
29. Yu, C.-H.; Qin, Z.; Buehler, M.J. Artificial Intelligence Design Algorithm for Nanocomposites Optimized for Shear Crack Resistance. *Nano Futures* **2019**, *3*, 035001. [CrossRef]
30. Gu, G.H.; Noh, J.; Kim, I.; Jung, Y. Machine Learning for Renewable Energy Materials. *J. Mater. Chem. A* **2019**, *7*, 17096–17117. [CrossRef]

31. Yang, K.; Xu, X.; Yang, B.; Cook, B.; Ramos, H.; Krishnan, N.M.A.; Smedskjaer, M.M.; Hoover, C.; Bauchy, M. Predicting the Young's Modulus of Silicate Glasses Using High-Throughput Molecular Dynamics Simulations and Machine Learning. *Sci. Rep.* **2019**, *9*, 8739. [CrossRef]
32. Venkatraman, V.; Evjen, S.; Knuutila, H.K.; Fiksdahl, A.; Alsberg, B.K. Predicting Ionic Liquid Melting Points Using Machine Learning. *J. Mol. Liq.* **2018**, *264*, 318–326. [CrossRef]
33. Qu, N.; Liu, Y.; Liao, M.; Lai, Z.; Zhou, F.; Cui, P.; Han, T.; Yang, D.; Zhu, J. Ultra-High Temperature Ceramics Melting Temperature Prediction via Machine Learning. *Ceram. Int.* **2019**, *45*, 18551–18555. [CrossRef]
34. Chen, L.; Tran, H.; Batra, R.; Kim, C.; Ramprasad, R. Machine Learning Models for the Lattice Thermal Conductivity Prediction of Inorganic Materials. *Comput. Mater. Sci.* **2019**, *170*, 109155. [CrossRef]
35. Luo, Y.; Li, M.; Yuan, H.; Liu, H.; Fang, Y. Predicting Lattice Thermal Conductivity via Machine Learning: A Mini Review. *NPJ Comput. Mater.* **2023**, *9*, 4. [CrossRef]
36. Chen, Z.; Andrejevic, N.; Drucker, N.C.; Nguyen, T.; Xian, R.P.; Smidt, T.; Wang, Y.; Ernstorfer, R.; Tennant, D.A.; Chan, M.; et al. Machine Learning on Neutron and X-Ray Scattering and Spectroscopies. *Chem. Phys. Rev.* **2021**, *2*, 031301. [CrossRef]
37. Anker, A.S.; Butler, K.T.; Selvan, R.; Jensen, K.M.Ø. Machine Learning for Analysis of Experimental Scattering and Spectroscopy Data in Materials Chemistry. *Chem. Sci.* **2023**, *14*, 14003–14019. [CrossRef]
38. Deagen, M.E.; Walsh, D.J.; Audus, D.J.; Kroenlein, K.; de Pablo, J.J.; Aou, K.; Chard, K.; Jensen, K.F.; Olsen, B.D. Networks and Interfaces as Catalysts for Polymer Materials Innovation. *Cell Rep. Phys. Sci.* **2022**, *3*, 101126. [CrossRef]
39. Lutz, J.-F.; Lehn, J.-M.; Meijer, E.W.; Matyjaszewski, K. From Precision Polymers to Complex Materials and Systems. *Nat. Rev. Mater.* **2016**, *1*, 16024. [CrossRef]
40. Meijer, H.E.H.; Govaert, L.E. Mechanical Performance of Polymer Systems: The Relation between Structure and Properties. *Prog. Polym. Sci.* **2005**, *30*, 915–938. [CrossRef]
41. Gurnani, R.; Kuenneth, C.; Toland, A.; Ramprasad, R. Polymer Informatics at Scale with Multitask Graph Neural Networks. *Chem. Mater.* **2023**, *35*, 1560–1567. [CrossRef] [PubMed]
42. Patra, T.K. Data-Driven Methods for Accelerating Polymer Design. *ACS Polym. Au* **2022**, *2*, 8–26. [CrossRef]
43. Tran, H.; Gurnani, R.; Kim, C.; Pilania, G.; Kwon, H.-K.; Lively, R.P.; Ramprasad, R. Design of Functional and Sustainable Polymers Assisted by Artificial Intelligence. *Nat. Rev. Mater.* **2024**, *9*, 866–886. [CrossRef]
44. Egrioglu, E.; Bas, E. A New Deep Neural Network for Forecasting: Deep Dendritic Artificial Neural Network. *Artif. Intell. Rev.* **2024**, *57*, 171. [CrossRef]
45. Zhou, J.; Cui, G.; Hu, S.; Zhang, Z.; Yang, C.; Liu, Z.; Wang, L.; Li, C.; Sun, M. Graph Neural Networks: A Review of Methods and Applications. *AI Open* **2020**, *1*, 57–81. [CrossRef]
46. Yoo, H.; Byun, H.E.; Han, D.; Lee, J.H. Reinforcement Learning for Batch Process Control: Review and Perspectives. *Annu. Rev. Control* **2021**, *52*, 108–119. [CrossRef]
47. Martin, T.B.; Audus, D.J. Emerging Trends in Machine Learning: A Polymer Perspective. *ACS Polym. Au* **2023**, *3*, 239–258. [CrossRef]
48. Liu, N.; Jafarzadeh, S.; Lattimer, B.Y.; Ni, S.; Lua, J.; Yu, Y. Large Language Models, Physics-Based Modeling, Experimental Measurements: The Trinity of Data-Scarce Learning of Polymer Properties. *arXiv* **2024**, arXiv:2407.02770. [CrossRef]
49. Schmid, F. Understanding and Modeling Polymers: The Challenge of Multiple Scales. *ACS Polym. Au* **2023**, *3*, 28–58. [CrossRef]
50. Xu, B.; Yang, G. Interpretability Research of Deep Learning: A Literature Survey. *Inf. Fusion.* **2025**, *115*, 102721. [CrossRef]
51. Pugliese, R.; Regondi, S.; Marini, R. Machine Learning-Based Approach: Global Trends, Research Directions, and Regulatory Standpoints. *Data Sci. Manag.* **2021**, *4*, 19–29. [CrossRef]
52. Unke, O.T.; Chmiela, S.; Sauceda, H.E.; Gastegger, M.; Poltavsky, I.; Schütt, K.T.; Tkatchenko, A.; Müller, K.-R. Machine Learning Force Fields. *Chem. Rev.* **2021**, *121*, 10142–10186. [CrossRef] [PubMed]
53. Xu, P.; Chen, H.; Li, M.; Lu, W. New Opportunity: Machine Learning for Polymer Materials Design and Discovery. *Adv. Theory Simul.* **2022**, *5*, 2100565. [CrossRef]
54. Badini, S.; Regondi, S.; Pugliese, R. Unleashing the Power of Artificial Intelligence in Materials Design. *Materials* **2023**, *16*, 5927. [CrossRef]
55. Jain, A.; Ong, S.P.; Hautier, G.; Chen, W.; Richards, W.D.; Dacek, S.; Cholia, S.; Gunter, D.; Skinner, D.; Ceder, G.; et al. Commentary: The Materials Project: A Materials Genome Approach to Accelerating Materials Innovation. *APL Mater.* **2013**, *1*, 011002. [CrossRef]
56. Curtarolo, S.; Setyawan, W.; Wang, S.; Xue, J.; Yang, K.; Taylor, R.H.; Nelson, L.J.; Hart, G.L.W.; Sanvito, S.; Buongiorno-Nardelli, M.; et al. AFLOWLIB.ORG: A Distributed Materials Properties Repository from High-Throughput *Ab. Initio* Calculations. *Comput. Mater. Sci.* **2012**, *58*, 227–235. [CrossRef]
57. Kirklin, S.; Saal, J.E.; Meredig, B.; Thompson, A.; Doak, J.W.; Aykol, M.; Rühl, S.; Wolverton, C. The Open Quantum Materials Database (OQMD): Assessing the Accuracy of DFT Formation Energies. *NPJ Comput. Mater.* **2015**, *1*, 15010. [CrossRef]
58. Chanussot, L.; Das, A.; Goyal, S.; Lavril, T.; Shuaibi, M.; Riviere, M.; Tran, K.; Heras-Domingo, J.; Ho, C.; Hu, W.; et al. Open Catalyst 2020 (OC20) Dataset and Community Challenges. *ACS Catal.* **2021**, *11*, 6059–6072. [CrossRef]

59. Otsuka, S.; Kuwajima, I.; Hosoya, J.; Xu, Y.; Yamazaki, M. PoLyInfo: Polymer Database for Polymeric Materials Design. In Proceedings of the 2011 International Conference on Emerging Intelligent Data and Web Technologies, Tirana, Albania, 7–9 September 2011; pp. 22–29. [CrossRef]
60. Graziano, G. Fingerprints of Molecular Reactivity. *Nat. Rev. Chem.* **2020**, *4*, 227. [CrossRef]
61. Zhao, Y.; Mulder, R.J.; Houshyar, S.; Le, T.C. A Review on the Application of Molecular Descriptors and Machine Learning in Polymer Design. *Polym. Chem.* **2023**, *14*, 3325–3346. [CrossRef]
62. Lee, D.E.; Song, S.-O.; Yoon, E.S. Weighted Support Vector Machine for Quality Estimation in Polymerization Processes. *Korean J. Chem. Eng.* **2004**, *21*, 1103–1107. [CrossRef]
63. Malashin, I.; Tynchenko, V.; Gantimurov, A.; Nelyub, V.; Borodulin, A. Support Vector Machines in Polymer Science: A Review. *Polymers* **2025**, *17*, 491. [CrossRef]
64. Savage, N. Neural Net Worth. *Commun. ACM* **2019**, *62*, 10–12. [CrossRef]
65. Breiman, L. Random Forests. *Mach. Learn.* **2001**, *45*, 5–32. [CrossRef]
66. Orsenigo, C.; Vercellis, C. Kernel Ridge Regression for Out-of-Sample Mapping in Supervised Manifold Learning. *Expert. Syst. Appl.* **2012**, *39*, 7757–7762. [CrossRef]
67. Wang, X.; Huang, F.; Cheng, Y. Computational Performance Optimization of Support Vector Machine Based on Support Vectors. *Neurocomputing* **2016**, *211*, 66–71. [CrossRef]
68. Tolles, J.; Meurer, W.J. Logistic Regression: Relating Patient Characteristics to Outcomes. *JAMA* **2016**, *316*, 533–534. [CrossRef]
69. Schulz, E.; Speekenbrink, M.; Krause, A. A Tutorial on Gaussian Process Regression: Modelling, Exploring, and Exploiting Functions. *J. Math. Psychol.* **2018**, *85*, 1–16. [CrossRef]
70. Jakob, P.; Madan, M.; Schmid-Schirling, T.; Valada, A. Multi-Perspective Anomaly Detection. *Sensors* **2021**, *21*, 5311. [CrossRef] [PubMed]
71. Ertefaie, A.; McKay, J.R.; Oslin, D.; Strawderman, R.L. Robust Q-Learning. *J. Am. Stat. Assoc.* **2021**, *116*, 368–381. [CrossRef]
72. Rodemann, J.; Augustin, T. Imprecise Bayesian Optimization. *Knowl.-Based Syst.* **2024**, *300*, 112186. [CrossRef]
73. Yu, C.; Yao, W.; Bai, X. Robust Linear Regression: A Review and Comparison. *arXiv* **2014**. [CrossRef]
74. He, Z.; Wu, Z.; Xu, G.; Liu, Y.; Zou, Q. Decision Tree for Sequences. *IEEE Trans. Knowl. Data Eng.* **2023**, *35*, 251–263. [CrossRef]
75. Safavian, S.R.; Landgrebe, D. A Survey of Decision Tree Classifier Methodology. *IEEE Trans. Syst. Man. Cybern.* **1991**, *21*, 660–674. [CrossRef]
76. Valkenborg, D.; Rousseau, A.-J.; Geubbelmans, M.; Burzykowski, T. Support Vector Machines. *Am. J. Orthod. Dentofac. Orthop.* **2023**, *164*, 754–757. [CrossRef]
77. Zhao, Z.; Fitzsimons, J.K.; Fitzsimons, J.F. Quantum-Assisted Gaussian Process Regression. *Phys. Rev. A* **2019**, *99*, 052331. [CrossRef]
78. Boloix-Tortosa, R.; Murillo-Fuentes, J.J.; Payán-Somet, F.J.; Pérez-Cruz, F. Complex Gaussian Processes for Regression. *IEEE Trans. Neural Netw. Learn. Syst.* **2018**, *29*, 5499–5511. [CrossRef]
79. Aldana-Bobadila, E.; Kuri-Morales, A.; Lopez-Arevalo, I.; Rios-Alvarado, A.B. An Unsupervised Learning Approach for Multilayer Perceptron Networks. *Soft Comput.* **2019**, *23*, 11001–11013. [CrossRef]
80. Chen, T.; Guestrin, C. XGBoost: A Scalable Tree Boosting System. In Proceedings of the 22nd ACM SIGKDD International Conference on Knowledge Discovery and Data Mining, San Francisco, CA, USA, 13–17 August 2016; pp. 785–794. [CrossRef]
81. Ke, G.; Meng, Q.; Finley, T.; Wang, T.; Chen, W.; Ma, W.; Ye, Q.; Liu, T.-Y. LightGBM: A Highly Efficient Gradient Boosting Decision Tree. In *Advances in Neural Information Processing Systems*; Curran Associates, Inc.: Red Hook, NY, USA, 2017; Volume 30.
82. Hancock, J.T.; Khoshgoftaar, T.M. CatBoost for Big Data: An Interdisciplinary Review. *J. Big Data* **2020**, *7*, 94. [CrossRef]
83. Arik, S.Ö.; Pfister, T. TabNet: Attentive Interpretable Tabular Learning. *Proc. AAAI Conf. Artif. Intell.* **2021**, *35*, 6679–6687. [CrossRef]
84. Tan, L.; Wu, S.; Zhou, W.; Huang, X. Weighted Neural Tangent Kernel: A Generalized and Improved Network-Induced Kernel. *Mach. Learn.* **2023**, *112*, 2871–2901. [CrossRef]
85. Yoder, J.; and Priebe, C.E. Semi-Supervised k-Means++. *J. Stat. Comput. Simul.* **2017**, *87*, 2597–2608. [CrossRef]
86. Cohen-addad, V.; Kanade, V.; Mallmann-trenn, F.; Mathieu, C. Hierarchical Clustering: Objective Functions and Algorithms. *J. ACM* **2019**, *66*, 1–42. [CrossRef]
87. Greenacre, M.; Groenen, P.J.F.; Hastie, T.; D'Enza, A.I.; Markos, A.; Tuzhilina, E. Principal Component Analysis. *Nat. Rev. Methods Primers* **2022**, *2*, 100. [CrossRef]
88. Sainburg, T.; McInnes, L.; Gentner, T.Q. Parametric UMAP Embeddings for Representation and Semisupervised Learning. *Neural Comput.* **2021**, *33*, 2881–2907. [CrossRef]
89. Wei, X.; Zhang, Z.; Huang, H.; Zhou, Y. An Overview on Deep Clustering. *Neurocomputing* **2024**, *590*, 127761. [CrossRef]
90. Geadah, V.; Barello, G.; Greenidge, D.; Charles, A.S.; Pillow, J.W. Sparse-Coding Variational Autoencoders. *Neural Comput.* **2024**, *36*, 2571–2601. [CrossRef]
91. Joo, W.; Lee, W.; Park, S.; Moon, I.-C. Dirichlet Variational Autoencoder. *Pattern Recognit.* **2020**, *107*, 107514. [CrossRef]

92. Wang, J.; Yang, J.; He, J.; Peng, D. Multi-Augmentation-Based Contrastive Learning for Semi-Supervised Learning. *Algorithms* **2024**, *17*, 91. [CrossRef]
93. He, K.; Fan, H.; Wu, Y.; Xie, S.; Girshick, R. Momentum Contrast for Unsupervised Visual Representation Learning. *arXiv* **2020**, arXiv:1911.05722. [CrossRef]
94. Amini, M.-R.; Feofanov, V.; Pauletto, L.; Hadjadj, L.; Devijver, É.; Maximov, Y. Self-Training: A Survey. *Neurocomputing* **2025**, *616*, 128904. [CrossRef]
95. Ning, X.; Wang, X.; Xu, S.; Cai, W.; Zhang, L.; Yu, L.; Li, W. A Review of Research on Co-Training. *Concurr. Comput. Pract. Exp.* **2023**, *35*, e6276. [CrossRef]
96. Zhou, D.; Bousquet, O.; Lal, T.; Weston, J.; Schölkopf, B. Learning with Local and Global Consistency. In *Advances in Neural Information Processing Systems*; MIT Press: Cambridge, MA, USA, 2003; Volume 16.
97. Sohn, K.; Berthelot, D.; Li, C.-L.; Zhang, Z.; Carlini, N.; Cubuk, E.D.; Kurakin, A.; Zhang, H.; Raffel, C. FixMatch: Simplifying Semi-Supervised Learning with Consistency and Confidence. *arXiv* **2020**, arXiv:2001.07685. [CrossRef]
98. Berthelot, D.; Carlini, N.; Goodfellow, I.; Papernot, N.; Oliver, A.; Raffel, C. MixMatch: A Holistic Approach to Semi-Supervised Learning. *arXiv* **2019**, arXiv:1905.02249. [CrossRef]
99. Lee, D.H. Pseudo-label: The simple and efficient semi-supervised learning method for deep neural networks. *Workshop Chall. Represent. Learn.* **2013**, *3*, 896.
100. Ratner, A.; Bach, S.H.; Ehrenberg, H.; Fries, J.; Wu, S.; Ré, C. Snorkel: Rapid Training Data Creation with Weak Supervision. *Proc. VLDB Endow.* **2017**, *11*, 269–282. [CrossRef] [PubMed]
101. Fan, Y.; Kukleva, A.; Schiele, B. Revisiting Consistency Regularization for Semi-Supervised Learning. *arXiv* **2021**, arXiv:2112.05825. [CrossRef]
102. Derry, A.; Krzywinski, M.; Altman, N. Convolutional Neural Networks. *Nat. Methods* **2023**, *20*, 1269–1270. [CrossRef]
103. Zaremba, W.; Sutskever, I.; Vinyals, O. Recurrent Neural Network Regularization. *arXiv* **2015**, arXiv:1409.2329. [CrossRef]
104. Van Houdt, G.; Mosquera, C.; Nápoles, G. A Review on the Long Short-Term Memory Model. *Artif. Intell. Rev.* **2020**, *53*, 5929–5955. [CrossRef]
105. Cai, Z.; Xiong, Z.; Xu, H.; Wang, P.; Li, W.; Pan, Y. Generative Adversarial Networks: A Survey Toward Private and Secure Applications. *ACM Comput. Surv.* **2021**, *54*, 1–38. [CrossRef]
106. Vaswani, A.; Shazeer, N.; Parmar, N.; Uszkoreit, J.; Jones, L.; Gomez, A.N.; Kaiser, L.; Polosukhin, I. Attention Is All You Need. *arXiv* **2023**, arXiv:1706.03762. [CrossRef]
107. Song, J.; Meng, C.; Ermon, S. Denoising Diffusion Implicit Models. *arXiv* **2020**, arXiv:2010.02502. [CrossRef]
108. Graph Neural Networks. *Nat. Rev. Methods Primers* **2024**, *4*, 18. [CrossRef]
109. Watkins, C.J.C.H.; Dayan, P. Q-Learning. *Mach. Learn.* **1992**, *8*, 279–292. [CrossRef]
110. Liu, Q.; Cui, C.; Fan, Q. Self-Adaptive Constrained Multi-Objective Differential Evolution Algorithm Based on the State–Action–Reward–State–Action Method. *Mathematics* **2022**, *10*, 813. [CrossRef]
111. Mnih, V.; Kavukcuoglu, K.; Silver, D.; Graves, A.; Antonoglou, I.; Wierstra, D.; Riedmiller, M. Playing Atari with Deep Reinforcement Learning. *arXiv* **2013**, arXiv:1312.5602. [CrossRef]
112. Betancourt, M. The Convergence of Markov Chain Monte Carlo Methods: From the Metropolis Method to Hamiltonian Monte Carlo. *Ann. Der Phys.* **2019**, *531*, 1700214. [CrossRef]
113. Okada, M.; Taniguchi, T. DreamingV2: Reinforcement Learning with Discrete World Models without Reconstruction. *arXiv* **2022**, arXiv:2203.00494. [CrossRef]
114. Liashchynskyi, P.; Liashchynskyi, P. Grid Search, Random Search, Genetic Algorithm: A Big Comparison for NAS. *arXiv* **2019**, arXiv:1912.06059. [CrossRef]
115. Wang, X.; Jin, Y.; Schmitt, S.; Olhofer, M. Recent Advances in Bayesian Optimization. *ACM Comput. Surv.* **2023**, *55*, 1–36. [CrossRef]
116. Baymurzina, D.; Golikov, E.; Burtsev, M. A Review of Neural Architecture Search. *Neurocomputing* **2022**, *474*, 82–93. [CrossRef]
117. Li, L.; Jamieson, K.; DeSalvo, G.; Rostamizadeh, A.; Talwalkar, A. Hyperband: Bandit-based configuration evaluation for hyperparameter optimization. *J. Mach. Learn. Res.* **2017**, *18*, 1–52.
118. Finn, C.; Abbeel, P.; Levine, S. Model-Agnostic Meta-Learning for Fast Adaptation of Deep Networks. *arXiv* **2017**, arXiv:1703.03400. [CrossRef]
119. Behl, H.S.; Baydin, A.G.; Torr, P.H.S. Alpha MAML: Adaptive Model-Agnostic Meta-Learning. *arXiv* **2019**, arXiv:1905.07435. [CrossRef]
120. Jose, S.T.; Simeone, O. Information-Theoretic Generalization Bounds for Meta-Learning and Applications. *Entropy* **2021**, *23*, 126. [CrossRef]
121. Long, T.; Li, J.; Wang, C.; Wang, H.; Cheng, X.; Lu, H.; Zhang, Y.; Zhou, C. Polymers Simulation Using Machine Learning Interatomic Potentials. *Polymer* **2024**, *308*, 127416. [CrossRef]

122. Patel, R.A.; Borca, C.H.; Webb, M.A. Featurization Strategies for Polymer Sequence or Composition Design by Machine Learning. *Mol. Syst. Des. Eng.* **2022**, *7*, 661–676. [CrossRef]
123. Zhang, B.; Li, X.; Zeng, M.; Cao, J. COMFO: Integrated Deep Learning Model Facilitates Discovery of Multifunctional Polyimide Materials. *Polymer* **2025**, *320*, 128081. [CrossRef]
124. del Rio, B.G.; Phan, B.; Ramprasad, R. A Deep Learning Framework to Emulate Density Functional Theory. *NPJ Comput. Mater.* **2023**, *9*, 158. [CrossRef]
125. Yao, K.; Herr, J.E.; Parkhill, J. The Many-Body Expansion Combined with Neural Networks. *J. Chem. Phys.* **2017**, *146*, 014106. [CrossRef] [PubMed]
126. Gautham, S.M.B.; Patra, T.K. Deep Learning Potential of Mean Force between Polymer Grafted Nanoparticles. *Soft Matter* **2022**, *18*, 7909–7916. [CrossRef]
127. Shi, J.; Quevillon, M.J.; Amorim Valença, P.H.; Whitmer, J.K. Predicting Adhesive Free Energies of Polymer–Surface Interactions with Machine Learning. *ACS Appl. Mater. Interfaces* **2022**, *14*, 37161–37169. [CrossRef] [PubMed]
128. Yang, S.; Zhao, C.; Ren, J.; Zheng, K.; Shao, Z.; Ling, S. Acquiring Structural and Mechanical Information of a Fibrous Network through Deep Learning. *Nanoscale* **2022**, *14*, 5044–5053. [CrossRef]
129. Fang, X.; Murphy, E.A.; Kohl, P.A.; Li, Y.; Hawker, C.J.; Bates, C.M.; Gu, M. Universal Phase Identification of Block Copolymers From Physics-Informed Machine Learning. *J. Polym. Sci.* **2025**, *63*, 1433–1440. [CrossRef]
130. Nguyen, D.; Tao, L.; Li, Y. Integration of Machine Learning and Coarse-Grained Molecular Simulations for Polymer Materials: Physical Understandings and Molecular Design. *Front. Chem.* **2022**, *9*, 820417. [CrossRef]
131. Li, H.; Collins, C.R.; Ribelli, T.G.; Matyjaszewski, K.; Gordon, G.J.; Kowalewski, T.; Yaron, D.J. Tuning the Molecular Weight Distribution from Atom Transfer Radical Polymerization Using Deep Reinforcement Learning. *Mol. Syst. Des. Eng.* **2018**, *3*, 496–508. [CrossRef]
132. Kranthiraja, K.; Saeki, A. Machine Learning: Experiment-Oriented Machine Learning of Polymer:Non-Fullerene Organic Solar Cells (Adv. Funct. Mater. 23/2021). *Adv. Funct. Mater.* **2021**, *31*, 2170168. [CrossRef]
133. Webb, M.A.; Jackson, N.E.; Gil, P.S.; de Pablo, J.J. Targeted Sequence Design Within the Coarse-Grained Polymer Genome | Science Advances. *Sci. Adv.* **2020**, *6*, eabc6216. [CrossRef]
134. Hiraide, K.; Oya, Y.; Suzuki, M.; Muramatsu, M. Inverse Design of Polymer Alloys Using Deep Learning Based on Self-Consistent Field Analysis and Finite Element Analysis. *Mater. Today Commun.* **2023**, *37*, 107233. [CrossRef]
135. Zhu, C.; Zhang, J. Inferential Estimation of Polymer Melt Index Using Deep Belief Networks. In Proceedings of the 2018 24th International Conference on Automation and Computing (ICAC), Newcastle Upon Tyne, UK, 6–7 September 2018; pp. 1–6. [CrossRef]
136. Sattari, K.; Xie, Y.; Lin, J. Data-Driven Algorithms for Inverse Design of Polymers. *Soft Matter* **2021**, *17*, 7607–7622. [CrossRef] [PubMed]
137. Queen, O.; McCarver, G.A.; Thatigotla, S.; Abolins, B.P.; Brown, C.L.; Maroulas, V.; Vogiatzis, K.D. Polymer Graph Neural Networks for Multitask Property Learning. *NPJ Comput. Mater.* **2023**, *9*, 90. [CrossRef]
138. Pilania, G.; Iverson, C.N.; Lookman, T.; Marrone, B.L. Machine-Learning-Based Predictive Modeling of Glass Transition Temperatures: A Case of Polyhydroxyalkanoate Homopolymers and Copolymers. *J. Chem. Inf. Model.* **2019**, *59*, 5013–5025. [CrossRef] [PubMed]
139. Lee, C.-K.; Lu, C.; Yu, Y.; Sun, Q.; Hsieh, C.-Y.; Zhang, S.; Liu, Q.; Shi, L. Transfer Learning with Graph Neural Networks for Optoelectronic Properties of Conjugated Oligomers. *J. Chem. Phys.* **2021**, *154*, 024906. [CrossRef]
140. Kuenneth, C.; Lalonde, J.; Marrone, B.L.; Iverson, C.N.; Ramprasad, R.; Pilania, G. Bioplastic Design Using Multitask Deep Neural Networks. *Commun. Mater.* **2022**, *3*, 96. [CrossRef]
141. Phan, B.K.; Shen, K.-H.; Gurnani, R.; Tran, H.; Lively, R.; Ramprasad, R. Gas Permeability, Diffusivity, and Solubility in Polymers: Simulation-Experiment Data Fusion and Multi-Task Machine Learning. *NPJ Comput. Mater.* **2024**, *10*, 186. [CrossRef]
142. Himanshu; Chakraborty, K.; Patra, T.K. Developing Efficient Deep Learning Model for Predicting Copolymer Properties. *Phys. Chem. Chem. Phys.* **2023**, *25*, 25166–25176. [CrossRef]
143. Liang, Z.; Tan, Z.; Hong, R.; Ouyang, W.; Yuan, J.; Zhang, C. Automatically Predicting Material Properties with Microscopic Images: Polymer Miscibility as an Example. *J. Chem. Inf. Model.* **2023**, *63*, 5971–5980. [CrossRef]
144. Kim, C.; Chandrasekaran, A.; Huan, T.D.; Das, D.; Ramprasad, R. Polymer Genome: A Data-Powered Polymer Informatics Platform for Property Predictions. *J. Phys. Chem. C* **2018**, *122*, 17575–17585. [CrossRef]
145. Upadhya, R.; Tamasi, M.; Di Mare, E.; Murthy, S.; Gormley, A. Data-Driven Design of Protein-Like Single-Chain Polymer Nanoparticles. *ChemRxiv* **2022**. [CrossRef]
146. Doan Tran, H.; Kim, C.; Chen, L.; Chandrasekaran, A.; Batra, R.; Venkatram, S.; Kamal, D.; Lightstone, J.P.; Gurnani, R.; Shetty, P.; et al. Machine-Learning Predictions of Polymer Properties with Polymer Genome. *J. Appl. Phys.* **2020**, *128*, 171104. [CrossRef]
147. Hong, X.; Yang, Q.; Liao, K.; Pei, J.; Chen, M.; Mo, F.; Lu, H.; Zhang, W.-B.; Zhou, H.; Chen, J.; et al. AI for Organic and Polymer Synthesis. *Sci. China Chem.* **2024**, *67*, 2461–2496. [CrossRef]

148. Khanzadeh, M.; Rao, P.; Jafari-Marandi, R.; Smith, B.K.; Tschopp, M.A.; Bian, L. Quantifying Geometric Accuracy With Unsupervised Machine Learning: Using Self-Organizing Map on Fused Filament Fabrication Additive Manufacturing Parts. *J. Manuf. Sci. Eng.-Trans. ASME* **2018**, *140*, 031011. [CrossRef]
149. Roy, M.; Wodo, O. Data-Driven Modeling of Thermal History in Additive Manufacturing. *Addit. Manuf.* **2020**, *32*, 101017. [CrossRef]
150. Zhu, Y.; Kwok, T.; Haug, J.C.; Guo, S.; Chen, X.; Xu, W.; Ravichandran, D.; Tchoukalova, Y.D.; Cornella, J.L.; Yi, J.; et al. 3D Printable Hydrogel with Tunable Degradability and Mechanical Properties as a Tissue Scaffold for Pelvic Organ Prolapse Treatment. *Adv. Mater. Technol.* **2023**, *8*, 2201421. [CrossRef]
151. Edlim, F.W.; Mafazy, M.M.; Ansori, D.B.; Klemm, S.F.; Purwitasari, D. Failure Detection on 3D Printing Images with Color-Based Data Augmentation. In Proceedings of the 2024 International Conference on Smart Computing, IoT and Machine Learning (SIML), Surakarta, Indonesia, 6–7 June 2024; pp. 68–73. [CrossRef]
152. Lee, X.Y.; Saha, S.K.; Sarkar, S.; Giera, B. Automated Detection of Part Quality during Two-Photon Lithography via Deep Learning. *Addit. Manuf.* **2020**, *36*, 101444. [CrossRef]
153. Sassaman, D.; Phillips, T.; Milroy, C.; Ide, M.; Beaman, J. A Method for Predicting Powder Flowability for Selective Laser Sintering. *JOM* **2022**, *74*, 1102–1110. [CrossRef]
154. Satterlee, N.; Torresani, E.; Olevsky, E.; Kang, J.S. Comparison of Machine Learning Methods for Automatic Classification of Porosities in Powder-Based Additive Manufactured Metal Parts. *Int. J. Adv. Manuf. Technol.* **2022**, *120*, 6761–6776. [CrossRef]
155. Nasrin, T.; Pourkamali-Anaraki, F.; Peterson, A.M. Application of Machine Learning in Polymer Additive Manufacturing: A Review. *J. Polym. Sci.* **2024**, *62*, 2639–2669. [CrossRef]
156. Chen, F.; Wang, J.; Guo, Z.; Jiang, F.; Ouyang, R.; Ding, P. Machine Learning and Structural Design to Optimize the Flame Retardancy of Polymer Nanocomposites with Graphene Oxide Hydrogen Bonded Zinc Hydroxystannate. *ACS Appl. Mater. Interfaces* **2021**, *13*, 53425–53438. [CrossRef]
157. Li, C.; Rubín de Celis Leal, D.; Rana, S.; Gupta, S.; Sutti, A.; Greenhill, S.; Slezak, T.; Height, M.; Venkatesh, S. Rapid Bayesian Optimisation for Synthesis of Short Polymer Fiber Materials. *Sci. Rep.* **2017**, *7*, 5683. [CrossRef] [PubMed]
158. Takasuka, S.; Ito, S.; Oikawa, S.; Harashima, Y.; Takayama, T.; Nag, A.; Wakiuchi, A.; Ando, T.; Sugawara, T.; Hatanaka, M.; et al. Bayesian Optimization of Radical Polymerization Reactions in a Flow Synthesis System. *ChemRxiv* **2024**, *4*, 2425178. [CrossRef]
159. Qi, Y.; Hu, D.; Jiang, Y.; Wu, Z.; Zheng, M.; Chen, E.X.; Liang, Y.; Sadi, M.A.; Zhang, K.; Chen, Y.P. Recent Progresses in Machine Learning Assisted Raman Spectroscopy. *Adv. Opt. Mater.* **2023**, *11*, 2203104. [CrossRef]
160. Ranjan, N.; Kumar, R.; Kumar, R.; Kaur, R.; Singh, S. Investigation of Fused Filament Fabrication-Based Manufacturing of ABS-Al Composite Structures: Prediction by Machine Learning and Optimization. *J. Mater. Eng. Perform.* **2023**, *32*, 4555–4574. [CrossRef]
161. Aizawa, T. New Design Method for Fabricating Multilayer Membranes Using CO2-Assisted Polymer Compression Process. *Molecules* **2020**, *25*, 5786. [CrossRef]
162. Barnett, J.W.; Bilchak, C.R.; Wang, Y.; Benicewicz, B.C.; Murdock, L.A.; Bereau, T.; Kumar, S.K. Designing Exceptional Gas-Separation Polymer Membranes Using Machine Learning. *Sci. Adv.* **2020**, *6*, eaaz4301. [CrossRef]
163. Sbrana, A.; de Almeida, A.G.; de Oliveira, A.M.; Neto, H.S.; Rimes, J.P.C.; Belli, M.C. Plastic Classification With NIR Hyperspectral Images and Deep Learning. *IEEE Sens. Lett.* **2023**, *7*, 1–4. [CrossRef]
164. Rizkin, B.A.; Shkolnik, A.S.; Ferraro, N.J.; Hartman, R.L. Combining Automated Microfluidic Experimentation with Machine Learning for Efficient Polymerization Design. *Nat. Mach. Intell.* **2020**, *2*, 200–209. [CrossRef]
165. Cao, L.; Russo, D.; Lapkin, A.A. Automated Robotic Platforms in Design and Development of Formulations. *AIChE J.* **2021**, *67*, e17248. [CrossRef]
166. Knox, S.T.; Parkinson, S.J.; Wilding, C.Y.P.; Bourne, R.A.; Warren, N.J. Autonomous Polymer Synthesis Delivered by Multi-Objective Closed-Loop Optimisation. *Polym. Chem.* **2022**, *13*, 1576–1585. [CrossRef]
167. Gong, Y.; Shao, H.; Luo, J.; Li, Z. A Deep Transfer Learning Model for Inclusion Defect Detection of Aeronautics Composite Materials. *Compos. Struct.* **2020**, *252*, 112681. [CrossRef]
168. Cheng, J.; Tan, W.; Yuan, Y.; Zhao, Z.; Cheng, Y. Research on Defect Detection Method for Composite Materials Based on Deep Learning Networks. *Appl. Sci.* **2024**, *14*, 4161. [CrossRef]
169. Gao, L.; Lin, J.; Wang, L.; Du, L. Machine Learning-Assisted Design of Advanced Polymeric Materials. *Acc. Mater. Res.* **2024**, *5*, 571–584. [CrossRef]
170. Han, L.; Peng, X.F.; Li, L.X. Entire-Region Constitutive Relation for Treloar's Data. *Rubber Chem. Technol.* **2021**, *95*, 119–127. [CrossRef]
171. Meyer, T.A.; Ramirez, C.; Tamasi, M.J.; Gormley, A.J. A User's Guide to Machine Learning for Polymeric Biomaterials. *ACS Polym. Au* **2023**, *3*, 141–157. [CrossRef]
172. Zhang, H.; Chen, W.; Rondinelli, J.M.; Chen, W. ET-AL: Entropy-Targeted Active Learning for Bias Mitigation in Materials Data. *Appl. Phys. Rev.* **2023**, *10*, 021403. [CrossRef]

173. Ma, X.; Chen, H.; He, R.; Yu, Z.; Prokhorenko, S.; Wen, Z.; Zhong, Z.; Íñiguez-González, J.; Bellaiche, L.; Wu, D.; et al. Active Learning of Effective Hamiltonian for Super-Large-Scale Atomic Structures. *NPJ Comput. Mater.* **2025**, *11*, 70. [CrossRef]
174. Yang, Z.; Li, S.; Li, S.; Yang, J.; Liu, D. A Two-Step Data Augmentation Method Based on Generative Adversarial Network for Hardness Prediction of High Entropy Alloy. *Comput. Mater. Sci.* **2023**, *220*, 112064. [CrossRef]
175. Chen, X.; Wang, T.; Guo, T.; Guo, K.; Zhou, J.; Li, H.; Song, Z.; Gao, X.; Zhang, X. Unveiling the Power of Language Models in Chemical Research Question Answering. *Commun. Chem.* **2025**, *8*, 4. [CrossRef]
176. Yan, R.; Jiang, X.; Wang, W.; Dang, D.; Su, Y. Materials Information Extraction via Automatically Generated Corpus. *Sci. Data* **2022**, *9*, 401. [CrossRef]
177. Hu, B. Data Curation of a Findable, Accessible, Interoperable, Reusable Polymer Nanocomposites Data Resource—MaterialsMine. Ph.D. Thesis, Duke University, Durham, NC, USA, 2022. Available online: https://www.proquest.com/docview/2773518516/abstract/E55EDA9E16E247CDPQ/1 (accessed on 25 September 2024).
178. Brinson, L.C.; Deagen, M.; Chen, W.; McCusker, J.; McGuinness, D.L.; Schadler, L.S.; Palmeri, M.; Ghumman, U.; Lin, A.; Hu, B. Polymer Nanocomposite Data: Curation, Frameworks, Access, and Potential for Discovery and Design. *ACS Macro Lett.* **2020**, *9*, 1086–1094. [CrossRef] [PubMed]
179. Yu, T.; Sun, Z.-B.; Sang, J.-P.; Huang, S.-Y.; Zou, X.-W. Structural Class Tendency of Polypeptide: A New Conception in Predicting Protein Structural Class. *Phys. A Stat. Mech. Its Appl.* **2007**, *386*, 581–589. [CrossRef]
180. Niu, J.; Miao, B.; Guo, J.; Ding, Z.; He, Y.; Chi, Z.; Wang, F.; Ma, X. Leveraging Deep Neural Networks for Estimating Vickers Hardness from Nanoindentation Hardness. *Materials* **2024**, *17*, 148. [CrossRef]
181. Lee, F.L.; Park, J.; Goyal, S.; Qaroush, Y.; Wang, S.; Yoon, H.; Rammohan, A.; Shim, Y. Comparison of Machine Learning Methods towards Developing Interpretable Polyamide Property Prediction. *Polymers* **2021**, *13*, 3653. [CrossRef]
182. Bahoria, B.V.; Pande, P.B.; Dhengare, S.W.; Raut, J.M.; Bhagat, R.M.; Shelke, N.M.; Uparkar, S.S.; Vairagade, V.S. Predictive Models for Properties of Hybrid Blended Modified Sustainable Concrete Incorporating Nano-Silica, Basalt Fibers, and Recycled Aggregates: Application of Advanced Artificial Intelligence Techniques. *Nano-Struct. Nano-Objects* **2024**, *40*, 101373. [CrossRef]
183. Tao, Q.; Yang, X.; Bao, L.; Zhou, Y.; Yang, T.; Zhao, Y.; Shi, R.; Yao, Z.; Liu, X. Transforming Machine Learning Model Knowledge into Material Insights for Multi-Principal-Element Superalloy Phase Design. *NPJ Comput. Mater.* **2025**, *11*, 99. [CrossRef]
184. Yang, Z.; Shan, X.; Zhang, W. Discovery of Knowledge of Wall-Bounded Turbulence via Symbolic Regression. *arXiv* **2024**, arXiv:2406.08950. [CrossRef]
185. Ouyang, R.; Curtarolo, S.; Ahmetcik, E.; Scheffler, M.; Ghiringhelli, L.M. SISSO: A Compressed-Sensing Method for Identifying the Best Low-Dimensional Descriptor in an Immensity of Offered Candidates. *Phys. Rev. Mater.* **2018**, *2*, 083802. [CrossRef]

MDPI AG
Grosspeteranlage 5
4052 Basel
Switzerland
Tel.: +41 61 683 77 34

Polymers Editorial Office
E-mail: polymers@mdpi.com
www.mdpi.com/journal/polymers

www.ingramcontent.com/pod-product-compliance
Lightning Source LLC
Chambersburg PA
CBHW040804250926
51850CB00007BA/1078
9783725858392